AF876645

Microsomes and Drug Oxidations

John French

Microsomes and Drug Oxidations

Proceedings of the 7th International Symposium

Adelaide, Australia
17 - 21 August, 1987

Edited by
J.O. Miners, D.J. Birkett, R. Drew,
B.K. May and M.E. McManus

Taylor & Francis
London*NewYork*Philadelphia

UK Taylor & Francis Ltd, 4 John St, London WC1N 2ET

USA Taylor & Francis Inc., 242 Cherry St, Philadelphia, PA 19106-1906

Copyright © Taylor & Francis 1988

All rights reserved. No part of this publication may be reproduced, stored in a retrieval system, or transmitted, i any form or by any means, electronic, electrostatic, magnetic tape, mechanical, photocopying, recording o otherwise, without the prior permission of the copyrigh owner.

British Library Cataloguing in Publication Data

Microsomes and drug oxidations.
1. Drug metabolism 2. Microsomes 3. Oxidation, Physiological
1. Miners, J.O.
615'.7 RM301.55

ISBN 0-85066-361-X

Printed in Great Britain by Taylor & Francis (Printers) Lt Basingstoke, Hants.

CONTENTS

4. HORMONAL REGULATION OF CYTOCHROME P-450

5. CHEMISTRY AND PHYSICS OF CYTOCHROME P-450

6. HUMAN CYTOCHROME P-450 POLYMORPHISMS AND FUNCTIONAL CORRELATES OF CYTOCHROME P-450 MULTIPLICITY IN MAN

7. UDP-GLUCURONOSYLTRANSFERASE

8. METABOLIC ACTIVATION: CELL TOXICITY AND CARCINOGENESIS

PREFACE

The Seventh International Symposium on Microsomes and Drug Oxidations was held in Adelaide, Australia in August, 1987. Despite the barriers of distance, the meeting attracted an outstanding attendance of participants from all parts of the world. A very strong programme was developed by the Organising Committee with substantial advice from an International Advisory Committee, whose assistance we acknowledge gratefully. "Poster Discussion Sessions" were a new feature introduced to give proper emphasis to the poster presentations and were a great success. We thank all participants in the meeting for the high and stimulating standard of presentation, discussion and debate which characterised all the scientific proceedings. It is only rarely that such a meeting can be held "downunder" and the Adelaide Symposium has undoubtedly been a stimulus to interest and research in xenobiotic metabolism in Australia and the Southeast Asian region generally. We thank all our friends and colleagues in the international scientific community for their generous support.

It was clear that there is a continuing place for a broadly based meeting which allows integration across enzyme systems and between disciplines ranging from molecular biology to toxicology and clinical pharmacology. We wish every success to the next Symposium to be held in Stockholm in 1990.

Don Birkett
Chairman, Organising Committee

7TH INTERNATIONAL SYMPOSIUM ON MICROSOMES AND DRUG OXIDATIONS

Organising Committee

D.J. Birkett	Chairman
J.O. Miners	Secretary
F. Bochner	
R. Drew	
B.K. May	
M.E. McManus	
A. Somogyi	
B. Stock	

International Advisory Committee

A.H. Conney	(USA)
M.J. Coon	(USA)
D.S. Davies	(UK)
R.W. Estabrook	(USA)
Y. Fujii-Kuriyama	(Japan)
J.R. Gillette	(USA)
F.P. Guengerich	(USA)
E.F. Johnson	(USA)
R. Kato	(Japan)
U.A. Meyer	(Switzerland)
F. Oesch	(FRG)
S. Orrenius	(Sweden)
R.L. Smith	(UK)

7TH INTERNATIONAL SYMPOSIUM ON MICROSOMES AND DRUG OXIDATIONS

Sponsors

Ansett Airlines
Abbott Diagnostics (Aust)
Anti-Cancer Foundation of the Universities of South Australia
Astra Pharmaceuticals (Aust) Pty Ltd
Beckman Instruments (Aust) Pty Ltd
Ciba-Geigy Australia Ltd
Cyanamid Pacific
F.H. Faulding Ltd
Glaxo Australia Pty Ltd
ICI Australia Operations Ltd
International Society for the Study of Xenobiotics
International Union of Biochemistry
IUPHAR - Drug Metabolism Section
May and Baker Australia Ltd
Millipore-Waters (Aust)
National Institutes of Health (USA)
Parke-Davis & Co Ltd (Aust)
Qantas Airways
Roche Products Pty Ltd (Aust)
6th International Symposium on Microsomes and Drug Oxidations (UK)
Smith, Kline and French Laboratories (Australia) Ltd
Utah Foundation
Wellcome Australia
3rd World Conference on Clinical Pharmacology and Therapeutics (Sweden)

These Proceedings are dedicated to Professor Ryo Sato in recognition of his contributions to research on cytochrome P-450.

INTER-RELATIONSHIPS AND DISTRIBUTION OF CYTOCHROMES P-450

Ethanol-inducible Rat and Human Liver Cytochrome P-450: Characterization and Role as High Affinity N-Nitrosodimethylamine Demethylases

DENE E. RYAN, PAUL E. THOMAS, STEVEN A. WRIGHTON [1], PHILIP S. GUZELIAN [1] and WAYNE LEVIN

Department of Molecular Oncology, Roche Institute of Molecular Biology, Nutley, NJ 07110, USA and [1] Division of Clinical Toxicology and Environmental Medicine, Medical College of Virginia, Richmond, VA 23298, USA

INTRODUCTION

Alcohol consumption by humans has been associated with an increased incidence of cancer at various sites including the esophagus, colon, rectum and liver (Lieber et al., 1981), although the basis of the increased suseptibility has not been defined. The influence of alcohol is apparently indirect since ethanol is not carcinogenic in laboratory animals (Gricuite et al., 1981). One plausible hypothesis has been that ethanol influences the incidence of human cancer by affecting the pharmacokinetics of nitrosamines (Swann, 1984) which are derived from a multitude of exogenous sources as well as endogenous synthesis in the gastrointestinal tract. Ethanol has also been shown to prevent the first pass clearance of N-nitrosodimethylamine (NDMA) in the rat thereby exposing extrahepatic organs to the carcinogen (Swann, 1984; Swann et al., 1984). In addition, the excretion of NDMA by humans is influenced by ethanol which suggests that first pass clearance is also prevented in humans (Swann et al., 1984; Spiegelhalder et al., 1982).

NDMA, a potent hepatocarcinogen in a variety of species, requires metabolic activation to exert its toxic and carcinogenic effects. This activation, which is cytochrome P450-mediated, leads to formation of formaldehyde and an electrophilic alkylating agent as the presumed ultimate carcinogen (Preussmann & Stewart, 1984). The influence of ethanol on NDMA metabolism is multifaceted: the presence of ethanol inhibits the *in vivo* and *in vitro* metabolism of NDMA (Khanna et al., 1980; Phillips et al., 1977; Schwarz et al., 1980; Peng et al., 1982); whereas chronic ethanol treatment yields enhanced metabolism in rat liver microsomes (Schwarz et al., 1980; Peng et al., 1982; Maling et al., 1975; Garro et al., 1981). Demethylation of NDMA by hepatic microsomes proceeds by both a high and low affinity enzyme (Lake et al., 1974; Arcos et al., 1977; Sipes et al., 1978; Tu et al., 1981), although only the low Km enzyme plays a significant role in vivo based on maximum concentrations of NDMA that can be obtained in liver (Argus & Arcos, 1978; Magee, 1980). The high affinity NDMA demethylase is inducible in animals by compounds such as ethanol, isoniazid, pyrazole, acetone and isopropanol (Peng et al., 1982; Garro et al., 1981; Sipes et al., 1978; Tu et al., 1981; Sipes et al., 1973; Evarts et al., 1982; Tu et al., 1983).

Ethanol-inducible rat liver cytochrome P450j[1] and its ortholog in rabbits, cytochrome P450 LM3a, have been shown to be efficient catalysts of NDMA metabolism at low substrate concentrations (Tu & Yang, 1985; Levin et al., 1986; Yang et al., 1985). Other preferential substrates for these enzymes apparently include acetone, acetaminophen, ethanol, and carbon tetrachloride (Koop & Casazza, 1985; Morgan et al., 1983; Morgan et al., 1982; Johansson & Ingelman-Sundberg, 1985). The conservation of a functionally related cytochrome P450 through evolution may be of significant relevance to human toxicology. For example, is the increased cancer incidence, the potentiation of acetaminophen hepatotoxicity (Seelf et al., 1986) and/or toxic responses to carbon tetrachloride (Folland et al., 1976) observed in humans exposed to ethanol mediated through the induction of a cytochrome P450 similar to that observed in animals? In addition, liver damage observed in some patients treated with the anti-tuberculosis drug, isoniazid (Nelson et al., 1976), an inducer of rat cytochrome P450j, could be related to the induction of an orthologous isozyme in man.

In this study, we describe the characterization of purified human cytochrome P450 (HLj) isolated from microsomes from an ethanol-intoxicated male patient and compare this enzyme to rat liver cytochrome P450j. The role of these enzymes in microsomal metabolism of NDMA is also evaluated using monospecific antibody to the rat isozyme. Furthermore, extrahepatic levels of cytochrome P450j and NDMA metabolism have been determined in order to

[1] According to a recommended nomenclature system for cytochromes P450, rat P450j, rabbit LM3a and HLj are encoded by the gene IIE1 (Nebert et al.,1987).

assess the organ specificity of expression of this isozyme.

RESULTS AND DISCUSSION

Cytochrome P450j has been purified from hepatic microsomes of ethanol- or isoniazid-treated adult male rats by chromatography on n-octylamine-Sepharose 4B, hydroxylapatite, DEAE-Sepharose, CM-Sepharose and phosphocellulose (Ryan et al., 1985). When human liver microsomes from an ethanol-intoxicated male patients were subjected to the same purification protocol with some modifications, purified cytochrome P450 HLj was obtained (Wrighton et al., 1987). The spectral characteristics of HLj are reminiscent of rat P450j and orthologous rabbit LM3a (Ryan et al., 1985; Koop et al., 1982) since the human isozyme has an absorbance maximum at 451 nm in the CO-reduced difference spectrum and is primarily a high spin ferric hemoprotein (data not shown). The electrophoretic homogeneity of rat P450j and HLj are shown in Fig. 1 since only one protein-stained band is observed in each sample. The minimum molecular weights were determined to be 51,500 and 54,000 for the rat and human isozymes, respectively. The NH_2-terminal amino acid sequence of HLj was determined to be Ser-Ala-Leu-Gly-Val-Thr-Val-Ala-Leu-Leu-Val-Trp-Ala-Ala-Phe-Leu-Leu-Leu-. The lack of any contaminating sequence in the HLj preparation also confirmed the purity of the sample. Eleven of the first eighteen amino acids of HLj are identical to rat P450j. Recently, Song et al. (1986) have sequenced cDNA's for rat and human P450j and calculated that the proteins share 78% overall structural homology based on their deduced amino acid sequences. Immunoblot analysis (Fig. 1) revealed that monospecific polyclonal antibody against rat P450j recognizes HLj and a protein of identical molecular weight in human liver microsomes. Further evidence that rat anti-P450j recognizes a single protein in human liver was obtained from the successful isolation of HLj from solubilized microsomes by immunoaffinity chromatography. That isolated protein also possessed a single NH_2-terminal (Wrighton et al., 1986).

Rat P450j and HLj apparently share a high degree of functional, as well as structural, relatedness. Both hemoproteins have no measurable activity toward the metabolism of benzo(a)pyrene nor do they participate in the hydroxylation of the steroids testosterone and progesterone. Low, but measurable, metabolism of benzphetamine and 7-ethoxycoumarin are catalyzed by

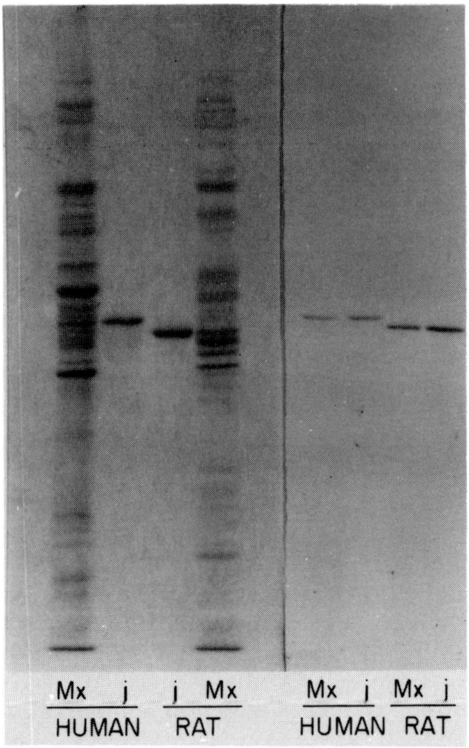

FIGURE 1. SDS gel and immunoblot of microsomes (Mx) and purified P450j and HLj. Left side, 8 ug of microsomal protein and 0.5 ug of each purified sample was stained for protein. Right side, 10 ug and 4 ug of microsomal protein from human No. 17 and ethanol-treated rats, respectively, were applied to the gel. Purified rat P450j (.05 ug) and HLj (0.2 ug) were electrophoresed in the wells as labeled. The blot was probed with monospecific polyclonal anti-rat P450j.

these isozymes (Wrighton et al., 1987). In contrast, these cytochromes effectively catalyze the p-hydroxylation of aniline, with turnover numbers of 13.6 and 17.8 nmol/min/nmol P450 for the human and rat isozymes, respectively. The parallel catalytic profiles of rat P450j and HLj also extends to N-demethylation of NDMA. Of eleven purified rat liver cytochrome P450 isozymes, only P450j has high catalytic activity toward this substrate (Levin et al., 1986). Kinetic constants for NDMA metabolism catalyzed by rat P450j and HLj were determined from Lineweaver Burk plots; Km's,app were calculated to be 3.5 and 1.3 mM and Vmax's were determined to be 23.9 and 7.1 nmol/min/nmol P450 for the rat and human isozymes, respectively. When cytochrome b_5 was added to either reconstituted system at a molar ratio of 4 to 1 with respect to cytochrome P450, similar effects on the kinetic constants were observed. The Vmax of the reaction was increased slightly while a substantial decrease (4-6 fold) in Km,app was observed. The mechanism by which cytochrome b_5 decreases the Km,app for NDMA demethylation is unknown but could reflect a conformational change in rat P450j and HLj as a result of complex formation between cytochromes P450 and b_5. The influence of cytochrome b_5 on the Km's,app for this reaction in the reconstituted system correlates reasonably well with the values obtained with microsomes from ethanol-treated rats (Levin et al., 1986).

The participation of rat P450j and HLj in overall microsomal metabolism of NDMA has been evaluated using monospecific antibody to the rat enzyme as shown in Table 1. Ethanol, isoniazid and pyrazole, which are inducers of rat P450j, also induce NDMA demethylation in rat liver microsomes (Thomas et al., 1987). A comparison of microsomal NDMA metabolism and levels of cytochrome P450j has revealed a strong correlation which is further supported by the results of antibody inhibition experiments. NDMA demethylation by rat liver microsomes is primarily cytochrome P450j-mediated since 85-90% of the activity was inhibited by the antibody. Microsomes from patients who had received ethanol (No. 17) or isoniazid (No. 22) exhibited elevated rates of NDMA metabolism as compared to the rates observed in microsomes from patients (No. 18, No. 20) who had not been exposed to known inducers of rat P450j. Patients No. 17 and No. 22 also had higher levels of microsomal HLj than patients No. 18 and No. 20 (Wrighton et al., 1986). The results of

TABLE 1. Effect of anti-rat P450j on microsomal metabolism of NDMA (1 mM).

Microsomes		NDMA Metabolism (nmol/min/mg protein)	
		control	+ anti-P450j
Rat	untreated	0.47	0.07 (85)
	ethanol	5.09	0.53 (90)
	isoniazid	3.54	0.40 (89)
	pyrazole	3.61	0.39 (90)
Human	No. 17 (ethanol)	2.18	0.49 (78)
	No. 18	0.56	0.16 (71)
	No. 20	0.30	0.09 (70)
	No. 22 (isoniazid)	1.13	0.26 (77)

from Thomas et al., 1987, Wrighton et al., 1986. Percent inhibition by anti-P450j is given in parentheses.

antibody inhibition in Table 1 indicate that HLj is responsible for at least 70% of NDMA metabolism in microsomes from these patients.

The residual NDMA demethylation catalyzed by rat and human liver microsomes that was not inhibited by the antibody could be due to additional cytochrome P450 isozymes capable of catalyzing this reaction or to the inaccessibility of a certain amount of rat P450j or HLj to the antibody. The latter possibility is more likely based on results of experiments changing the order of addition of the antibody and components of the reconstituted system (Thomas _et al_., 1987; Wrighton _et al_., 1987). These results have indicated that the interaction of P450j or HLj with NADPH-cytochrome P450 reductase may prevent complete access of the antibody to the isozyme. Therefore, P450j and HLj are clearly the primary, and perhaps sole, microsomal catalysts of NDMA demethylation at substrate concentrations ($\leq$ 1 mM) relevant to carcinogenesis induced by NDMA.

Since the organ specificity of NDMA-induced carcinogenesis is dose-dependent in the rat (Magee & Barnes, 1962), P450j content and NDMA metabolism were measured in extrahepatic tissues of the rat. In microsomes from untreated male rats, P450j content is highest in liver but measurable amounts are also found in kidney. Lung and testes, however, contain barely detectable levels of this isozyme. An induction of approximately 3-fold in P450j levels was observed in these extrahepatic tissues following isoniazid treatment. Extrahepatic NDMA metabolism is clearly cytochrome P450j-mediated since 80-90% of demethylation was inhibited by antibody to this isozyme (Thomas _et al_., 1987). The above results on the distribution of rat P450j in liver and extrahepatic tissues in conjunction with the potent antibody inhibition of NDMA metabolism strongly indicate that cytochrome P450j is the critical _in vivo_ catalyst responsible for the activation of NDMA to an ultimate carcinogen. A similar role for the human ortholog, HLj, in NDMA metabolism is also indicated from the data presented; therefore, the rat appears to be an accurate model for this enzyme and substrate in humans.

ACKNOWLEDGMENT

The authors thank Ms. Cathy Michaud for her excellent secretarial assistance.

REFERENCES

Arcos, J.C., Davies, D.L., Brown, C.E.L., and Argos, M.F. 1977, Z. Krebsforsch. 89, 181.

Argus, M.F., & Arcos, J.C., 1978, Cancer Research, 38, 226.

Evarts, R.P., Haliday, E., Negishi, M., and Hjelmeland, L.M., 1982, Biochemical Pharmacology, 31,1245.

Folland, D.S., Schaffner, W., Ginn, H.E., Crofford, O.B., and McMurray, D.R., 1976, Journal American Medical Association, 236,1853.

Garro, A.J., Seitz, H.K., and Lieber, C.S., 1981, Cancer Research, 41,120.

Gricuite, L., Castegnaro, M., and Bereziat, J.C., 1981, Cancer Letters,13, 345.

Johansson, I., and Ingelman-Sundberg, M., 1985, FEBS Letters, 183,265.

Khanna, J.M., Chung, S., Ho, G., and Shak, G., 1980, In Currents in Alcoholism Vol. VII, p.93, edited by M. Galenter (New York: Greene and Strattan).

Koop, D.R., and Casazza, J.P., 1985, Journal of Biological Chemistry,260,13607.

Koop, D.R., Morgan, E.T., Tarr, G.E., and Coon,M.J. 1982, Journal of Biological Chemistry, 257, 8472.

Lake, B.G., Heading, C.E., Phillips, J.C., Gangoli, S.D.,and Lloyd, A.G., 1974 Biochemical Society Transactions, 2,10.

Levin, W., Thomas, P.E., Oldfield, N., and Ryan, D.E., 1986, Archives Biochemistry and Biophysics,248,158.

Lieber, C.S., Seitz, H.K., Garro, A.J., and Worner, T.M., 1981, In Frontiers in Liver Disease p. 320, edited by P.D. Berk and T. C. Chalmers (New York: Thieme-Stratton).

Magee, P., 1980, In Microsomes, Drug Oxidations and Chemical Carcinogenesis Vol. II, p. 1081 , edited by M.J. Coon, A.H. Conney, R.W. Estabrook, H.V. Gelboin, J.R. Gillette, and P.J. O'Brien (New York: Academic Press).

Magee, P.N., & Barnes, J.M.,1962, Journal of Pathology & Bacteriology, 84,19.

Maling, H.M., Stripp, B., Sipes, I.G., Highman, B., Saul, W., and Williams,M.A., 1975, Toxicology and Applied Pharmacology,33, 291.

Morgan, E.T., Koop, D. R., and Coon, M.J., 1983, Biochemical and Biophysical Research Communications, 112, 8.

Morgan, E.T., Koop, D.R., and Coon, M.J., 1982, Journal of Biological Chemistry 257, 13951.

Nebert, D.W., Adesnik, M., Coon, M.J., Estabrook, R.W., Gonzalez, F.J., Guengerich, F.P., Gunsalus, I.C., Johnson, E.F., Kemper, B., Levin, W., Phillips, I.R., Sato, R., and Waterman, M.R., 1987, DNA, 6,1.

Nelson, S.D., Mitchell, J.R., Timbrell, J.A., Snodgrass, W.R., and Corcoran, G.B., 1976, Science, 193,901.

Peng, R., Tu, Y.Y., and Yang, C.S., 1982, Carcinogenesis, 3, 1457.

Phillips, J.C., Lake, B.G., Gangolli, S.D., Grasso, P., and Lloyd, A.G.,1977, Journal of National Cancer Institute, 58,629.

Preussmann, R., and Stewart, B.W., 1984, In Chemical Carcinogens Vol. 2,p.643, edited by C.E. Searle (Washington D.C.: ACS Monograph 182, American Chemical Society)

Ryan, D.E., Ramanathan, L., Iida, S., Thomas, P.E., Haniu, M., Shively, J.E., Lieber,C.S., and Levin, W., 1985, Journal of Biological Chemistry, 260, 6385.

Schwarz, M., Appel, K.E., Schrenk, D., and Kunz, W., 1980, Journal of Cancer Research and Clinical Oncology, 97,233.

Seelf, L.B., Cuccherini, B.A., Zimmerman, H.J., Alder, E., and Benjamin,S.B., 1986, Annuals of Internal Medicine, 104,399.

Sipes, I.G., Slocumb, M.L., and Holtzman, G., 1978, Chemical-Biological Interactions, 21, 155.

Sipes, I.G., Stripp, B., Krishna, G., Malling, H.M., and Gillette, J., 1973, Proceedings Society Experimental Biology and Medicine, 142, 237.

Song, B.J., Gelboin, H.V., Park, S.S., Yang, C.S., and Gonzalez, F.J., 1986, Journal of Biological Chemistry, 261, 16689.

Spiegelhalder, B., Eisenbrand, G., and Preussmann, R. 1982, In N-Nitroso Compounds: Occurrence and Biological Effects p.443, edited by H. Bartsch, I.K. O'Neill, M. Castegnaro and M. Okada (Lyon: IARC Scientific Publications)

Swann, P.F.,1984, In N-Nitroso Compounds: Occurrence, Biological Effects and Relevance to Human Cancer p.501 edited by I.K. O'Neill, R.C. Borstel, C.T. Miller, J.Long, and H. Bartsch (Lyon: IARC Scientific Publications).

Swann, P.F., Coe, A.M., and Mace, R., 1984, Carcinogenesis, 5, 1337.

Thomas, P.E., Bandiera, S., Maines, S.L., Ryan, D.E., and Levin, W., 1987, Biochemistry, 26, 2280.

Tu., Y.Y., Peng, R.X., Cheng, Z.F., and Yang, C.S., 1983, Chemical-Biological Interactions, 44, 247.
Tu., Y.Y., Sonnenberg,J., Lewis, K.F., and Yang, C.S., 1981, Biochemical & Biophysical Research Communications, 103, 905.
Tu, Y.Y., and Yang, C.S., 1985, Archives Biochemistry & Biophysics, 242, 32.
Wrighton, S.A., Thomas, P.E., Molowa, D.T., Haniu, M., Shively, J.E., Maines, S.L., Watkins, P.B., Parker, G., Mendez-Picon, G., Levin, W., and Guzelian, P.S., 1986, Biochemistry, 25, 6731.
Wrighton, S.A., Thomas, P.E., Ryan, D.E., and Levin, W., 1987, Archives Biochemistry & Biophysics, in press.
Yang, C.S., Tu, Y.Y., Koop, D.R., and Coon, M.J., 1985, Cancer Research, 45, 1140.

Structure and Function of Homologous Pulmonary and Hepatic Isozymes of Cytochrome P-450

RUDOLFO GASSER and RICHARD M. PHILPOT

Laboratory of Pharmacology, National Institute of Environmental Health Sciences, National Institutes of Health, PO Box 12233, Research Triangle Park, NC 27709, USA

INTRODUCTION

Three isozymes of cytochrome P-450 (P-450) are known to be present in rabbit lung. The majority (>90%) of the pulmonary P-450 is comprised of two isozymes (2 and 5) that are relatively minor components of the hepatic P-450 system (Domin et al., 1986). The levels of isozymes 2 and 5 increase dramatically in liver, but not in lung, following treatment of rabbits with phenobarbital. The pulmonary and hepatic forms of these isozymes have been purified, and no differences between them have been noted (Parandoosh et al., 1987). Isozyme 6 is also present in rabbit lung, but accounts for only 1-2% of the total P-450 (Domin and Philpot, 1986). Both the pulmonary and hepatic concentrations of isozyme 6 increase following treatment of rabbits with 3-methylcholanthrene-type inducers. The same inducers increase the hepatic concentration of isozyme 4, a form of P-450 not detected in lung. Homologs of isozymes 4 and 6 in rats, mice, guinea pigs, and hamsters exhibit similar patterns of pulmonary and hepatic expression (Philpot et al., 1985 and unpublished).

Although a significant amount of information about pulmonary P-450 systems is now available, a number of important details have not been resolved. For example, the primary structures of the P-450 isozymes expressed in lung are not known, the extent to which the rabbit pulmonary system is representative of other species is not clear, and the mechanisms responsible for the different responses of lung and liver to various P-450 inducers have not been elucidated. We have now made some progress in these areas, and the results of our studies on the distribution and structure of isozyme 5, and on the structure, multiplicity, and tissue-specific expression of isozyme 2 are presented.

METHODS

The methods for immunoblotting and analysis of the metabolism of 2-aminofluorene are listed by Vanderslice et al. (1987). mRNA was isolated by a modification of the methods of Chirgwin et al. (1979) and Glisin et al. (1974). Libraries of cDNA derived from pulmonary and hepatic mRNA were constructed in lambda gt11 (Young and Davis, 1983). Libraries were screened by plaque hybridization (Benton et al., 1977) with nick translated cDNA or by immunodetection with polyclonal antibodies. Positive recombinants were subcloned into plasmid pGEM-2 (Promega Biotech, Madison, WI). Recombinant M13 clones were generated from restriction fragments or by sequential deletion (Dale et al., 1985) in E. coli JM103 (Messing, 1983; Schreier and Cortese, 1979). Sequences were determined by the dideoxy chain termination method (Sanger et al., 1977; Biggin et al., 1983).

RESULTS AND DISCUSSION

Isozyme 5

Determination of distribution and induction by immunochemical methods. Homologs of rabbit P-450 isozyme 5 in lungs and livers of rats, mice, guinea pigs, and hamsters have been identified by immunoblotting and inhibition of enzymatic activity with antibodies to isozyme 5 (Vanderslice et al., 1987). These results are summarized in Table 1.

Table 1. Expression of homologs of isozyme 5 in lungs and livers of several species

					Induction[c]			
	Immunoblots[a]		Inhibition[b]		TCDD		PB	
Species	Lung	Liver	Lung	Liver	Lung	Liver	Lung	Liver
Rabbit	+	+	78%	60%	-	-	-	+
Rat	+	-	82%	10%	-	-	-	-
Guinea pig	+	-	80%	8%	-	-	-	-
Mouse	+	-	75%	5%	-	-	-	-
Hamster	+	+	85%	42%	-	-	-	-

a. The results show whether or not a pulmonary or hepatic microsomal protein was detected by antibodies to isozyme 5.
b. The results show the extent of inhibition by antibodies to isozyme 5 of the pulmonary and hepatic microsomal metabolism of 2-aminofluorene.
c. Induction was determined both by immunoblotting and activity determinations; a negative sign indicates that the change was less that ±1.5-fold.

A pulmonary microsomal protein in samples from all species examined is detected with antibodies to isozyme 5. In contrast, detection of a protein in hepatic microsomal samples is observed only with rabbits and hamsters. The same pattern is obtained by inhibition of the metabolism of 2-aminofluorene by antibodies to isozyme 5; marked inhibition (>75%) is observed with all pulmonary samples but inhibition of hepatic activity occurs only with rabbits and hamsters. A second major species difference is seen with the responses to treatment with phenobarbital (PB). Treatment of rabbits with PB increases the hepatic concentration of isozyme 5 by about 10-fold (Robertson et al., 1983), but this effect is not observed with the other species examined.

The primary structure of isozyme 5. Pulmonary and hepatic mRNA was isolated and used for the construction of cDNA libraries. A hepatic cDNA clone for isozyme 5 was identified by immunodetection of the recombinant protein. This probe was used for subsequent screening. The identities of cDNAs encoding isozyme 5 were established by comparing the derived amino acid sequences with the known N-terminal sequence (Parandoosh et al., 1987) (Table 2).

Table 2. Sequences from protein and cDNA for the N-terminal region of isozyme.5[a]

CGCTGCCGCTCCAGCCACG					ATG	CTC	GGC	TTC	CTC	TCC	CGC	CTG	GGC	CTG
					MET	LEU	GLY	PHE	LEU	SER	ARG	LEU	GLY	LEU
TGG	GCT	TCC	GGA	CTG	ATC	TTG	ATC	CTA	GGC	TTC	CTC	AAG	CTC	CTC
TRP	ALA	SER	GLY	LEU	ILE	LEU	ILE	LEU	GLY	PHE	leu	lys	leu	leu

a. The N-terminal sequence reported for isozyme 5 extends to residue 21 and is shown in upper case.

The complete coding region of isozyme 5 cDNA contains 1,515 bases encoding 505 amino acids. The molecular weight of isozyme 5 is 58,018 daltons. Sequences of cDNAs isolated from pulmonary and hepatic libraries show no differences. The homology of isozyme 5 with other mammalian isozymes of P-450 is between 30 and 45%. Alignment of the cysteinyl peptide thought to provide the proximal thiolate heme ligand with the homologous sequences of other selected P-450 isozymes is shown in Table 3. Homologies in this region between isozyme 5 and either isozyme 2 or 6 are about the same as with the complete coding region (<40%). Homologies with the cysteinyl peptide of two isozymes from rat, P-450PCN and P-450LA$_{\omega}$, are >50%.

Table 3. Comparison of cysteinyl peptide sequences of selected isozymes of cytochrome P-450.[a]

Isozyme	Cysteinyl Peptide	References
Rabbit 5	FSAGPRNCIGQQFAMNEMKVV	
Rat LA_{ω}	--G-A-----L----S--L-I	Hardwick et al. (1987)
Rat PCN	-GN-------MR--LMN--LA	Gonzalez et al. (1985)
Rabbit 2	--L-K-I-L-EGI-RT-LFLF	Tarr et al. (1983)
Rabbit 6	-GL-K-R-L-EIL-RL-VFLF	Okino et al. (1985)

a. Amino acids not designated correspond to those shown for rabbit isozyme 5. The cysteine of isozyme 5 is residue 447.

Expression of mRNA for isozyme 5 in rabbit. Hybridization of cDNA for isozyme 5 and mRNA from rabbit lung and liver is shown in Figure 1. Hybridization occurred only with the polyadenylated fraction from either tissue. The greater amount of hybridization with pulmonary mRNA is consistent with the difference between the isozyme 5 contents of the two tissues.

Figure 1. Expression of mRNA for isozyme 5 in rabbit lung and liver. Samples of polyadenylated RNA from lung (lane 2, 10μg; lane 3, 5μg) and liver (lane 5, 10μg; lane 6, 5μg) and non-polyadenylated RNA from lung (lane 1, 10μg) and liver (lane 4, 10μg) were probed with cDNA for isozyme 5.

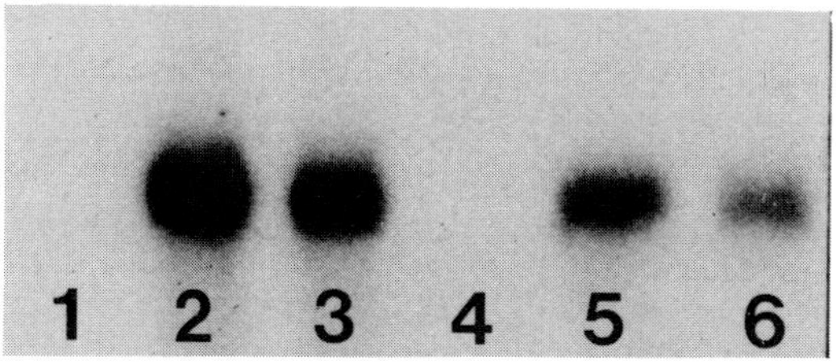

Treatment of rabbits with PB increases the amount of mRNA for isozyme 5 in liver by about 10-fold, but has no effect on the content of isozyme 5 mRNA in lung (not shown).

Expression of mRNA for homologs of isozyme 5 in several species. The cDNA probe for isozyme 5 hybridizes to the same extent with mRNA from hamster lung or liver, shows much greater hybridization with mRNA from rat lung than from rat liver, and hybridizes with mRNA from guinea pig lung but not with mRNA from guinea pig liver. These results are entirely consistent with those obtained by immunoblotting.

Isozyme 2

Structures of multiple forms of isozyme 2. cDNAs encoding isozyme 2 were isolated from three libraries (Lg, Lv-1, and Lv-4). The Lg library was constructed with pulmonary mRNA, and hepatic libraries were constructed with mRNA from individual rabbits treated once with PB (Lv-1) or treated daily for 4 days (Lv-4). Three types of cDNAs were identified by restriction with Bst EII (B0 and B1 from Lg, B1 and B2 from Lv-1, and B0 from Lv-4), and the sequences of nine cDNAs were determined (Table 4).

Table 4. Isolation and characterization of cDNAs encoding cytochrome P-450 isozyme 2.

Library	cDNAs Isolated			cDNAs Sequenced (base pairs)		
	B0	B1	B2	B0	B1	B2
Lg	12	2	0	2044, 860	1091, 1074	
Lv-1	0	9	18		2063, 1850	2077, 720
Lv-4	3	0	0	1735		

All sequences for a given type were found to be identical, and the differences between the sequences are: B0 vs. B1, 56 bases (6 amino acids); B0 vs. B2, 86 bases (11 amino acids); B1 vs. B2, 82 bases (15 amino acids). All three cDNA types encode 491 amino acids with differences at a total of 16 positions (Table 5).

Table 5. Amino acid substitutions for forms of isozyme 2.

		Residue Number							
Type	Tissue	35	39	57	114	120	174	248	286
B0	Liver (Lv-4), Lung	SER	VAL	ARG	ILE	ARG	ile	SER	GLN
B1	Liver (Lv-1), Lung	SER	ile	gln	ILE	ARG	VAL	SER	GLN
B2	Liver (Lv-1)	pro	VAL	ARG	phe	his	VAL	thr	arg
		290	294	314	363	367	370	417	420
B0	Liver (Lv-4), Lung	LEU	SER	MET	ILE	VAL	THR	ASN	LEU
B1	Liver (Lv-1), Lung	ile	SER	leu	ILE	VAL	THR	ASN	met
B2	Liver (Lv-1)	LEU	thr	MET	val	ala	met	asp	LEU

The primary sequence derived from the B0 cDNA is identical to that reported for isozyme 2 by Tarr et al. (1983), but none of the differences among the three types correspond to the variances between the sequences reported by Tarr et al. (1983) and Heinemann and Ozols (1983).

Results obtained with a number of additional mRNA samples are summarized in Table 7.

Table 7. Detection of B0, B1, and B2 transcripts in pulmonary and hepatic samples.

							Phenobarbital Treated												
		Untreated					1x12h			1x20h				4x4 days					
		1	2	3	4	5	1	2	3	1	2	3	4	1	2	3	4	5	6
Liver	B0	-	-	-	-	-	-	+	+	+	-	-	+	+	-	+	+	+	-
	B1	-	-	-	-	-	+	-	-	+	+	+	+	-	+	+	-	-	+
	B2	-	+	+	+	+	+	-	-	+	+	+	+	-	+	+	-	-	+
Lung	B0	+	-	+	+	-				+	-	-	+	+	-	+	+	+	-
	B1	-	+	+	+	+				+	+	+	+	-	+	+	-	-	+
	B2	-	-	-	-	-				-	-	-	-	-	-	-	-	-	-

The B0 and B1 probes show the same patterns of detection with pulmonary and hepatic mRNA from rabbits treated with PB; B0 and/or B1 transcripts are present. Pulmonary mRNA from untreated rabbits also contains B0 and/or B1 transcripts, but neither is detected in hepatic mRNA from the same animals. Detection of the B2 transcript in hepatic mRNA from rabbits treated with PB coincides with the presence of the B1 transcript, but no B2 mRNA is detected in any pulmonary sample. Hepatic mRNA from four untreated rabbits contained B2 mRNA. The absence of a specific transcript in mRNA from one untreated animal, which contained mRNA for isozyme 2 as determined by hybridization with a full length probe, indicates that at least one isozyme 2 gene product in addition to B0, B1, and B2 can be expressed.

SUMMARY

Rabbit P-450 isozyme 5 and its homologs in other species appear to comprise a unique type of P-450 that exhibits unusual tissue-specific and species-specific patterns of expression.

Rabbit isozyme 2 comprises a minimum of three distinct gene products, two of which are expressed constitutively in lung but are detected in liver only following treatment with PB. A third form is expressed constitutively in liver but is not present in lung. The concentrations of all three forms are increased in liver by PB, an inducer that has no effect on the levels of the forms present in lung. Analysis of the distribution of specific transcripts in hepatic mRNA from individual rabbits indicates that at least one form of isozyme 2 remains to be characterized.

<u>Development of specific probes for B0, B1, and B2.</u> Analysis of mRNA content with full length cDNA probes for isozyme 2 shows the expected results, lung contains about twice as much total mRNA for isozyme 2 than does liver, and treatment of rabbits with PB increases hepatic mRNA for isozyme 2 by about 10-fold but has no effect on pulmonary mRNA. The distribution of B0, B1, and B2 transcripts was assessed with oligonucleotide probes complementary to highly variable sequences present in the 3'-flanking regions near the stop codons (Table 6). The specificity of these probes was demonstrated by hybridization to specific cDNAs.

Table 6. Alignment of 3'-flanking regions showing the sequences complementary to the oligonucleotides (lower case).

	STOP	
B0	TGA	AGGGGCCACAGGgaccc-------------------tggtcattaggTGTCC
B1	TGA	AGGGGCCACAGGGACccccccagccca-----------gtcgttagGTGTCC
B2	TGA	AGGGGCCACAGGGACACCCCCCCCCaccgcaccgccctggtcgtaggtGTCC

<u>Northern analysis of rabbit hepatic and pulmonary mRNA with specific probes for B0, B1, and B2.</u> An illustration of results obtained by hybridization of the B0 and B1 probes and hepatic mRNA is shown in Figure 2. Similar autoradiograms were obtained with the B2 probe, and with pulmonary mRNA. All three probes and full length cDNAs detected mRNA of approximately 2.4Kb.

Figure 2. Northern analysis of hepatic mRNA with B0 and B1 probes. mRNA (5μg) from livers of rabbits treated 4 times with PB (lanes 1-4), or once with PB (lanes 5-7, 20 hours; lanes 8 and 9, 12 hours) were examined.

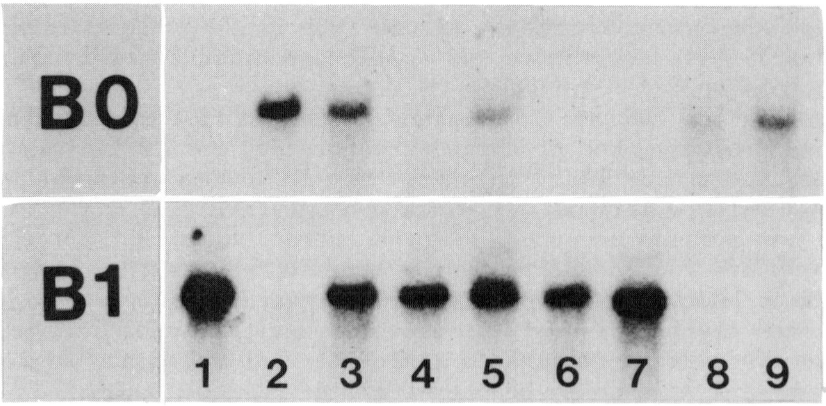

REFERENCES

Benton, W. D. & Davis, R. W., 1977, Science, 196, 180.

Biggin, M. D., Gibson, T. J. & Hong, G. F., 1983, Proceedings of the National Academy of Science, USA, 80, 3963.

Chirgwin, J. M., Przybyla, A. E., MacDonald, J. R. & Rutter, W. J., 1979, Biochemistry, 18, 5294.

Dale, R. M. K., McClure, B. A. & Houchins, J. P., 1985, Plasmid, 13, 31.

Domin, B. A., Devereux, T. R. & Philpot, R. M., 1986, Molecular Pharmacology, 30, 296.

Glisin, V., Crkvenjakov, R. & Byus, C., 1974, Biochemistry, 13, 2633.

Gonzalez, F. J., Nebert, D. W., Hardwick, J. P. & Kaspar, C. B., 1985, Journal of Biological Chemistry, 260, 7435.

Hardwick, J. P., Song, B. J., Huberman, E. & Gonzalez, F. J., 1987, Journal of Biological Chemistry, 262, 801.

Heinemann, F. S. & Ozols, J., 1983, Journal of Biological Chemistry, 258, 4195.

Messing, J., 1983, New M13 vectors for cloning. In Methods in Enzymology, Vol. 101, edited by R. Wu, L. Grossman & K. Moldave, (Academic Press), p. 20.

Okino, S. T., Quattrochi, L. C., Barnes, H. J., Osanto, S., Griffin, K. J., Johnson, E. F. & Tukey, R. H., 1985, Proceedings of the National Academy of Science, USA, 82, 5310.

Parandoosh, Z., Fujita, V. S., Coon, M. J. & Philpot, R. M., 1987, Drug Metabolism and Disposition, 15, 59.

Philpot, R. M., Domin, B. A., Devereux, T. R., Harris, C., Anderson, M. W., Fouts, J. R. & Bend, J. R., 1985, Cytochrome P-450-dependent monooxygenase systems of lung: relationships to pulmonary toxicity. In Microsomes and Drug Oxidations, edited by A. R. Boobis, J. Caldwell, F. Matteis & C. R. Elcombe, (Taylor and Francis), p. 248.

Robertson, I. C. G., Serabjit-Singh, C. J., Croft, J. E. & Philpot, R. M., 1983, Molecular Pharmacology, 24, 156.

Sanger, F., Nicklen, S. & Coulson, A. R., 1977, Proceedings of the National Academy of Science, USA, 74, 5463.

Schreier, P. H. & Cortese, H., 1979, Journal of Molecular Biology, 129, 169.

Tarr, G. W., Black, S. D., Fujita, V. S. & Coon, M. J., 1983, Proceedings of the National Academy of Science, USA, 80, 6552.

Vanderslice, R. A., Domin, B. A., Carver, G. T. & Philpot, R. M., 1987, Molecular Pharmacology, 31, 320.

Young, R. A. & Davis, R. W., 1983, Proceedings of the National Academy of Science, USA, 80, 1194.

Immunohistochemical Localization and Distribution of NADPH-Cytochrome P-450 Reductase in Human Tissues

MICHAEL E. McMANUS, PAULINE de la HALL, IEVA STUPANS, WENDY BURGESS, JIM BRENNAN and DON J. BIRKETT

Departments of Clinical Pharmacology and Histopathology, Flinders Medical Centre, Bedford Park, Adelaide, South Australia 5042, Australia

INTRODUCTION

The first event in the initiation stage of chemical carcinogenesis involves the metabolic activation of a chemical to a highly reactive intermediate that is capable of modifying tissue macromolecules (Gelboin 1980; Thorgeirsson et al, 1983). The major enzyme system shown to be involved in this process is the cytochrome P450 monooxygenase. This enzyme system is composed of a flavoprotein, NADPH-cytochrome P450 reductase (EC 1.6.2.4), and a family of cytochrome P450 isozymes. Indications are that specific isozymes of cytochrome P450 are involved in the metabolic activation of chemical carcinogens. For example, when the capacity of six highly purified forms of rabbit cytochrome P450 to metabolically activate (N-hydroxylate) the model carcinogen 2-acetylaminofluorene (AAF) was assessed, only Form 4, the major polycyclic hydrocarbon inducible form in the adult was capable of mediating this reaction (Johnson et al, 1980; McManus et al, 1984). In addition, the regulation of cytochrome P450 isozymes in hepatic and extrahepatic tissue are under different control mechanisms (Schwab & Johnson, 1987; Kimura et al, 1986). These observations have provided the basis for the notion that the relative composition of cytochrome P450 isozymes may be a major determinant in individual susceptibility to cancer as well as target organ selection on exposure to chemical carcinogens.

While unequivocal evidence for the multiplicity of the cytochrome P450 system in man has been provided by protein purification and DNA sequencing studies (Guengerich et al, 1986; Nebert & Gonzalez, 1987), only one form of NADPH-cytochrome P450 reductase has been identified (Guengerich et al, 1981; McManus et al, 1987).

NADPH-cytochrome P450 reductase catalyses electron transfer from NADPH to cytochrome P450 during catalysis and, together with this hemoprotein is absolutely essential for reconstitution of xenobiotic metabolising activity (Lu and West, 1978). Except mitochondrial forms of cytochrome P450 that are involved in endogenous substrate metabolism and utilize adrenodoxin reductase and adrenodoxin to provide reducing equivalents from NADPH (Hanukoglu and Hanukogen, 1986), the requirement for NADPH-cytochrome P450 reductase is common to all forms of cytochrome P450. These data indicate that an antibody to NADPH-cytochrome P450 reductase can be used as a immunohistochemical probe for studying the tissue distribution of the cytochrome P450 system involved in exogenous substrate metabolism. This approach has been used in the present study to map the distribution and cellular localization of the reductase and hence, cytochrome P450 system in the human liver, lung, kidney, gastrointestinal tract, adrenal gland and pancreas.

Materials and Methods

The purification of human liver NADPH-cytochrome P450 reductase and the raising of an antibody against this protein were carried out as previously described by McManus et al (1987a). Immunoperoxidase staining for NADPH-cytochrome P450 reductase on human tissue sections was performed as outlined by McManus et al (1987b). The human organs selected for this study were liver, lung, kidney, stomach, colon, rectum, pancreas and adrenal gland. Only tissues that were found to be 'normal' on histological examination were used in this study. The use of such tissue in these studies had local Research Ethics Committee approval and, where appropriate, Coroner's permission.

Results and Discussion

A polyclonal antibody raised against the human liver NADPH-cytochrome P450 reductase has been used to study the tissue distribution and cellular localisation of the cytochrome P450 system in man. On Western blots this antibody has been shown to recognise a protein in human liver microsomes of identical molecular weight to the purified NADPH-cytochrome P450 reductase (McManus et al, 1987b). Employing anti-reductase IgG, reductase staining was detected in all zones of the human liver acinus, with maximal staining in zone 3 hepatocytes adjacent to the terminal hepatic venules. This pattern of distribution is consistent with zone 3 hepatocytes being a primary target of drug induced toxicity. It is also consonant with that reported in control rat liver (Baron et al, 1978; Redick et al, 1980; Taira et al, 1980a & b; Dees et al, 1980).

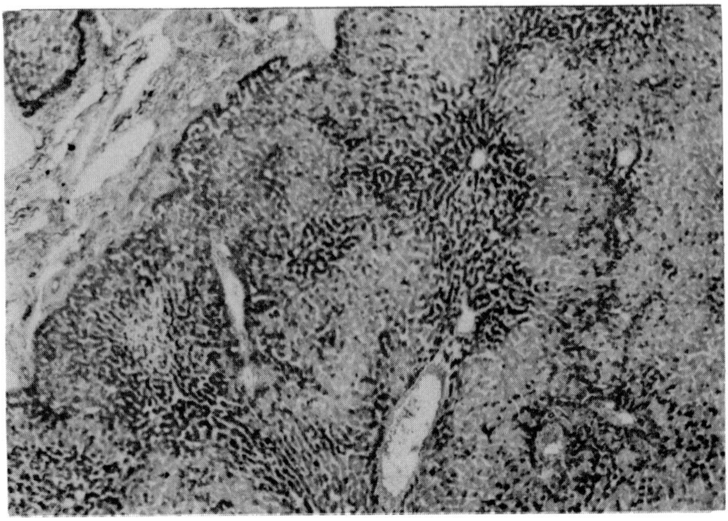

Figure 1. Normal liver, all hepatocytes stain positively for NADPH-cytochrome P450 reductase, with maximal staining around terminal hepatic venules and large veins. Immunoperoxidase x 60.

The lung is a major portal of entry of xenobiotics and has been shown to metabolise a wide range of drugs and chemical carcinogens (McManus et al, 1980; Minchin & Boyd, 1983). In the human lung strong immunoreactive NADPH-cytochrome P450 reductase staining was detected in the bronchial epithelial cells (Figure 2). Clara cells and alveolar lining cells also stained positively for the reductase (data not shown). The distribution of the reductase in human lung is similar to that reported by Dees et al (1980) for rat and minipig lung, where positive staining was reported in the bronchi and bronchiols and in an additional cell type in the lung parenchyma. It is also similar to the collective distribution of cytochromes P450 Form 2 ($P450_1$) and Form 5 ($P450_{11}$) in rabbit lung (Serabjit-Singh et al, 1979; Devereux et al, 1981). The strong immunoreactive NADPH-cytochrome P450 reductase staining in the bronchial epithelial cells is consistent with the majority of lung tumours arising from these cells.

In contrast to the lung, the pattern of reductase staining in the human kidney was considerably different to that reported in animal species. The cortical structures of the human kidney such as the glomeruli, proximal and distal convoluted tubules were negative (Figure 3a) whereas the medullary structures, namely the loop of Henle, collecting ducts and transitional epithelium of the renal pelvis were positive (Figure 3b). By contrast, in laboratory animals the proximal tubules have been shown to stain positively for NADPH-cytochrome P450 reductase (Dees et al, 1980) and specific forms of cytochrome P450 (Dees et al, 1982), whereas other structures are essentially negative. The pattern of reductase staining is consistent with the renal medulla and pelvis being the major sites of chemical induced injury in man.

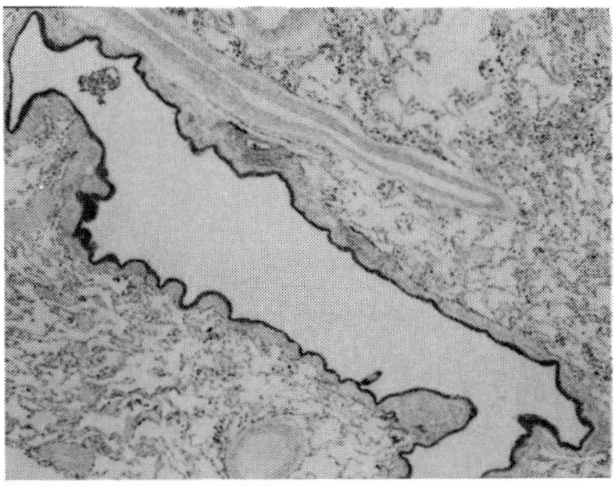

Figure 2. Lung showing strong positive staining for reductase in bronchiolar lining cells, the alveolar lining cells are weakly positive. Immunoperoxidase x 20.

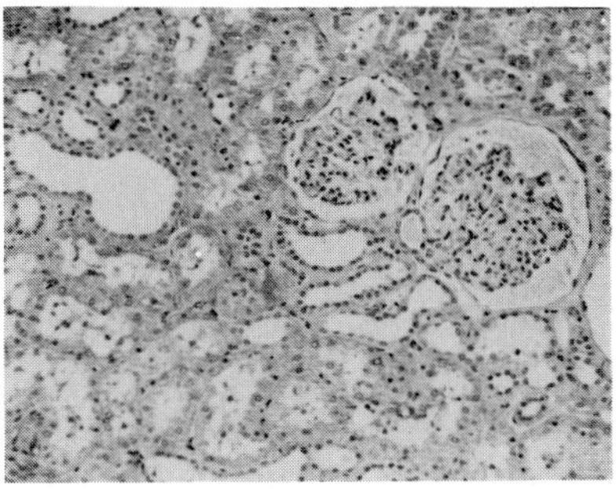

Figure 3a) Renal cortex, the glomeruli and proximal tubules do not show staining for reductase. Immunoperoxidase x 90.

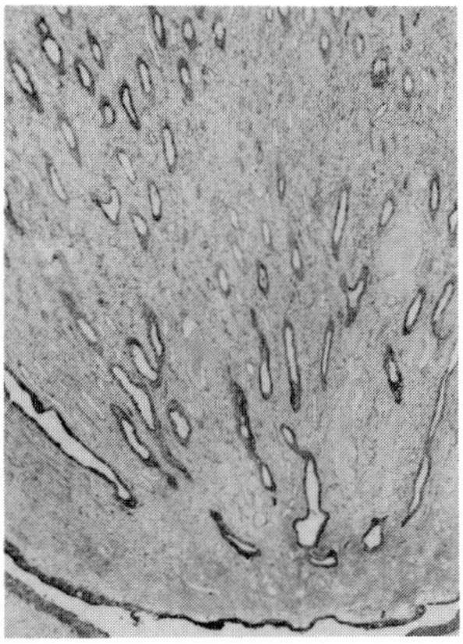

Figure 3b) Renal medulla, the large collecting ducts and thin limbs of the loops of Henle stain positively for reductase. Immunoperoxidase x 20.

In different sections of the human gastrointestinal tract considerable variation in the intensity of the reductase staining was observed. The most intense staining was found in the entrocytes of the small intestine, with maximal staining at the tips of the villi (Figure 4a). In the rat Foster et al (1986) have also shown these cells to stain intensely for cytochrome P450c, but only following beta-naphthoflavone treatment. The surface epithelium of the human stomach (Figure 4b) also stained positively for the reductase whereas colonic epithelial cells were usually weakly positive, while the rectum was negative (data not shown). The variation in the intensity of reductase staining in different segments of the gastrointestinal tract may reflect the differential induction of the cytochrome P450 system in this tissue. However, regardless of the effect of inducing agents the pattern of reductase staining does not correlate with the colon and rectum having the highest frequency of carcinomas in the gastrointestinal tract.

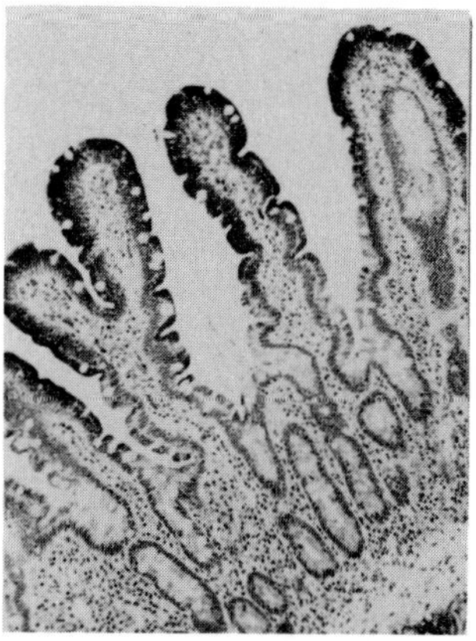

Figure 4a) Small intestine, all enterocytes stain positively for reductase, with maximal staining at the tips of the villi. Immunoperoxidase x 70.

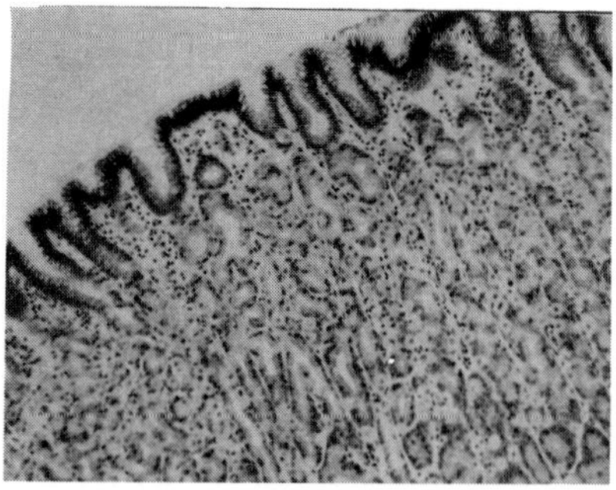

Figure 4b) Stomach, the surface columnar epithelial cells show positive staining for reductase, while the zymogenic glands and parietal cells are negative. Immunoperoxidase x 70.

NADPH-cytochrome P450 reductase staining was positive in all components of the pancreatic ductal system, with centroacinar cells staining strongly positive for the reductase (Figure 5). Peripheral acinar cells were negative and the islets of Langerhans were weakly positive. This distribution of reductase staining is consistent with the majority of pancreatic tumours arising from ductal cells.

In the adrenal gland positive staining for the reductase was found throughout the cortex whereas the medulla was negative (Figure 6). This pattern of reductase staining in the human adrenals is consistent with the localization of adrenodoxin and adrenodoxin reductase (Mitani, et al, 1979) cytochrome $P450_{11\beta}$ and cytochrome $P450_{scc}$ (Mitani et al, 1982; Ginze et al, 1987) in the same parenchymal cells in the bovine adrenals. Human foetal adrenal glands have also been shown to possess NADPH-cytochrome P450 reductase activity and to be capable of metabolising numerous xenobiotics (Vaino and Hietanen, 1980; Pelkonen, 1977).

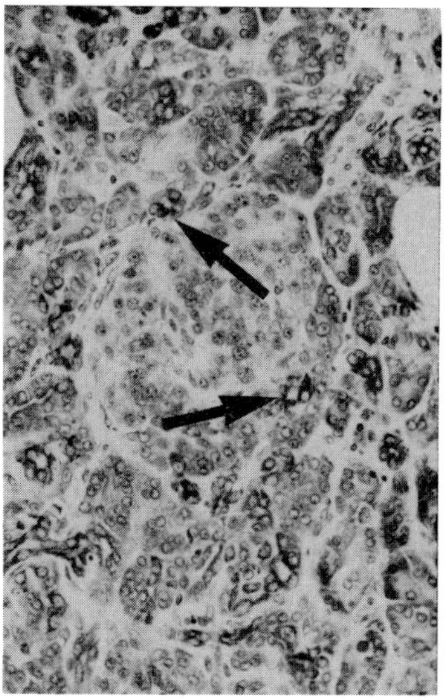

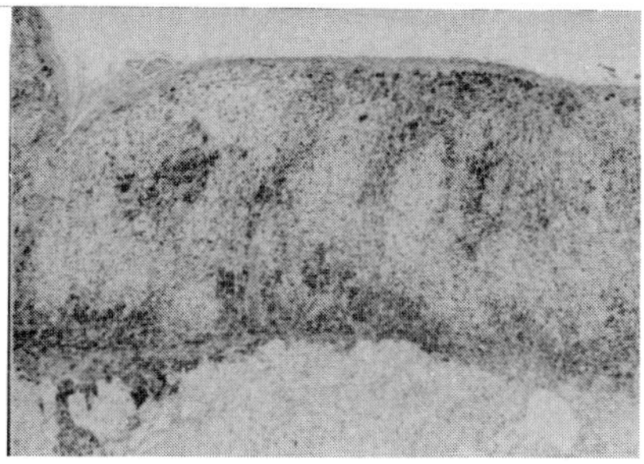

Figure 6. Adrenal, the cortical cells stain positively for reductase with maximal staining in the zona glomerulosa and zona reticularis, the medulla is negative - Immunoperoxidase x 26.

Figure 5. Pancreas showing strong positive staining for reductase in the centroacinar cells, weak staining in the islet cells (arrows) and no staining in the peripheral acinar cells. Immunoperoxidase x 180.

In summary, except for the kidney, the distribution and localization of NADPH-cytochrome P450 and hence the cytochrome P450 system, is similar in human liver, lung and gastrointestinal tracts to that reported for laboratory animals. No comparable animal data exists to compare the distribution of NADPH-cytochrome P450 reductase in the adrenal gland and pancreas. The differential cellular localization of NADPH-cytochrome P450 reductase in the lung, kidney and pancreas is consistent with the major sites of tumour formation and chemical injury. However, in the gastrointestinal tract, the colorectal region (the major site of tumour formation) was either weakly positive or negative. This may infer that other enzymes involved in carcinogen metabolism, or the appropriate cytochrome P450 inducing agent may be important in tumour formation at this site.

REFERENCES

Baron, J., Redick, J.A., Greenspan, P. & Taira, Y., 1978, Life Sciences, 22, 1097.

Dees, J.H., Coe, L.D., Yasukochi, Y. and Masters, B.S., 1980, Science (Washington DC) 208, 1473.

Dees, J.H., Masters, B.S.S., Muller-Eberhard, U. & Johnson, E.F., 1982, Cancer Research, 42, 1423.

Devereux, T.R., Serabjit-Singh, C.J., Slaughter, S.R., Wolf, C.R., Philpot, R.M. & Fouts, J.R., 1981, Experimental Lung Research, 2, 221.

Foster, J.R., Elcombe, C.R., Boobis, A.R., Davies, D.S., Sesardic, D., McQuade, J., Robson, R.T., Hayward, C. & Lock, E.A., 1986, Biochemical Phamacology, 35, 4543.

Gelboin, H.V., 1980, Physiological Reviews, 60, 1107.

Geuze, H.J., Slot, J.W., Yanagebashi, K., McCracken, J.A., Schwartz, A.L. & Hall, P.F., 1987, Histochemistry, 86, 551.

Guengerich, F.P., Distlerath, L.M., Reilly, P.E.B., Wolff, T., Shimada, T., Umbenhauer, D.R. & Martin, M.V., 1986, Xenobiotica, 16, 367.

Guengerich, F.P., Wang, P. and Mason, P.S., 1981, Biochemistry, 20, 2379.

Hanukoglu, I. & Hanukoglu, Z., 1986, European Journal of Biochemistry, 157, 27.

Johnson, E.F., Levitt, D.S., Muller-Eberhard, U. & Thorgeirsson, S.S., 1980, Cancer Research, 40, 4456.

Kimura, S., Gonzalez, F.J. & Nebert, D.W., 1986, Molecular Cell Biology, 6, 1471.

Lu, A.Y.H. & West, S.B., Pharmacology and Therapeutics, Part A. Chemotherapy, Toxicology, 1978, Metabolism and Inhibitors, 2, 337.
McManus, M.E., Boobis, A.R., Pacifici, G.M., Frempong, R.Y., Brodie, M.J., Kahn, G.C., Whyte, C. & Davies, D.S., 1980, Life Sciences, 26, 481.
McManus, M.E., Hall, P. de la M., Stupans, I., Brennan, J., Burgess, W., Robson, R. & Birkett, D.J. 1987b, Molecular Pharmacology, 32, 189.
McManus, M.E., Minchin, R.F., Sanderson, N., Schwartz, D., Johnson, E.F. & Thorgersson, S.S., 1984, Carcinogenesis, 5, 1717.
McManus, M.E., Stupans, I., Burgess, W., Koenig, J.A., Hall, P. de la M. & Birkett, D.J., 1987a, Drug Metabolism and Disposition, 15, 256.
Minchin, R.F. & Boyd, M.R., 1983, Annual Review of Pharmacology and Toxicology, 23, 217.
Mitani, F., Ishimura, Y., Izumi, S. & Watanabe, K., 1979, Acta Endocrinology, 90, 317.
Mitani, F., Shimizu, T., Ueno, R., Ishimura, Y., Izumi, S., Komatsu, N. & Watanable, K., 1982, Journal of Histochemistry and Cytochemistry, 30, 1066.
Nebert, D.W. & Gonzalez, F.J., 1987, Annual Reviews of Biochemistry, 56, 945.
Pelkonen, O., 1977. Transplacental transfer of foreign compounds and their metabolism by the foetus. In Progress in Drug Metabolism, Vol 2, edited by J.W. Bridges & L.F. Chasseaud (John Wiley & Sons, London), p. 119.
Redick, J.A., Kawabata, T.T., Guengerich, F.P., Krieter, P.A., Shires, T.K. & Baron, J., 1980, Life Sciences, 27, 2465.
Schwab, G.E. & Johnson, E.F., Enzymology of rabbit cytochromes P450. In Mammalian cytochromes P450, edited by F.P. Guengerich (CRC Press Boca Raton, Florida) (in press).
Serabjit-Singh, C.J., Wolf, C.R., Philpot, R.M. & Plopper, C.G., 1979, Science (Washington DC), 207, 1469.
Taira, Y., Greenspan, P., Kapke, G.F., Redick, J.A. & Baron, J., 1980a, Molecular Pharmacology 18, 304.
Taira, Y., Redick, J.A. & Baron, J., 1980b, Molecular Pharmacology, 17, 374.
Thorgeirsson, S.S., Glowinski, I.B. & McManus, M.E., 1983, Reviews of Biochemical Toxicology, 5, 349.
Vaino, H. & Hietanen, E., 1980. Role of extrahepatic metabolism, Part A, edited by P. Jenner & B. Testa (Marcel Dekker Incorporated, New York), p. 251.

THE POLYCYCLIC AROMATIC HYDROCARBON-INDUCIBLE AND PHENOBARBITONE-INDUCIBLE CYTOCHROME P-450 GENE FAMILIES IN ANIMALS AND IN MAN

The Ah Receptor: First Step in the Mechanism of Response to 3-Methylcholanthrene-type Chemicals in Rodents and in Man

ALAN B. OKEY, MICHAEL S. DENISON, PATRICIA A. HARPER and REBECCA D. PROKIPCAK

Division of Clinical Pharmacology and Toxicology, The Hospital for Sick Children, Toronto, Ontario M5G 1X8, Canada

INTRODUCTION

Techniques of biochemistry and molecular biology have revealed, in great detail, the transcriptional and post-transcriptional events involved in microsomal enzyme induction. In contrast, little is known about the early steps in enzyme induction --- especially how target cells initially recognize the presence of specific xenobiotic chemicals that induce microsomal enzymes.

This brief review will focus on a specific recognition site ("receptor") for one category of inducer, i.e. "3-methylcholanthrene(MC)-type" compounds. We will concentrate on data from our own laboratory. Citations will be made to recent review articles and to selected original research papers that will provide detailed experimental information and an entry to the expanding literature on *Ah* receptor.

Figure 1 presents a simplified model of enzyme regulation by the *Ah* receptor. Evidence supporting the model is presented in the following sections.

FIGURE 1

General Scheme Illustrating the Role of the Ah *receptor in Mediating Cellular Responses to "MC-type" Xenobiotic Chemicals.*

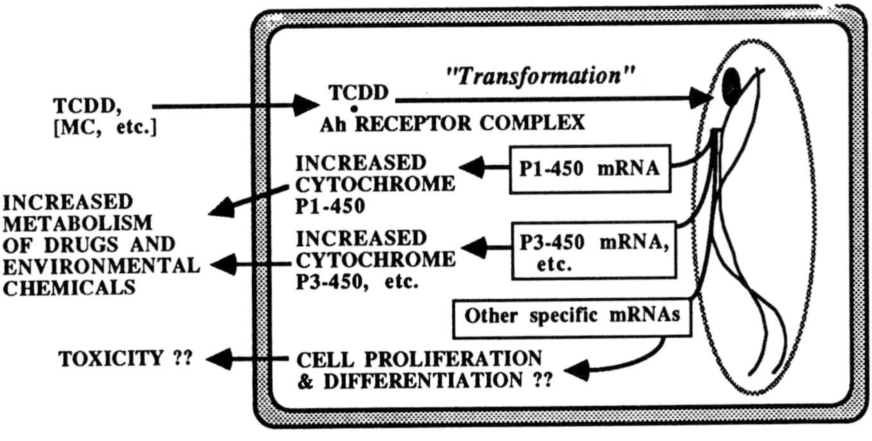

EVIDENCE THAT THE *Ah* RECEPTOR MEDIATES INDUCTION OF CYTOCHROME P_1-450.

Quantitative Structure-Activity Relationships (QSAR).

Cytochrome P_1-450 (aryl hydrocarbon hydroxylase, AHH) is induced by MC and by several other carcinogenic polycyclic aromatic hydrocarbons (PAHs) such as benzo[a]pyrene (BP), benz[a]anthracene (BA), and dibenz[a,h]anthracene (DB[a,h]A). The most potent P_1-450 inducers are specific halogenated aromatic hydrocarbons (HAHs) such as 2,3,7,8-tetrachlorodibenzo-*p*-dioxin (TCDD) and 2,3,7,8-tetrachlorodibenzofuran (TCDF).

Virtually all P_1-450 inducers are highly hydrophobic compounds that bind extensively to many cellular proteins and other macromolecules. Evidence that a particular binding protein, the *Ah* (aromatic hydrocarbon) receptor serves to mediate P_1-450 induction first was provided by Poland and coworkers (*Poland* et al. 1976; *Poland & Knutson* 1982); they demonstrated that the affinity with which several halogenated congeners of dibenzo-*p*-dioxins and dibenzofurans bind to *Ah* receptor is well correlated with the potencies of these congeners as P_1-450 inducers. Subsequent extensive QSAR studies by *Safe and coworkers* indicate that within each family of polychlorinated compounds (dibenzo-*p*-dioxins, dibenzofurans, biphenyls) the congeners that are the most toxic and the most potent as P_1-450 inducers are those compounds that bind with the highest affinity to the *Ah* receptor (*Safe et al.* 1985; *Safe* 1986).

The most potent halogenated inducers are highly planar molecules (dibenzo-*p*-dioxins & dibenzofurans) or can assume a coplanar conformation (polychlorinated biphenyls that lack ortho substitutent groups). Binding affinity and potency for P_1-450 induction are highest in those HAHs which have halogens in all four lateral positions (*Poland & Knutson* 1982; *Safe* 1986); the optimal two-dimensional size is a rectangle 3 Å x 10 Å (*Poland & Knutson* 1982) or approximately 7 Å by 14 Å when van der Waals radii are taken into account for halogenated and nonhalogenated ligands (*Poellinger et al.* 1985).

Although quantitative structure-activity relationships hold well within each class of halogenated *Ah* receptor ligand, there are certain discrepancies in cross-class comparisons. For example, the affinity with which the *Ah* receptor binds the nonhalogenated ligand, MC, *in vitro* is nearly equal to the affinity with which the receptor binds TCDD (*Okey et al.* 1984). Yet, *in vivo* TCDD is about 30,000 times more potent than MC as an inducer of P_1-450 (*Poland & Knutson* 1982). Thus the affinity with which a compound binds to the *Ah* receptor may not always predict the efficacy of that compound as an enzyme inducer; the reasons for such discrepancies are not fully known.

Mouse Genetics In Vivo

Strains of inbred mice such as C57BL/6 are termed "responsive" since they exhibit high P_1-450 induction in liver when adults are treated with nonhalogenated PAHs such as MC, BP, BA, etc. "Nonresponsive" strains (prototype: DBA/2N) show no significant increase in hepatic P_1-450 in adults even when treated with high doses of nonhalogenated PAHs. "Nonresponsive" mice do exhibit P_1-450 induction when treated with potent HAHs such as 2,3,7,8-TCDD (*Poland & Knutson* 1982), but the dose required for half-maximal induction by TCDD is approximately 15-fold higher in DBA/2N mice than in C57BL/6N mice *(Eisen et al.* 1983).

In genetic crosses and backcrosses in mice, "responsiveness" for P_1-450 induction by PAHs segregates with the presence of a specific high-affinity binding site for [^{3}H]TCDD, i.e. the *Ah* receptor (*Poland & Knutson* 1982; *Eisen et al.* 1983). Thus the *Ah* locus governing P_1-450 inducibility appears to be a regulatory gene coding for some essential component of the *Ah* receptor.

In early experiments the *Ah* receptor was completely undetectable in cytosol from "nonresponsive" mice, even using the potent agonist [^{3}H]TCDD as the receptor ligand (*Poland et al.* 1976; *Okey et al.* 1979). Since "nonresponsive" mice do respond (albeit with reduced sensitivity) to TCDD and since the *Ah* receptor is postulated to regulate P_1-450 induction it would be expected that [^{3}H]TCDD should bind *Ah* receptor in "nonresponsive" strains. Hypotheses as to the nature of the genetically-based receptor "defect" in "nonresponsive" strains have existed for several years (*Poland & Knutson* 1982; *Eisen et al.* 1983) and are illustrated in Figure 2. Recent work in our laboratory (*Okey et al.*, in preparation) reveals that hepatic cytosols from adult DBA/2 mice and from other "nonresponsive" strains *do* contain measurable amounts of *Ah* receptor. However, the *Ah* receptor from "nonresponsive "mice is present in lower quantity than in "responsive" strains and has an apparent affinity for [^{3}H]TCDD that is about one log unit weaker than the affinity in "responsive" mice. The ability to directly measure a "low-affinity" receptor in "nonresponsive" mice further strengthens evidence that the *Ah* receptor regulates induction of P_1-450 and related "MC-induced" enzymes.

FIGURE 2

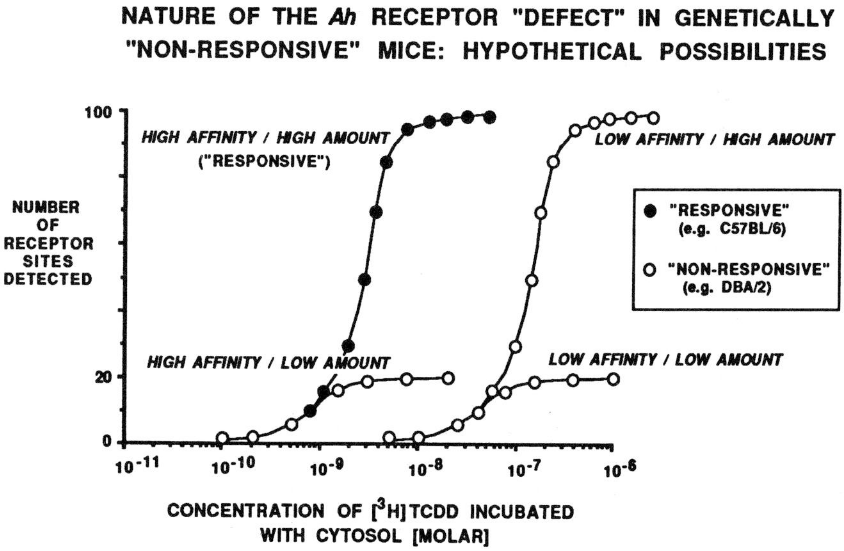

Somatic Cell Genetics and Molecular Genetics

Hankinson and coworkers (1985) and *Whitlock and coworkers* (1986) selected and characterized mutant cell lines derived from Hepa-1 mouse hepatoma cells. Genetic complementation studies in the mutant lines indicate that several genes are required for P_1-450 induction. P_1-450 inducibility is greatly reduced or absent in mutant cell lines that have either decreased cytosolic *Ah* receptor or defective nuclear translocation of the TCDD•*Ah* receptor complex.

The level of P_1-450 mRNA in livers of mice treated *in vivo* with TCDD is highly correlated with the concentration of TCDD•*Ah* receptor complexes in the nucleus (*Eisen et al.* 1983); the increased level of P_1-450 mRNA in cells exposed to "MC-type inducers" is a consequence of increased rates of transcription (*Whitlock* 1986; *Nebert & Gonzalez* 1987). A number of laboratories have shown that specific regions flanking the 5' end of the P_1-450 gene appear to be sites at which TCDD•*Ah* receptor complexes bind to DNA and thus stimulate increased transcription of P_1-450 mRNA. The molecular biology of P_1-450 regulation has been extensively discussed in several excellent recent reviews and research papers (*Adesnik & Atchison* 1985; *Whitlock* 1986; *Sogawa et al.* 1986; *Nebert & Gonzalez* 1987).

Taken together, quantitative structure-activity relationships, mouse genetics, and molecular genetics provide strong evidence that the *Ah* receptor is the prime mediator of P_1-450 induction. Other enzymes that also may be regulated by the *Ah* receptor are described in the review by *Nebert & Gonzalez* (1987). The postulated role of the *Ah* receptor in regulating events related to cell proliferation, cell differentiation, and toxicity (Figure 1) is discussed in reviews by *Poland & Knutson* (1982) and *Greenlee et al.* (1987).

MOLECULAR CHARACTERISTICS OF CYTOSOLIC AND NUCLEAR FORMS OF *Ah* RECEPTOR

Physicochemical properties of the *Ah* receptor and its general mechanism of gene regulation are very similar to those of receptors for steroid hormones (*Eisen et al.* 1983; *Poellinger et al.* 1985; *Denison et al.* 1986a). It is likely that the *Ah* receptor is evolutionarily related to steroid receptors; however, the *Ah* receptor is a distinct protein that does not interact specifically with any known steroid.

Prior to ligand binding the *Ah* receptor appears to be a cytosolic protein (*Denison et al.* 1986b; *Nebert & Gonzalez* 1987), but it also is possible that the unoccupied receptor is loosely associated with the cell nucleus (*Whitlock* 1986). In conditions of low ionic strength the cytosolic *Ah* receptors from rat liver, mouse liver, and mouse Hepa-1 cells have relative molecular masses (M_r) ≈ 260,000 - 300,000 (*Denison et al.* 1986a; *Cuthill et al.* 1987; *Prokipcak et al.*, in preparation). Under conditions of high ionic strength (0.4 M KCl), the cytosolic *Ah* receptor from rat liver is readily converted into a ligand-binding subunit of M_r ≈ 100,000 - 120,000 (*Denison et al.* 1986a; *Cuthill et al.* 1987). In our laboratory (*Denison et al.* 1986a; *Prokipcak et al.*, in preparation) cytosolic *Ah* receptors from both mouse liver and mouse Hepa-1 cells have been highly resistant to salt-induced dissociation into the M_r ≈ 100,000 subunit, whereas *Cuthill et al.* (1987)

observed resistance to salt-induced dissociation in Hepa-1 cytosol, but not in mouse liver cytosol.

An important step in P_1-450 regulation by the *Ah* receptor is the "transformation " of *Ah* receptor from the high-M_r cytosolic form into the form which binds specifically to DNA and stimulates transcription of P_1-450 mRNA. "Transformation" appears to involve changes both in the M_r of the receptor and in the ionic properties of the receptor. "Transformed" *Ah* receptor can be most readily obtained by extraction from the nuclei of cells that have been exposed in culture to [^{3}H]TCDD at physiologic temperatures. The nuclear *Ah* receptor from Hepa-1 cells has an apparent $M_r \approx$ 175,000, and thus corresponds neither to "native" $M_r \approx$ 250,000 - 300,000 cytosolic form nor to an apparently monomeric $M_r \approx$ 105,000 ligand-binding subunit which can be generated by exposing Hepa-1 cytosol to 0.5 M KCl for extended periods of time (*Prokipcak et al.*, in preparation). The nature of the subunits (if any) that make up the M_r 175,000 nuclear form of *Ah* receptor in Hepa-1 cells is unknown.

Although it initially had been difficult to "transform" cytosolic *Ah* receptor *in vitro*, *Hannah et al.* (1986) and *Gasiewicz & Bauman* (1987) recently have shown that heat-activation of cytosolic [^{3}H]TCDD•*Ah* receptor complexes under extracellular conditions *in vitro* converts a portion of the complexes to forms that bind DNA immobilized on chromatography columns. It remains to be determined whether these *in vitro* products correspond to the nuclear form of receptor obtained from whole cells incubated with [^{3}H]TCDD in culture.

Purification of the *Ah* receptor has proceeded slowly. *Poland et al.* (1986), using photoaffinity labeling techniques, have identified a peptide which has an apparent $M_r \approx$ 95,000 on denaturing gel electrophoresis. The labeled peptide, from hepatic cytosol of C57BL/6J mice, exhibits the ligand-specificity expected of the *Ah* receptor and represents, to date, the most significant achievement in purification of the *Ah* receptor. Given the strong resemblance of the *Ah* receptor to the receptors for steroid hormones, it is not unlikely that the gene(s) coding for the *Ah* receptor (or its subunits) will be identified and sequenced before purification of nondenatured *Ah* receptor protein has been accomplished.

HUMAN *Ah* RECEPTOR

Many human cell lines in culture as well as some human tissues *in vivo* are known to exhibit P_1-450 induction when exposed to "MC-type" compounds (*Okey et al.* 1986). For example, AHH (P_1-450) is highly induced in human placentas from smokers or from women exposed to polychlorinated biphenyls and dibenzofurans (*Wong et al.* 1986). In addition, human cell lines, such as the MCF-7 breast carcinoma, show significant P_1-450 induction when treated with TCDD or BA in culture (*Jaiswal et al.* 1985). Although in each of these cases the human tissues and cells clearly responded to "MC-type" compounds, the *Ah* receptor was not detectable (*Wong et al.*, 1986; *Jaiswal et al.* 1985), leading the authors to suggest that P_1-450 induction in human tissues might be mediated via mechanisms other than the *Ah* receptor.

Initially our laboratory also found that *Ah* receptor either was absent from human tissues or was present only at low concentrations (*Roberts et al.* 1985).

However, with improved assay procedures we now detect *Ah* receptor in many human tissues and cells and find that human placenta has *Ah* receptor concentrations equivalent to those in mouse or rat liver *(Manchester et al.*, in the press).

FIGURE 3

Saturation Analysis of [^{3}H]TCDD Binding to Cytosolic Ah Receptor & Log-Dose Response Curves for AHH Induction by TCDD: Comparison of Human A431 Cells with Mouse Hepa-1 Cells.

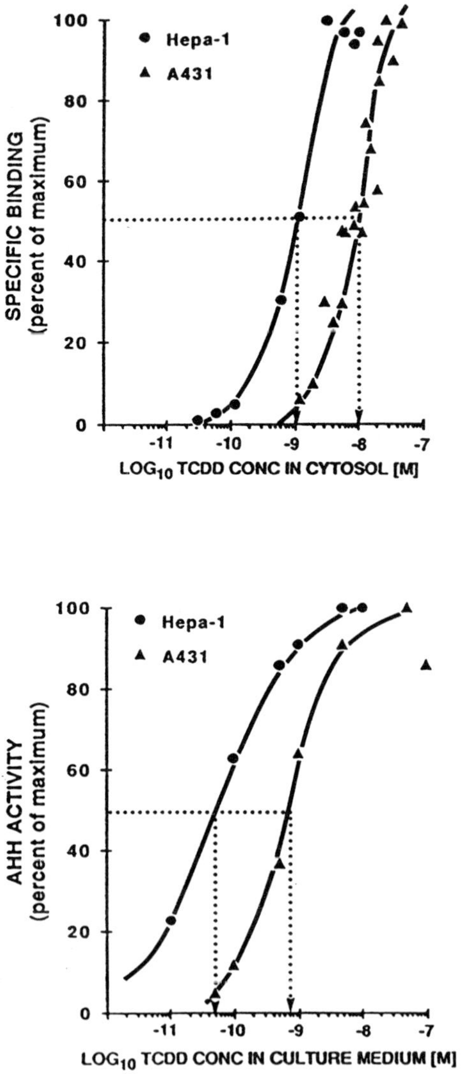

<u>Receptor Saturation (upper panel)</u>: cytosols were incubated with concentrations of [^{3}H]TCDD from 5 x 10^{-11} M to 5 x 10^{-8} M, then specific binding was determined by density gradient centrifugation. The ordinate scale is normalized to show fractional binding as a percent of the maximum binding observed in the cytosol for each cell type. The actual maximum specific binding (fmol/mg cytosol protein) was 740 for Hepa-1 and 200 for A431. Dotted lines indicate the concentration of [^{3}H]TCDD required to saturate half the receptor sites in cytosol from each cell line. <u>AHH Induction (lower panel)</u>: Hepa-1 and A431 cells were incubated for 24 hr with TCDD in the culture medium at the concentrations indicated. AHH activity, determined on whole cells, was normalized to show the percent of maximum response. The actual maximum was 6.3 pmol/min/10^6 cells for Hepa-1 and 1.3 pmol/min/10^6 cells for A431. (From: *Harper et al.*, in preparation).

Although several human cell lines and tissues have substantial concentrations of *Ah* receptor, in each case which we have studied the affinity with which [^{3}H]TCDD binds to human *Ah* receptor has been lower than with rodent *Ah* receptors. In human placentas the apparent equilibrium dissociation constants (K_d) for [^{3}H]TCDD binding to *Ah* receptor were 5 to 8 nM, compared with 1-2 nM in rat or mouse livers (*Manchester et al.*, in the press).

As might be expected, the lower affinity with which human *Ah* receptor binds TCDD alters the sensitivity of human cells to P_1-450 induction by TCDD. As shown in Figure 3 (top panel), the saturation curve for *Ah* receptor occupancy by [^{3}H]TCDD in human A431 cells is shifted about one log unit to the right of the curve for mouse Hepa-1 cells. There is a proportional shift in the dose-response curve for AHH induction by TCDD (Figure 3, bottom panel). In essence, the affinity of [^{3}H]TCDD binding to *Ah* receptor in human A431 cells is about ten-fold weaker than in mouse Hepa-1 cells and, probably as a consequence of the lower affinity, the concentration of TCDD required for AHH induction in human A431 cells is about ten-fold higher than in mouse Hepa-1 cells.

In addition to human placenta, we also have detected *Ah* receptor in samples of human liver, lung, thymus, fetal lung, and fetal kidney (*Roberts et al.* 1985). In addition to the A431 squamous cell carcinoma line characterized in our laboratory (*Harper et al.* in preparation), several other human squamous cell carcinoma lines have been shown by *Greenlee et al.* (1987) to contain *Ah* receptor.

Although human *Ah* receptor appears generally to have significantly lower affinity for [^{3}H]TCDD than do receptors from rats or mice, the mechanism of P_1-450 induction seems to be fundamentally the same in humans as in rodents (*Manchester et al.*, in the press; *Kawajiri et al.* 1986; *Harper et al.*, in preparation).

THE KNOWN AND THE UNKNOWN

Research contributions from several laboratories have provided considerable progress in understanding the structure and function of the *Ah* receptor during the decade since the receptor originally was discovered by *Poland et al.* (1976). The *Ah* receptor, like P_1-450 inducibility, has a broad phylogenetic distribution and appears to be present in most mammalian tissues, including human tissues. However, there appears to be some structural heterogeneity among receptors from different animal species; genetically-based structural differences and differences in receptor affinity and ligand specificity may allow considerable latitude in the degree and nature of receptor-mediated responses in different organisms exposed to the same environmental agents.

The major unknown concerning the *Ah* receptor is whether there exists any endogenous (physiologic) ligand for this receptor. Other than the induction of specific drug-metabolizing enzymes, the biochemical responses that are proven to be receptor-mediated also remain poorly defined (*Poland & Knutson* 1982; *Greenlee et al.* 1987).

Despite these significant gaps in understanding, to date the *Ah* receptor constitutes the only well-established recognition site known to mediate induction of specific forms of P-450 and associated enzymes. The challenge remains to identify the means by which cells recognize and respond to other specific categories of microsomal enzyme inducers.

REFERENCES

Adesnik, M. & Atchison, M., 1985, *CRC Critical Reviews in Biochemistry* , **19** , 247.

Cuthill, S., Poellinger, L. & Gustafsson, J.-Å., 1987, *The Journal of Biological Chemistry* , **262**, 3477.

Denison, M.S., Vella, L.M. & Okey, A.B., 1986a, *The Journal of Biological Chemistry* , **261**, 3987.

Denison, M.S., Harper, P.A. & Okey, A.B., 1986b, *European Journal of Biochemistry* , **155**, 223.

Eisen, H.J., Hannah, R.R., Legraverend, C., Okey, A.B., & Nebert, D.W., 1983, The *Ah* receptor: controlling factor in the induction of drug-metabolizing enzymes by certain chemical carcinogens and other environmental pollutants. In: *Biochemical Actions of Hormones* , v. X, edited by G. Litwack, Academic Press, New York, p. 227.

Gasiewicz, T.A. & Bauman, P.A., 1987, *The Journal of Biological Chemistry* , **262**, 2116.

Greenlee, W.F., Osborne, R. & Dold, K.M., 1987, Altered regulation of epidermal cell proliferation and differentiation by 2,3,7,8-tetrachlorodibenzo-p-dioxin (TCDD). In: *Reviews in Biochemical Toxicology* , v. 8, edited by E. Hodgson, J.R. Bend & R.M. Philpot, Elsevier, New York, p. 1.

Hannah, R.R., Lund, J., Poellinger, L., Gillner, M. & Gustafsson, J.-Å., 1986, *European Journal of Biochemistry*, **156**, 237.

Hankinson, O., Andersen, R.D., Birren, B.W., Sander, F., Negishi, M. & Nebert, D.W., 1985, *The Journal of Biological Chemistry*, **260**, 1790.

Jaiswal, A.K., Nebert, D.W. & Eisen, H.J., 1985, *Biochemical Pharmacology* , **34** , 2721.

Kawajiri, K., Watanabe, J., Gotoh, O., Tagashira, Y., Sogawa, K. & Fujii-Kuriyama, Y., 1986, *European Journal of Biochemistry*, **159**, 219.

Manchester, D.K., Gordon, S.K., Golas, C.L., Roberts, E.A. & Okey, A.B., 1987, *Cancer Research* (in the press).

Nebert, D.W. & Gonzalez, F.J., 1987, *Annual Review of Biochemistry* , **56** , 945.

Okey, A.B., Bondy, G.P., Mason, M.E., Kahl, G.F., Eisen, H.J., Guenthner, T.M. & Nebert, D.W., 1979, *The Journal of Biological Chemistry*, **254**, 11636.

Okey, A.B., Dube, A.W. & Vella, L.M., 1984, *Cancer Research*, **44**, 1426.

Okey, A.B., Roberts, E.A., Harper, P.A. & Denison, M.S., 1986, *Clinical Biochemistry* , **19**, 132.

Poellinger, L., Lund, J., Gillner, M. & Gustafsson, J.-Å., 1985, The receptor for 2,3,7,8-tetrachlorodibenzo-p-dioxin: similarities and dissimilarities with steroid hormone receptors. In: *Molecular Mechanism of Steroid Hormone Action*, edited by V.K. Moudgil, Walter de Gruyter & co., Berlin, p. 755.

Poland, A., Glover, E. & Kende, A.S., 1976, *The Journal of Biological Chemistry* , **251**, 4936.

Poland, A. & Knutson, J.C., 1982, *Annual Review of Pharmacology and Toxicology*, **22**, 517.

Poland, A., Glover, E., Ebetino, F.H. & Kende, A.S., 1986, *The Journal of Biological Chemistry* , **261**, 6352.

Roberts, E.A., Shear, N.H., Okey, A.B. & Manchester, D.K., 1985, *Chemosphere*, **14**, 661.

Safe, S., 1986, *Annual Review of Pharmacology and Toxicology*, **26**, 371.

Safe, S., Bandiera, S., Sawyer, T., Robertson, L., Safe, L., Parkinson, A., Thomas, P.E., Ryan, D.E., Reik, L.M., Levin, W., Denomme, M.A. & Fujita, T., 1985, *Environmental Health Perspectives*, **60**, 47.

Sogawa, K., Fujisawa-Sehara, A., Yamane, M. & Fujii-Kuriyama, Y., 1986, *Proceedings of the National Academy of Sciences (U.S.A.)*, **83**, 8044.

Whitlock, J.P., Jr., 1986, *Annual Review of Pharmacology and Toxicology*, **26**, 333.

Wong, T.K., Sloop, T. & Lucier, G.W., 1986, *Toxicology and Applied Pharmacology* , **85**, 60.

Genetic Analysis of the Ah Receptor and Cytochrome P_1-450 Expression in the Mouse Hepatoma Cell Line, Hepa-1

OLIVER HANKINSON and ROY M. BANNISTER

Department of Pathology and Laboratory of Biomedical and Environmental Sciences, University of California, Los Angeles, CA 90024, USA

General Introduction

Certain polycyclic aromatic hydrocarbons (PAH's), such as benzo(a)pyrene (BP), and certain halogenated aromatic hydrocarbons (HAH's), such as 2,3,7,8-tetrachlorodibenzo-p-dioxin (TCDD), are important environmental pollutants. Many of these compounds are extremely toxic and carcinogenic. These same compounds also bind the soluble Ah receptor and are thereby capable of inducing cytochrome P_1-450. Cytochrome P_1-450-dependent aryl hydrocarbon hydroxylase (AHH) activity in turn plays a central role in metabolism of PAH's to their ultimate carcinogenic derivatives. Pathogenesis by HAH's also depends upon action of the Ah receptor. However in the case of these compounds, the HAH's themselves, rather than metabolites, are the pathogenic agents, and P_1-450 activity is not involved.

The 300 kd mouse Ah receptor appears to give rise by dissociation to a 100 kd ligand-binding polypeptide upon treatment with high salt (Denison et al, 1986). This suggests that the receptor is polymeric but does not shed light on whether it contains one, or more than one, type of subunit.

Besides their intimate involvement in chemical carcinogenesis, cytochrome P_1-450 and the Ah receptor are very interesting to study for at least two other reasons. i) They provides a good model for studying gene regulation in mammalian cells, since P_1-450 is highly inducible, and is expressed to widely different degrees in different tissues and at different times in development. ii) The Ah receptor appears to be involved in the regulation of cell division and cell differentiation in certain tissues (Reviewed by Poland, 1984; Nebert and Gonzalez, 1987).

We summarize here our studies on this system and provide new data concerning expression of the glucocorticoid receptor in mutant cells deficient in Ah receptor functioning.

Isolation of mutants deficient in induction of cytochrome P_1-450

Most of our research has utilized the cultured mouse hepatoma line, Hepa-1. Hepa-1 is highly inducible for AHH activity by PAH's and HAH's. (Mutations in the cytochrome P_1-450 structural gene abolish AHH activity in Hepa-1 cells and therefore all AHH activity in the line is ascribable to this form of cytochrome P-450 (Hankinson et al, 1985).) Moreover inducible AHH activity remained stable in Hepa-1 cells during many months in culture, and there was no detectable heterogeneity in inducible AHH activity among subclones of the line (Hankinson, 1979; 1981). This stability of AHH activity augured well for a genetic analysis of P_1-450 induction in the line.

Hepa-1 was found to be very sensitive to BP toxicity and a single-step selection procedure for isolating BP-resistant (BP^r) clones was designed (Hankinson, 1979). We have isolated over 100 such clones and all have proven to be non-inducible for AHH activity. We investigated whether the BP^r clones originate by a mutational process (i.e. by a change in DNA sequence), because we and others have obtained convincing evidence that loss in AHH activity can arise in some hepatoma cell lines by an epigenetic process(es) (Whitlock et al, 1976; Gudas and Hankinson, 1987). If a mutational event is responsible for loss of AHH activity, one can conclude that all the changes in phenotype probably result from a change in a single gene. Table 1 summarizes the evidence that we have obtained that our BP^r clones are bona fide mutants.

In order to determine whether AHH-deficiency was dominant or recessive, somatic cell hybrids were constructed between individual BP^r mutants and wild-type, Hepa-1 cells. 5 of the BP^r mutants proved to be dominant, while 65 were recessive. Hybridizations performed between the recessive mutants demonstrated that they belonged to four complementation groups, named A to D (Hankinson, 1983; Karenlampi et al, 1987). Whitlock and coworkers have utilized our BP selection procedure to isolate two complementation groups of AHH-deficient variants of Hepa-1 which resemble our B^- and C^- mutants (Miller et al, 1983).

Table 1. Evidence that the BP^r, AHH-deficient clones of Hepa-1 are mutational in origin.

1. They arise at the low spontaneous frequency of 2×10^{-7} events/cell generation (Hankinson, 1979)
2. Their phenotypes are stable (Hankinson, 1979, 1981)
3. Their frequency is increased by the application of known mutagens (Hankinson, 1981)
4. Their spontaneous reversion frequencies are low and increased by mutagens (not demonstrated for the D^- mutant) (Van Gurp and Hankinson, 1984)
5. They are not reverted by 5-azacytidine, which leads to demethylation of cytidine residues in DNA (not demonstrated for the D^- mutant) (Van Gurp and Hankinson, 1984)
6. Their defects are complemented by purified DNA (see below)
7. In the case of two A^- mutants, nucleotide substitution mutations have been shown to be responsible for loss of AHH activity (Kimura et al, 1987)

Biochemical characterization of the mutants

The results obtained with representative mutants are summarized in Table 2 (Legraverend et al, 1982; Hankinson et al, 1985; Karenlampi et al, 1987). AHH activities correlate with the levels of P_1-450 mRNA in all mutants, except some of the A^- class. The A^- mutants are heterogeneous. A^- subgroup II mutants lack P_1-450 mRNA, while subgroup IV mutants possess high levels of P_1-450 mRNA, although they totally lack inducible AHH activity. These observations provided the initial evidence that gene A is the P_1-450 structural gene. We also found that hybrids formed between individual dominant mutants and the wild-type lacked inducible P_1-450 mRNA as well as AHH activity (Hankinson et al, 1985). This suggests that the dominant mutants synthesize a repressor of P_1-450 transcription. One possibility is that the dominant mutants arose by the activation of a gene that normally regulates tissue and/or developmental expression of P_1-450.

There is some controversy as to whether the unoccupied Ah receptor, when not bound to ligand, is cytosolic or nuclear in location (Whitlock and Galeazzi, 1984; Denison et al, 1986). We enucleated uninduced Hepa-1 cells with cytochalasin B. With this procedure nuclear proteins do not appear to leak into the cytosol. We found Ah receptor in the cytosolic fraction, but not in the nuclei (Gudas et

Table 2. Properties of the mutants.

Type of Strain	AHH specific activity in TCDD-induced cells	P_1-450 mRNA levels in TCDD-induced cells	Ah receptor levels after *in vivo* treatment with [^{3}H]TCDD in:		Basis
			cytosol	nucleus	
Hepa-1	100	100	100	100	
A^-, subgroup II	0	0	100	100	P_1-450 structural
A^-, subgroup IV	0	100	100	100	gene
B^-	2	2	2	2	Ah
C^-	0	0	120	0	receptor
D^-	3	3	20	3	genes?
Dominant	1	1	100	100	Repressor

Each parameter is presented as a percentage of the same parameter in Hepa-1.

al, 1986; Gudas and Hankinson, 1986). We therefore believe that the unoccupied Ah receptor is cytoplasmic and we have interpreted our Ah receptor assays on the mutants accordingly. The A^- and dominant mutants have normal levels and functionality of the Ah receptor. B^- mutants have much reduced levels of the Ah receptor, which is nevertheless able to translocate to the nucleus on binding ligand. C^- mutants have normal levels of the Ah receptor but are totally defective in nuclear translocation. The D^- mutant has diminished levels of the Ah receptor which is also reduced in its ability to translocate to the nucleus. Thus mutations in three complementation groups can affect functioning of the Ah receptor. This is in marked contrast to the situation with mutants of the apparently similar glucorticoid receptor. Mutants analogous to our B^- and C^- classes have been isolated for this receptor (Gehring et al, 1977). However the two classes belong to the same complementation group, which corresponds to the 86 kd ligand- (and DNA-) binding subunit of the glucocorticoid receptor (Miesfeld et al, 1984). It is conceivable that our B^-, C^- and D^- mutants all carry lesions in the same gene and show intragenic complementation. However we think that it is much more likely that the B, C and D complementation groups each corresponds to a different gene, and thus three genes affect functioning of the receptor.

We found that the mutants are defective only in those

enzymes studied that are inducible by PAH's and HAH's. Thus the mutants possess normal activities of cytochrome P-450 reductase, epoxide hydrolase and ornithine decarboxylase (Duthu and Hankinson, 1983), but do manifest a defect in induction of UDP-glucuronosyl transferase (the D^- and dominant mutants were not tested) (Robertson et al, 1987). Furthermore the mutants also retain two liver specific functions possessed by the parental Hepa-1 line, namely albumin secretion and transferrin secretion (Duthu and Hankinson, 1983). Thus the loss of AHH inducibility in the mutants does not reflect a change in the general state of differentiation of the cells. We show here that all the classes of mutant also retain normal levels of the glucocorticoid receptor and have undiminished ability to concentrate the glucocorticoid receptor-ligand complex in the nucleus (Table 3). Thus the B^-, C^- and D^- mutations do not have a general effect on soluble steroid or steroid-like receptors.

Table 3. Glucocorticoid receptor levels and translocation in the mutants.

Strain	Genetic Class	Glucocorticoid receptor levels after <u>in vivo</u> treatment with [^{3}H]-dexamethasone (fmol/mg protein) $\pm$ S.E.M. in:		Number of independent experiments
		cytosol	nucleus	
Hepa-1	Wild-type	146$\pm$32	375$\pm$42	10
c38	Group A	121	272	2
c12	Group B	152$\pm$44	236$\pm$20	3
c40T	Group C	219	389	2
c35-1-T1-01	Group D	184$\pm$20	525$\pm$146	4
c31	Dominant	296$\pm$43*	710$\pm$244	3

* Significantly different from Hepa-1 cytosol ($P<0{\cdot}05$)

2.5 x 10^7 cells of each strain were treated for one hour at 37°C in suspension with 20 nM [^{3}H]-dexamethasone (45·8 Ci/mmol, NEN Research Products, Boston, Mass.) in the presence or absence of a 200-fold molar excess of radioinert dexamethasone. Nuclear and cytosolic fractions were then prepared by standard techniques (Gudas and Hankinson, 1986). 100 µl aliquots of the cytosolic extracts and 200 µl aliquots of the nuclear extracts were each added to 100 µl of a suspension of dextran-coated charcoal (5·5 mg charcoal and mg dextran per ml) and extracted for 10 minutes at 0°C. Radioactivity in the supernatants was then counted. Specific binding of [^{3}H]-dexamethasone was calculated as the difference between the amount of protein-bound radioactivity with and without non-radioactive dexamethasone.

<u>Development of a reverse selection procedure and complementation of the defects in the A^- mutants by transfection with genomic DNA.</u>

We developed an efficient procedure for selecting for AHH-inducible cells growing in the presence of a vast excess of AHH-deficient cells (Van Gurp and Hankinson, 1983). This procedure utilizes the PAH, benzo[ghi]perylene (B[ghi]P). Unlike benzo(a)pyrene, the metabolites of B[ghi]P are not toxic. However B[ghi]P itself is phototoxic. In the selection procedure, preinduced cells at 5 x 10^5 per dish are treated with B[ghi]P for two hours; the medium is then replaced with one lacking B[ghi]P; the cells are then incubated for a further two hours before being illuminated with near ultraviolet light. Cells which lack AHH activity are unable to eliminate B[ghi]P during the four hour incubation, and are killed on illumination, while AHH-positive cells survive.

We first used the reverse selection procedure to isolate revertants from representative mutants of each class (Van Gurp and Hankinson, 1984). However our major use of the method has been to isolate genomic DNA-derived transfectants of the mutants in which the genetic defect is complemented by the introduced DNA. In these experiments cells of a particular recessive mutant are treated with a calcium phosphate precipitate of genomic DNA from the rat or human, and the AHH-positive transfectant clones subsequently selected with our reverse selection.

Our first successful transfection experiments were obtained with the A^- mutants. We isolated both rat and human DNA-derived transfectants. Furthermore the rat transfectants that were tested all contained the rat P_1-450 structural gene, but little rat DNA overall (Montisano and Hankinson, 1985). These results provided additional evidence that the A gene is the P_1-450 structural gene, but equally importantly, indicated the feasibility of complementing our other mutants by transfection.

Present and future research

We obtained primary transfectants of a C^- mutant using human DNA donor material. Secondary transfectants were then obtained from the primaries. The secondaries contain certain unique human DNA sequences in common. We have cloned 18 kb of the common human DNA sequence from one of the secondaries. This 18 kb of DNA undoubtedly corresponds either to a part of the C gene or closely linked sequences (F.-F. Chu, F. Sander and O. Hankinson, unpublished). We have also obtained primary transfectants of a B^- mutant (A.J. Watson and O. Hankinson, unpublished) and of a D^- mutant (C. Legraverend, J. Nanthur, and O. Hankinson, unpublished). When secondaries are obtained we will clone these genes in the same fashion as we used for the C gene. We have also successfully transfected the dominant gene by treating wild-type Hepa-1 cells with DNA from a dominant mutant (A.J. Watson and O. Hankinson, unpublished). We propose to clone the dominant gene by a marker rescue strategy.

Success in cloning the B, C, D and dominant genes should lead to important insights into the structure, role and mechanism of action of the Ah receptor; the mechanism of interaction of the Ah receptor and the dominant repressor with the P_1-450 gene; the possible role of the Ah receptor in cell division and differentiation; and the possible involvement of one or more of the genes in determining individual differences in susceptibility to cigarette-induced cancer in the human population.

Acknowledgments

Supported by NCI grant 28868, DOE contract DE-AM03-76-SF00012 and a postdoctoral fellowship from the American Cancer Society and the California Institute for Cancer Research to R.M.B.

References

Denison, M.S., Harper, P.A., and Okey, A.B., 1986, European Journal of Biochemistry, 155, 223,229.

Denison, M.S., Vella, L.M. and Okey, A.B., 1986, Journal of Biological Chemistry, 261, 3987-3995.

Duthu, G.S. and Hankinson, O., 1983, Cancer Letters, 20, 249-254.

Gehring, U., Yamamoto, K.R. and Tomkins, G.M., 1977, Research in Steroids, 7, 43-48

Gudas, J.M. and Hankinson, O., 1987 (In press), Somatic Cell and Molecular Genetics.

Gudas, J.M. and Hankinson, O., 1986, Journal of Cell

Physiology, 128, 449-456.
Gudas, J.M., Karenlampi, S.O. and Hankinson, O., 1986, Journal of Cell Physiology, 128, 441-448.
Hankinson, O., 1983, Somatic Cell Genetics, 9, 497-514.
Hankinson, O., 1979, Proceedings of the National Academy of Sciences of the United States of America, 76, 373-376.
Hankinson, O., Andersen, R.D., Birren, B.W., Sander, F., Negishi, M., and Nebert, D.W., 1985, Journal of Biological Chemistry, 260, 1790-1795.
Karenlampi, S.O., Legraverend, C., Gudas, J.M., Carramanzana, N., and Hankinson, O., 1987, Submitted to Journal of Biological Chemistry.
Karenlampi, S.O., Montisano, D.F., Gudas, J.M., and Hankinson, O., 1986, Archives of Toxicology, 9, 159-162.
Kimura, S., Hänkinson, O., and Nebert, D.W., 1987, EMBO Journal.
Legraverend, C., Hannah, R., Eisen, H., Owens, O., Nebert, D., and Hankinson, O., 1982, Journal of Biological Chemistry, 257, 6502-6507.
Miesfeld, R., Okret, S., Wikström, A.-C., Wrange, Ö., Gustaffson, J.-A., and Yamamoto, K.R., 1984, Nature, 312, 779-781.
Miller, A.G., Israel, D., and Whitlock, J.P.Jr., 1983, Journal of Biological Chemistry, 258, 3523-3527.
Montisano, D.F. and Hankinson, O., 1985, Molecular and Cellular Biology, 5, 698-704.
Nebert, D.W. and Gonzalez, F.J., 1987, Annual Review of Biochemistry, 56, 945-993.
Poland, A., 1984, In Banbury Report No. 18, edited by Poland, A. and Kimbrough, R.D., (Cold Spring Harbor Laboratory), pp. 109-117.
Robertson, J.A., Hankinson, O., and Nebert, D.W., 1987 (In press), Chemica Scripta.
Van Gurp, J.R. and Hankinson, O., 1984, Molecular and Cellular Biology, 4, 1597-1604.
Van Gurp, J.R., and Hankinson, O., 1983, Cancer Research, 43, 6031-6038.
Whitlock, J.P.Jr. and Galeazzi, D.R., 1984, Journal of Biological Chemistry, 259, 980-985.
Whitlock, J.P., Gelboin, H.V., and Coon, 1976, Journal of Cell Biology, 70, 217-225.

Mechanism of Induction of the Rat Cytochrome P-450c Gene by Methylcholanthrene

Y. FUJII-KURIYAMA, A. FUJISAWA-SEHARA, K. SOGAWA, M. YAMANE and T. HIROMASA

Department of Biochemistry, Cancer Institute, Japanese Foundation for Cancer Research, Tokyo 170, Japan

SUMMARY

Regulatory mechanisms of P-450c gene expression were investigated by reverse genetics using the isolated gene DNA. cis-Acting DNA elements responsible for the inducible expression were localized in the 5' flanking region of the gene. The sequence of the DNA elements is composed of about 15 nucleotides and shows a significant homology to that of glucocorticoid responsive elements (GRE). We designate the DNA elements XRE(Xenobiotic responsive elements).

INTRODUCTION

Rat cytochrome P-450c is a member of cytochrome P-450 super-family which catalises a monooxygenase reaction of a wide variety of lipophilic xenobiotics. Nowaday, this hemoprotein has evoked general interest, because of its activity which is directed toward detoxication of xenobiotic substances on one hand and metabolic activation of carcinogenic polycyclic aromatic hydrocarbons(benzo(a) pyrene, 3-methylcholanthrene and so on) on the other(1, 2).

The expression of the P-450c gene is known to be regulated by exogenous inducers such as 3-methylcholanthrene (3-MC) and TCDD(2, 3, 7, 8-tetrachlorodibenzo-dioxin)(3,4). It is suggested that modulation of the gene expression by the inducers is mediated by intracellular receptor protein (s) which associates with a certain regulatory region of the gene as a complex with an inducer molecule(2, 5).

In order to investigate cis-acting DNA elements responsive to the inducers, we constructed a chimeric gene by ligating the 5' flanking sequence of P-450c gene to the structural gene for chloramphenicol acetyltransferase and transfected it into various cultured cells for assay of the CAT expression.

The fusion gene was found to express the CAT activity in response to added 3-MC in Hepa-1, cells derived from mouse hepatoma(7).

Transfection experiments using series of external deletion mutants could localize a short fragment responsible for the inducibility-conferring activity(7, 8). Subsequent subdivisions of the responsible fragment placed on the heterologous SV40 promoter allow us to define the essential DNA sequence for response to xenobiotic inducers(9).

We report here that two DNA elements of about 15 nucleotides localized in the responsible fragment act as tissue-specific inducible enhancers and that their effect largely depends upon normal function of the putative receptor for the inducers.

The two DNA elements are homologous and are inversely oriented with each other. Their homologues are distributed several times in the 5' flanking sequence of the P-450c gene at the positions whose deletion caused marked reduction in the induced activity. This homologous DNA sequence is also conserved at the equivalent positions of the human and mouse gene and shows some similarity to the glucocorticoid regulatory elements.

MATERIALS AND METHOD

Construction of Plasmids

An original fusion gene (pMC 6.3k) consisting of the 5' flanking sequence(6.3kb) was constructed as described previously(7). External deletion mutants in the 5' flanking sequence were produced from the original fusion gene pMC6.3k by digestion with BAL 31 exonuclease or appropriate restriction enzymes(8). Subdivided fragments of a part of the 5' flanking sequence were ligated to a heterologous promoter of SV40 as described(9). The correct constructions of these hybrid genes were confirmed by either sequencing or restriction mapping of the amplified plasimid DNAs.

Transfection and Induction

DNA transfections were performed as described previously (10). The cells were exposed to the media with or without 1μM 3-MC and them were harvested for the CAT assay 40 hours after treatment with inducers.

CAT Assay

The CAT activity was assayed as described(7, 11).

RESULTS AND DISCUSSION

Core Sequence of Inducible Enhancer

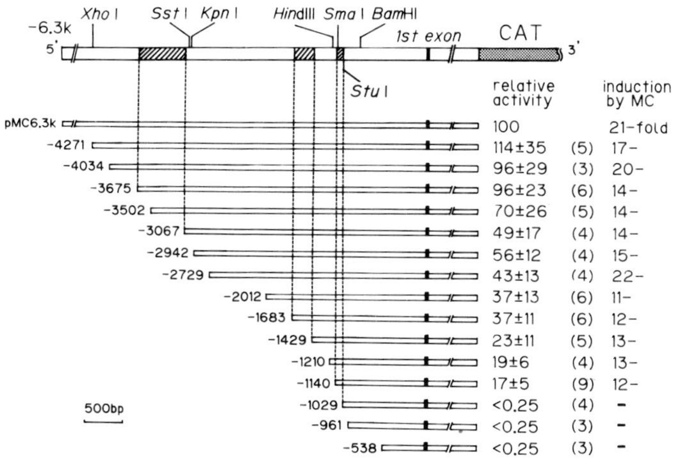

Fig. 1, External deletion mutants in the 5' flanking region expressed CAT activities of P-450-CAT fusion genes. Negative figures on the left indicate the 5' end of the sequences counted from the transcription start site. The column immediately to the right of the bars gives the CAT activity induced by MC as average values ± SD. The data are expressed as a percentage of the activity of pMC6.3k. Numbers in parentheses show numbers of separate experiments with each deletion mutant. Third column shows the ratios of induced to non-induced activity.

A chimeric gene consisting of the 5' flanking sequence of P-450c gene and the CAT structural gene expressed the CAT activity in response to 3-MC in Hepa-1 cells. Modes of the CAT gene expression in this transient expression system appear to mimic faithfully those of endogenous P-450c gene in livers and cultured cell lines in all respects so far tested (7). To define functional regions essential for the inducibility in the 5' flanking sequence of the P-450c gene, a set of hybrid genes with various external deletions were constructed and assayed for the inducibility of CAT activity in the transient expression system. The inducible expression of the CAT activity was decreased concomittantly with progressive deletions starting from the 5' end of the upstream flanking sequence of the P-450c gene. It was found that there were at least three regions whose deletions caused marked decrease in the induced CAT activity(Fig.1). Of these three regions, a fragment -1139 to -1029 appear to be the most important, because the inducibility and the level of the expressed CAT activity was almost lost by deletion of

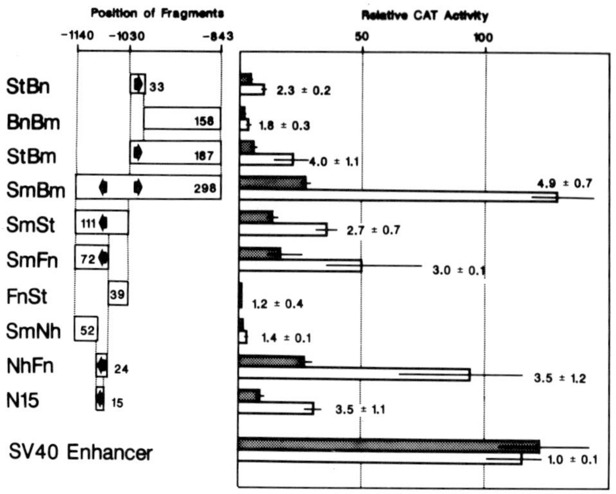

Fig. 2, Inducible enhancer activities of subfragments of SmBm on a heterologous promoter. Open boxes on the right side to the name of fragments indicate their lengths and positions in the original P-450c gene. Arrows in the boxes indicate the position of XREs. Open and stippled bars represent the corrected CAT activities in the presence and absence of 3-MC, respectively, and the figure shows the inducibility of each subfragment as a ratio of the induced to non-induced CAT activity with standard deviation. The corrected CAT activities are calculated as the ratios of the activity of subfragment-ligated fusion genes to that of the fusion gene consisting of the CAT structure gene and SV40 promoter without the enhancer sequence.

this fragment. We further pursued a key sequence for the inducibility in this fragment. The regulatory fragment(-844 to -1140bp) was subdivided into various subfragments and then these subfragments were placed separately on the 5' end of the SV40 promoter without its enhancer sequence which was fused to the CAT gene. The two halves of the regulatory fragment(SmSt) and (StBm) retained the enhancer activity in response to 3-MC. These two halves were finally reduced to their respective smallest fragments of 15 and 33 nucleotides with the inducible enhancer activity(Fig.2). Comparison of the DNA sequence between the two active subfragments(StBn and NhFn) in either orientation revealed the presence of a homologous sequence with a short alternate purine / pyrimidine stretch of CACGC(A)(Fig.3). The two homologous sequences are inversely oriented with each other in the original regulatory fragment(SmBm). These observations lead us to designate the homologous sequences as xenobiotic responsive element(XRE), that is, sequences derived from StBn and NhFn fragments are designated XRE1 and XRE2, respectively as shown in Fig.3. The XRE sequences exerted the enhancing

```
ggatcCCTCCAGGCTCTTCTCACGCAACTCCGGGGCACgatc   XRE1 (-1029/-997)
           |    |  | ||||||| |
gatcCGGGTCCCAGTGCTCTCACGCTAGgatccccgggtacc   XRE2 (-1069/-1092)
```

Fig. 3. comparison of DNA sequences of XRE1 and XRE2.
XRE1 and XRE2 correspond to the restriction fragments, StBn and NhFn, respectively. Sequences of the plus and minus strands of XRE1 and XRE2, respectively, were compared. Small letters indicate flanking polylinker sequences. Vertical lines represent matched bases between the two. The underline indicates the sequence of the fragment N15(XRE2').

effect in a manner relatively independent of their orientation or distance in relation to the promoter. Interestingly, tandem repeats(2 and 4 folds) of the XRE sequences, placed on the 5' end of the heterologous SV40 promoter exhibited much higher enhancing activity than a single XRE. A fully induced CAT activity of 4 fold tandem repeat of XRE was about 16 times as high as that of a single XRE(9).

Distribution of XRE Homologues

Close survey of the 5' flanking sequence of the P-450c gene reveals the presence of several XRE homologues in this sequence. Interestingly, their distribution appears to concur with that of the region whose deletion reduced markedly the induced CAT activity. The result may suggest inducible enhancer activities in these XRE homologues, though their functional roles have to be determined experimentally. It is reasonable to infer, therefore, that the strong expression of the P-450c gene in animal livers in response to 3-MC result from the concerted effects of the XREs and their homologues(Fig.4).

Sequence Similarity between XRE and GRE.

From comparison of XRE sequence with the regulatory sequences of various genes so far reported, the motif of XRE was fo nd to show some similarity with the sequence of DNA binding sites(GRE) for glucocorticoid receptor(12-16).
It has been recently reported that a trans-acting factor or 3-MC(or TCDD) receptor which is believed to interact with XRE sequences has many physicochemical and biochemical properties in common with the glucocorticoid receptor(5, 17-20). From the investigations of the polyamionic, proteolytic and hydrodynamic properties of the two receptors, Gustafsson and his coworkers concluded that the TCDD(or 3-MC) receptor and the glucocorticoid receptor appear to be structurally and functionally closely related proteins. All these results,

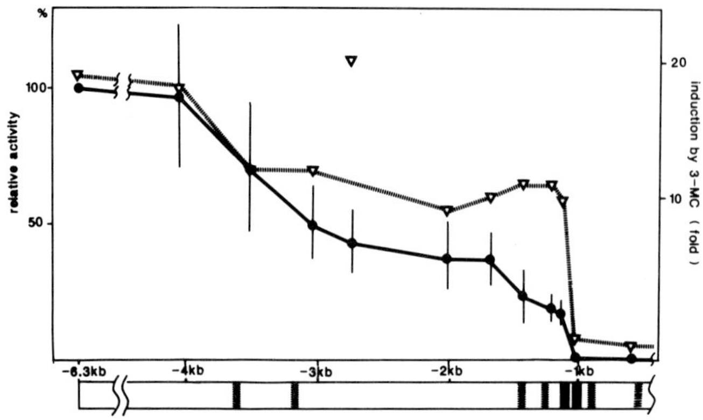

Fig. 4. The CAT activity of external deletion mutants and the distribution of XRE along the 5' flanking sequence. Relative CAT activities of various deletion mutants are shown by solid lines and the ratios (fold) of the induced to non-induced activity are represented by broken lines. These values are taken from Fig.1. XREs are indicated by filled blocks and their homologues are by hatched bars.

therefore, lead us to a notion that the inducible system of XRE and 3-MC(or TCDD) receptor is evolutionarily related to that of GRE and glucocorticoid receptor.

Transacting Factor

Recently, Hankinson isolated mutant cells(C4 and C12) of mutant cells with defective receptor function for 3-MC or TCDD(21). C4 mutant is shown to be defective in translocation of the receptor from cytoplasm to nucleus and C12 mutant show extremely low expression of the receptor. When XRE1 or XRE2 fusion gene was transfected into the mutant cells, practically no enhanced expression of the CAT activity was observed in response to 3-MC, suggesting that the XRE is a principle which interacts with the receptor-inducer complex for the inducer-specific expression. By protein-DNA mobility shift assay, we found a factor, which retarded mobility of the XRE in gel electrophoresis, in nuclear extracts from 3-MC-treated Hepa-1 cells. The interaction between the factor and the XRE was found to be specific for the XRE sequence and the species of cells. Purification of the XRE-binding factor from nuclear extracts of 3MC-treated cells are in progress in our laboratory.

ACKNOWLEDGEMENT

This work was supported in part by Grants-in-Aid for Scientific Research from the Ministry of Education, Science and Culture of Japan, By a grant from National Center for Nervous, Mental and Muscular Disorders of the Ministry of

Health and Welfare, Japan, and by funds obtained under the Life Science Project from the Institute of Physical and Chemical Research, Japan

REFERENCES

1) Conney, A.H. 1982, Cancer Research 42, 4875.
2) Adesnik, M. & Atchison, M. 1985, Critical Reviews in Biochemistry, 19, 247.
3) Kawajiri, K., Gotoh, O., Tagashira, Y., Sogawa, K. & Fujii-Kuriyama, Y. 1984, Journal of Biological Chemistry, 259, 10145.
4) Morville, A.L., Thomas, P., Levin, W., Reik, L., Ryan, D.E., Raphael, C. & Adesnik, M. 1983, Journal of Biological Chemistry, 258, 3901.
5) Okey, A.B., Bondy, G.P., Mason, M.E., Nebert, D.W., Forster-Gibson, C.J., Muncan, J. & Dufresne, M.J. 1980, Journal of Biological Chemistry, 255, 11415.
6) Eisen, J.J. 1986, in Cytochrome P-450; Structure, Mechanism and Biochemistry, P.R. Ortiz de Montellano ed. PP. 315, Plenum Press, New York and London.
7) Fujisawa-Sehara, A., Sogawa, K., Nishi, C. & Fujii-Kuriyama, Y, 1986, Nucleic Acid Research, 14, 1465.
8) Sogawa, K., Fujisawa-Sehara, A., Yamane, M. & Fujii-Kuriyama, Y. 1986, Proceedings of National Academy of Science U.S.A., 83, 8044.
9) Fujisawa-Sehara, A., Sogawa, K., Yamane, M. & Fujii-Kuriyama, Y. 1987, Nucleic Acids Research, 15, 4179.
10) Wigler, M., Sweet, R., Sim, G,K., Wold, B., Pellicer, A., Lacy, E., Maniatis, T., Silverstein, S. & Axel, R. 1979, Cell, 16, 777.
11) Gorman, C.M., Moffat, L.F. & Haward, B.H. 1982, Molecular and Cellular Biology, 2, 1044.
12) Karin, M., Haslinger, A., Holtgreve, H., Richards, R.I., Krauter, P., Westphal, H.M. & Beato, M. 1984, Nature, 308, 513.
13) Poyvar, F., de Franco, D., Firestone, G.L., Edger, G., Wrange, O., Okret, S., Gustafsson, J.-A. & Yamanoto, K.R. Cell, 35, 381.
14) Miksicek, R., Heber, A., Schmid, W., Danesch, V., Possechert, Beato, M. & Schutz, G. 1986, Cell, 46, 283.
15) Renkawitz, R., Schutz, G., Von der Ahe, D. & Beato, M. 1984, Cell, 37, 503.
16) Scheidereit, C. & Beato, M. 1984, Proceedings of National Academy of Science U.S.A., 81, 3029.
17) Willhelmsson, A., Wikstrom, A.- C. & Paellinger, L. 1986, Journal of Biological Chemistry, 261, 13456.
18) Poland, A., Glover, E., Ebetino, G.H. & Kende, A.S.

1986, Journal of Biological Chemistry, 261, 6352.
19) Whitlock, Jr., P. & Galeazzi, D.R. 1984, Journal of Biological Chemistry, 259, 980.
20) Cuthill, S., Poellinger, L. & Gustaffson, J.- A. 1987, Journal of Biological Chemistry, 262, 3477.
21) Hankinson, O. 1981, Somatic Cell Genetics, 7, 373.

Differential Regulation of Human Cytochrome P-450 Genes

ROBERT H. TUKEY, STEVE OKINO, USHA R. PENDURTHI, ERIC F. JOHNSON [1] and LINDA QUATTROCHI

Cancer Center, Department of Medicine and Department of Pharmacology, University of California, La Jolla, CA 92039, USA and [1] Research Institute of Scripps Clinic, La Jolla, CA 92037, USA

Introduction

The cytochrome P-450 monooxygenases are a group of proteins responsible for the oxidation of drugs, environmental pollutants and endogenous compounds such as steroids and fatty acids. Substrate oxidation reactions can be substantially accelerated resulting from the ability of foreign compounds to selectively induce certain forms of P-450, while many of the individual forms are present constitutively and are not susceptible to alterations in their steady state levels. In addition, it is known that the expression of certain forms of P-450 are influenced by unknown factors as a function of age, while others are sexually differentiated. The combination of environmental factors, temporal control, endocrine regulation, the sex of the animal, and the multiple forms adds tremendous flexibility to the P-450-dependent hydroxylations of such a wide range of endogenous and exogenous substrates.

Another mechanism that appears to have evolved during evolution which adds to the diversification of the cytochrome P-450 "super family" is the generation of multiple, yet highly homologous genes. For example, in rabbits, we have identified and characterized a liver cDNA clone that encodes for cytochrome P-450 1 (Tukey et al, 1985), a microsomal protein which is expressed constitutively. One of the major enzyme activities of P-450 1 is the 21-hydroxylation of progesterone to the mineralcorticoid, deoxycorticosterone, an activity which is also localized to the adrenal. Interestingly, comparison of the DNA sequence and its predicted translation product demonstrates that P-450 1 is unrelated to the human or pig adrenal 21-hydroxylase (White et al, 1984; Chung et al, 1985; Chaplin et al, 1986; White et al, 1986). However, Southern blot analysis with the P-450 1 cDNA shows that this gene is part of a complex multigene family (Tukey et al, 1985). Preliminary results also indicate that the structure of these genes are similar with regard to the placement of introns and exons, and that the gene transcripts are highly related to P-450 1. In addition, we have shown that the constitutive expression of P-450 1 and a highly related transcript are regulated differentially in various strains of rabbits (Johnson et al, 1987). These results indicate that evolution has coupled a mechanism by which genes can be altered and duplicated, while diversifying those events that control the overall expression of the homologous genes. This combination of gene duplication and differential expression adds additional flexibility to the functional role of the cytochrome P-450s.

To understand how gene duplication and evolution may influence the regulation of different human P-450 genes, we have been interested in contrasting the modes of regulation of human cytochrome P-450 genes from different gene families. The first family represents the highly conserved subfamily of human P-450 genes that are homologous to the polycyçlic aromatic hydrocarbon (PAH) inducible rabbit P-450 4 and P-450 6. We have identified and characterized the human P-450 4 gene (Quattrochi et al, 1986) as well as the P-450 6 gene, and are determining if these two genes are regulated in a similar fashion in the presence of PAHs. The second represents the subfamily which is represented by the human equivalent to rabbit P-450 1, which unlike human P-450 4 and 6, is encoded by a gene that is part of a much larger gene family. Evidence will be presented which indicates that during the expression of the human P-450 1, the nuclear transcripts are differentially processed into two forms of mRNA.

RESULTS AND DISCUSSION

Identification of the human P-450 4 and P-450 6 genes.

In rabbits, P-450 4 and P-450 6 are expressed constitutively, but are also susceptible to induction with PAHs and TCDD (Okino et al, 1985). To understand how this gene family has evolved and what mechanisms control its expression in humans, the mRNAs and genes that encode this family of P-450s were be characterized. Using rabbit cDNA clones to P-450 4 and P-450 6 (Okino et al, 1985), the cDNAs that encode the human equivalents were identified (Quattrochi et al, 1985; Quattrochi et al, 1986). These clones were in turn used to characterize the human P-450 4 structural gene (Quattrochi et al, 1986), as well as the human P-450 6 gene. Partial DNA sequence and restriction enzyme analysis of the human P-450 6 gene demonstrated that it was identical to the recently reported human P-450c gene (Kawajiri et al, 1986). Both genes structurally are spanned by 7 exons and 6 introns, similar to that observed in other species (Sogawa et al, 1985; Sogawa et al, 1984; Gonzalez et al, 1985). The recombinant clones that encode human P-450 4 and P-450 6 also contain substantial 5′ flanking sequence.

Expression of human P-450 4 and P-450 6 in cell culture.

Considerable progress has been made in understanding the mechanisms by which TCDD initiates the induction of P-450s. In mice, TCDD has been shown to initiate the induction of P_1-450 and P_3-450 by association with a regulatory gene product, the *Ah*-receptor, followed by activation of transcription of the structural genes (Tukey et al, 1982). In cell culture, the induction of the mouse P_1-450 gene (Jones et al, 1985; Jones et al, 1986; Gonzalez & Nebert, 1985) and its homolog in the rat, the P-450c gene (Fujisawa-Sehara et al, 1986; Sogawa et al, 1986), by TCDD and PAHs, has been shown to occur by the interaction of a *trans*-acting factor to *cis*-acting enhancer elements. Detailed analysis of the structure of these enhancer sequences, termed drug regulatory elements or DREs (Sogawa et al, 1986), has resulted in the identity of a consensus sequence that serves to confer inducibility in the presence of PAHs.

To determine if the human P-450 4 and P-450 6 genes were responsive to PAHs in culture, human Hep-G2 cells were treated with 3-methylcholanthrene and the mRNA analyzed for the presence of the two gene transcripts by hybridization with the human cDNA clones. In the absence of PAHs, there was no detectable human P-450 4 or P-450 6, suggesting that in cell culture neither gene is constitutively controlled. However, in the presence of PAHs, the 2600 kb human P-450 6 transcript (Quattrochi et al, 1986) is enhanced. Interestingly, the 3100 kb human P-450 4 transcript (Quattrochi et al, 1986) was not induced. Coupled with the observations that the rabbit P-450 4 gene is expressed only in liver (Liem et al, 1980), the lack of expression of the human P-450 4 in cell culture indicates that certain tissue specific factors are involved in the expression of this gene in combination with the *Ah*-receptor.

To define the potential *cis*-acting regulatory elements on the human genes, various portions of the 5′ regulatory regions were fused to the chloramphenicol acetyltransferase gene, and the constructs transfected into Hep-G2 cells. Preliminary experiments have led to the identification of a 1600 bp region, 5′ of the cap site in the P-450 6 gene that is responsive to PAHs (Figure 1). Analysis of the DNA

Figure 1. The expression of chloramphenicol acetyl transferase activity (CAT) in HepG2 cells following transfection with various 5′-flanking human P-450 4 (4) and P-450 6 (6) gene and CAT constructs. Activities were measured in extracts from 3-methylcholanthrene treated cells.

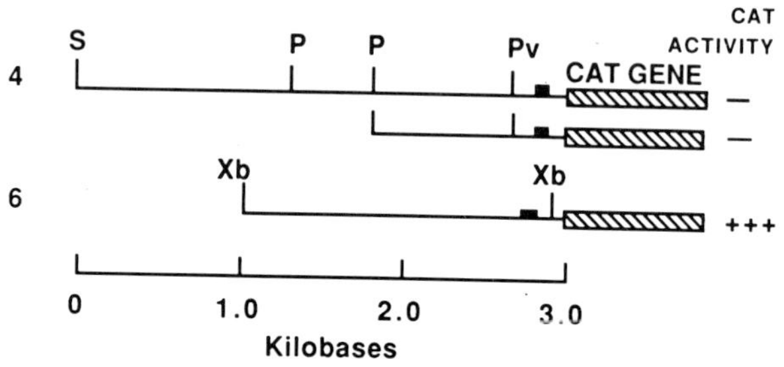

sequence in this region indicates that there are three DRE elements (Sogawa et al, 1986), located at bases -221 (CCCAGCCCTC), -521 (CAGGGCTGGG) and -1410 (GTTAGCTGGG) from the cap site. In contrast, DNA fragments as far as 3000 bp from the cap site of the human P-450 4 gene were not capable of initiating transcription in the presence of the human P-450 4 promoter. In addition, there were no consensus DRE elements in the 1700 bases 5′ from the cap site on the human P-450 4.

Although preliminary, these results suggest that the human P-450 4 and P-450 6 genes are differentially regulated. The ability of the rabbit P-450 4 gene to be expressed constitutively (Okino et al, 1985) and regulated primarily in the liver (Liem et al, 1980), indicates that tissue specific or other humoral factors play an important role in the expression of this gene. It would also appear that this "constitutive", tissue specific *trans*-acting element, which is absent in tissue culture, may play a critical role in coupling the *Ah*-receptor to the induction of the P-450 4 gene. In addition, the absence of consensus enhancer elements in proximity to the P-450 4 promoter, unlike that observed with the human P-450 6 gene, indicates that conservation of the regulatory regions of these two genes has not been maintained.

Identification of human P-450 1-like cDNA clones

Northern blot analysis with the rabbit P-450 1 cDNA clone, pl-8, against human liver mRNA showed that the rabbit clone hybridized to a human mRNA of approximately 2000 bases (Okino et al, 1987). To isolate the corresponding human cDNA, the 5′ coding region of the rabbit cDNA clone pl-8 was used as a probe to screen human liver cDNA libraries that had been generated in either pBR322 (Quattrochi et al, 1985)or λ-gt11 (Quattrochi et al, 1986). Over twenty positive clones from the two libraries were identified and purified, and characterized by partial restriction enzyme analysis. Close examination of the location of the restriction enzyme sites indicated that there were two populations of clones. Restriction enzyme analysis of the largest of the clones from each population, called Hpl-1 and Hpl-2, demonstrated that both inserts were identical, except clone Hpl-2 contained a segment of DNA that was not present in clone Hpl-1. Figure 2 shows a partial restriction enzyme map and the additional DNA sequence that was unique to clone Hpl-2.

Figure 2. Partial restriction enzyme digest and DNA sequence of the two human P-450 1 cDNA clones, termed Hpl-1 (the top cDNA clone), and Hpl-2 (the bottom cDNA clone). Note the two stop codons (underlined), and the absence of a 5′ GT consensus splice site.

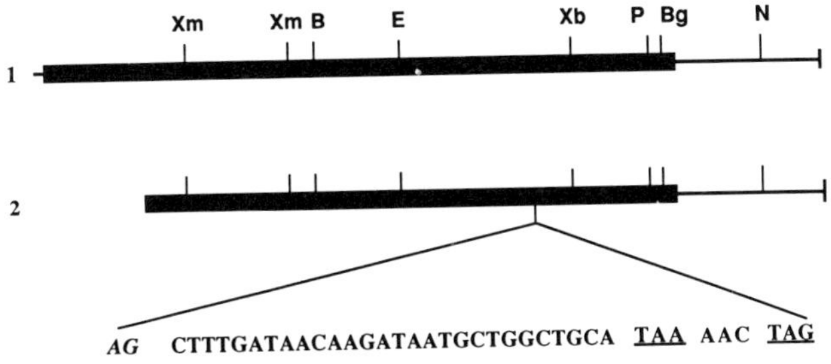

Characterization of Clone Hpl-1.

DNA sequence of clone Hpl-1 demonstrated that this cDNA contained an open reading frame that encoded a protein of 490 amino acids (Okino et al, 1987). When the predicted translation products of the human and rabbit forms were compared for similarity, they were shown to be 70% homologous, with the homology increasing to 81% when conserved amino acid changes were taken into consideration. When the predicted amino acid sequence of the human P-450 1 was compared to other known human forms of P-450, it was shown to be 80% homologous to the predicted translation products encoded by the human P-450 MP8 clone (Umbenhauer et al, 1987), which is related to the human microsomal P-450 s-mephenytoin 4-hydroxylase (Shimada et al, 1986).

The close homology between human P-450 1 and the s-mephenytoin 4-hydroxylase indicates that P-450 1 may be part of a larger gene family, as we had observed in rabbits (Tukey et al, 1985). To determine the genetic diversity of the human P-450 1 gene family, Southern blots were performed with the 5′ and 3′ portions of the Hp1-1 cDNA clone against human liver DNA. The 5′ portion of clone Hp1-1 generated a complex hybridization pattern, in contrast to the pattern generated with the 3′ non-coding portion of the clone. Since there exists a high degree of homology between the human P-450 1 and the s-mephenytoin 4-hydroxylase cDNA, some of the gene fragments that are hybridizing with the human P-450 1 cDNA most likely represent genes which encode the s-mephenytoin 4-hydroxylase P-450.

The human P-450 1 transcript is processed by alternate RNA splicing into two forms of mRNA.

The DNA sequence of clone Hp1-2 demonstrated that it was identical to clone Hp1-1, except Hp1-2 contains an additional 39 bases that are not present in Hp1-2 (Figure 2). When the DNA sequence of Hp1-2 was analyzed for an open reading frame, two stop codons, disrupting the reading frame, were found in the 39 bases. Analysis of the DNA sequence indicates that the 39 bases is not an intact intron since it contains the consensus 3′ splice dinucleotides AG, yet lacks the 5′ consensus GT dinucleotides present in the introns of all eukaryotic genes (Breathnach & Chambon, 1981). The presence of the 39 bases, which is located in the same position as the 7th intron of the P-450b gene (Suwa et al, 1985), combined with the lack of a 5′ consensus splice sequence, indicates that the Hp1-2 insert may represent part of an unspliced intron that was not completely removed during the processing of the nuclear RNA.

If we assume that there are several elements required for efficient RNA splicing, such as the 5′ and 3′ splice sites, the RNA branch point (Ruskin & Green, 1985; Padgett et al, 1984), and possibly exon sequences (Reed & Maniatis, 1986), a model for RNA splicing can be proposed. At the start of intron 7, the human P-450 1 RNA contains a 5′ donor splice site, followed by an intron of unidentified length that contains two 3′ splice sites. During removal of the intron, the formation of the 5′ lariat is directed to either a cryptic 3′ branch point, resulting in the formation of the Hp1-2 mRNA, or to a second branch point that would result in the generation of a functional mRNA (Hp1-1).

The use of cryptic 3′ branch points has been demonstrated *in vitro* in the splicing of human β-globin introns (Ruskin et al, 1985), while differential RNA splicing and the use of multiple 3′ splice sites serves a regulatory function for numerous other genes (Goldman et al, 1987; Voelkel-Meiman et al, 1987). It is intriguing to think that in humans, differential RNA splicing may serve as an additional mode of regulation for the P-450 gene families.

ACKNOWLEDGMENT

The work presented in this manuscript was supported by USPHS grants CA37139 and GM36590 and an ACS Faculty Research Award (R.H.T), and GM31005 (E.F.J). This research was conducted in part by the Clayton Foundation for Research-California Division. R.H.T. is a Clayton Foundation Investigator.

REFERENCES

Breathnach, R. & Chambon, P. 1981, Ann. Rev. Biochem., 50,349.

Chaplin, D.D., Galbraith, L.J., Seidman, J.G., White, P.C. & Parker, K.L. 1986, Proc. Natl. Acad. Sci. USA, 83,9601.

Chung, B-C., Matteson, K.J. & Miller, W.L. 1985, DNA, 4,211.

Fujisawa-Sehara, A., Sogawa, K., Nishi, C. & Fujii-Kuriyama, Y. 1986, Nucleic Acids Res., 14,1465.

Goldman, D., Deneris, E., Luyten, W., Kochhar, A., Patrick, J. & Heinemann, S. 1987, Cell, 48,965.

Gonzalez, F.J. & Nebert, D.W. 1985, Nucleic Acids Res., 13,7269.

Gonzalez, F.J., Kimura, S. & Nebert, D.W. 1985, J. Biol. Chem., 260,5040.

Johnson, E.F., Barnes, H.J., Griffin, K.J., Okino, S. & Tukey, R.H. 1987, J. Biol. Chem., 262,5918.

Jones, P.B.C., Galeazzi, D.R., Fisher, J.M. & Whitlock Jr.,

J.P. 1985, Science, 227,1499.
Jones, P.B.C., Durrin, L.K., Galeazzi, D.R. & Whitlock Jr., J.P. 1986, Proc. Natl. Acad. Sci. USA, 83,2802.
Kawajiri, K., Watanabe, J., Gotoh, O., Tagashira, Y., Sogawa, K. & Fujii-Kuriyama, Y. 1986, Eur. J. Biochem., 159,219.
Liem, H.H., Muller-Eberhard, U. & Johnson, E.F. 1980, Mol. Pharmacol., 18,565.
Okino, S.T., Quattrochi, L.C., Barnes, H.J., Osanto, S., Griffin, K.J., Johnson, E.F. & Tukey, R.H. 1985, Proc. Natl. Acad. Sci. USA, 82,5310.
Okino, S.T., Quattrochi, L.C., Pendurthi, U.R., McBride, O.W. & Tukey, R.H. 1987, J. Biol. Chem. In Press.
Padgett, R.A., Konarska, M.M., Grabowski, P.J., Hardy, S.F. & Sharp, P.A. 1984, Science, 225,898.
Quattrochi, L.C., Okino, S.T., Pendurthi, U.R. & Tukey, R.H. 1985, DNA, 4,395.
Quattrochi, L.C., Pendurthi, U.R., Okino, S.T., Potenza, C. & Tukey, R.H. 1986, Proc. Natl. Acad. Sci. USA, 83,6731.
Reed, R. & Maniatis, T. 1986, Cell, 46,681.
Ruskin, B. & Green, M.R. 1985, Nature, 317,732.
Ruskin, B., Greene, J.M. & Green, M.R. 1985, Cell, 41,833.
Shimada, T., Misono, K.S. & Guengerich, F.P. 1986, J. Biol. Chem., 261,909.
Sogawa, K., Gotoh, O., Kawajiri, K. & Fujii-Kuriyama, Y. 1984, Proc. Natl. Acad. Sci. USA, 81,5066.
Sogawa, K., Gotoh, O., Kawajiri, K., Harada, T. & Fujii-Kuriyama, Y. 1985, J. Biol. Chem., 260,5026.
Sogawa, K., Fujisawa-Sehara, A., Yamane, M. & Fujii-Kuriyama, Y. 1986, Proc. Natl. Acad. Sci. USA, 83,8044.
Suwa, Y., Mizukami, Y., Sogawa, K. & Fujii-Kuriyama, Y. 1985, J. Biol. Chem., 260,7980.
Tukey, R.H., Hannah, R.R., Negishi, M., Nebert, D.W. & Eisen, H.J. 1982, Cell, 31,275.
Tukey, R.H., Okino, S., Barnes, H., Griffin, K.J. & Johnson, E.F. 1985, J. Biol. Chem., 260,13347.
Umbenhauer, D.R., Martin, M.V., Lloyd, R.S. & Guengerich, F.P. 1987, Biochemistry, 26,1094.
Voelkel-Meiman, K., Keil, R.L. & Roeder, G.S. 1987, Cell, 48,1071.
White, P.C., New, M.I. & Dupont, B. 1984, Proc. Natl. Acad. Sci. USA, 81,1986.
White, P.C., New, M.I. & Dupont, B. 1986, Proc. Natl. Acad. Sci. USA, 83,5111.

Structure of Rabbit Cytochrome P-450 IIc Genes and Regulation by Phenobarbital

B. KEMPER, P. BELL, G. CHAN, S. GOVIND and J. ZHAO

Department of Physiology and Biophysics and the College of Medicine at Urbana-Champaign, University of Illinois, Urbana, IL 61801, USA

INTRODUCTION

The cytochrome P450 (P450) dependent monooxygenase activity in the liver microsomes is responsible for the metabolism of a large number of endogenous and foreign compounds. More than fifty P450s have now been characterized by protein sequencing or by molecular cloning of cDNAs and genes and can be classified into at least 8 families which together comprise a superfamily (Nebert et al., 1986). The levels of subsets of the P450s can be modulated or induced by a variety of hormones and foreign compounds while other P450s are constituitively expressed. An analysis of the differences in genes of these related P450s that respond selectively to different inducers should provide information about the molecular mechanisms of induction.

One of the classical inducers of microsomal P450s is phenobarbital. The phenobarbital-inducible P450s are primarily in the P450IIB and P450IIC subfamilies. The rabbit P450IIB subfamily contains only one member which is the major microsomal P450 after induction with phenobarbital (Haugen & Coon, 1976). In contrast the rabbit P450IIC subfamily contains at least 5 members (Leighton et al., 1984; Tukey et al., 1985; Zhao et al., 1987, Johnson et al., 1987), which respond selectively to phenobarbital. Comparative analysis of the genes in this subfamily is likely to provide considerable information about the mechanism and evolution of phenobarbital responsiveness in the P450s. In this report, our progress on the isolation and characterization of the rabbit P450IIC genes is reviewed.

RESULTS AND DISCUSSION

P450IIC cDNAs

We have cloned and characterized near full length cDNAs for P450IIC1, P450IIC2 and P450IIC3 (Leighton et al.,1984) and a partial cDNA for P450IIC4 (Zhao et al., 1987). Full length cDNAs for P450IIC4 and P450IIC5 have been characterized by Tukey et al., 1985 and Johnson et al., 1987. In the protein coding region of the cDNA sequences, P450IIC3 is about 75% homologous to the other four members. Within the subfamily, P450IIC1 and P450IIC2 form one closely related subgroup (88% homology) and P450IIC4 and P450IIC5 form a second subgroup (98% homology). The two subgroups are about 80% homologous to each other. In each case homology in the 3' noncoding region is considerably less than that in the protein coding region. At least two genes closely related to the C4/C5 subgroup have been detected by Southern analysis but expressed products for these genes have not yet been characterized (Tukey et al., 1985).

Dot blot hybridization or Northern blot analysis has been used to characterize the mRNA's for the P450IIC members (Leighton et al., 1984; Zhao et al., 1987, Tukey et al., 1985). P450IIC1 was not detected in livers of control animals but was induced by phenobarbital to levels similar to those of P450IIC2 in the treated animals (Table 1). P450IIC2 and P450IIC4 mRNA levels were increased 3 to 5 fold

Table 1. Summary of sizes, phenobarbital-inducibility and tissue specific expression of P450IIC mRNAs

	mRNA		Liver		Kidney	
	Species	Size(Kb)	C	PB	C	PB
C1	3-4	2.3-4.0	-	++++	-	-
C2	2	2.3-3.0	+	++++	+	++++
C3	1	2.0	++++	++++	-	-
C4	1	2.3	+	++++	-	-
C5	1	1.9	++++	++++	-	-

Summarized from Leighton and Kemper, 1984, for C1, C2 and C3; Zhao et al., 1987, for C4; and Dieter et al., 1982, and Tukey et al., 1985, for C5.

by phenobarbital treatment and P450IIC3 mRNA levels were not affected by the treatment. P450IIC5 is also expressed at high levels in untreated animals (Dieter et al., 1982). Tissue specific expression is also exhibited by these isozymes, since only P450IIC2 mRNA was detected in the kidney. P450IIC3, P450IIC4 and P450IIC5 each have one major mRNA species of about 2000 nucleotides that hybridizes to their cDNA. P450IIC1 and P450IIC2 also have a 2000 nucleotide species but contain additional larger species which probably result from alternate polyadenylation sites.

As noted above P450IIC2 is induced by phenobarbital in both the liver and kidney. This isozyme is the same as P450K (Finlayson et al., 1986) based on phenobarbital inducibility, immunoreactivity, gel mobility, tissue specific expression, and N-terminal sequence. The constituitive P450IIC3 gene is probably the same as P450(form 3b) on the basis of 98% homology of amino acid sequence. The nature of the isozyme corresponding to P450IIC4 is not known. P450IIC5 or P450(form 1) has been extensively characterized by Johnson and coworkers (Dieter et al., 1982).

Effect of phenobarbital on nuclear transcription

The genetic basis for the different responses of the P450IIC genes to phenobarbital is not known. Phenobarbital treatment increases the transcription rate of the rat P450IIB1/2 genes (Atchinson and Adesnik, 1983, Hardwick et al., 1983; Pike et al., 1985), but induction of glutathione transferase involves a posttranscriptional mechanism (Francis et al., 1986). In principle, the increases in mRNA for the responsive P450IIC genes could be due to either increased transcription of the gene or to altered metabolism of either pre-mRNA or mRNA. The lack of response in constituitive genes could result either from a loss of the phenobarbital-responsive elements or from the gain of a phenobarbital-dependent mechanism that compensates for induction of the gene, for example an increased degradation of mRNA in response to phenobarbital.

To distinguish between these possiblities we have estimated the rate of transcription by "run off" RNA synthesis in isolated nuclei. RNA from each incubation was hybridized to plasmid DNA containing cDNA inserts of P450IIC1, P450IIC2, and P450IIC3. The transcription rate is expressed as the fraction of the total radioactive nuclear RNA that binds specifically to each cDNA in parts per million (ppm). To obtain reliable estimates of the transcription rates, full length cDNA probes had to be used which results in cross-hybridization between P450IIC1 and P450IIC2 but little cross-hybridization of P450IIC3 with the other two.

Table 2. Relative increases in nuclear transcription rates and mRNA levels of P450IIC genes after phenobarbital treatment.

	IIC1		IIC2			IIC3		
Time	Transcript.		Transcript.		mRNA	Transcript.		mRNA
hr	ppm	fold	ppm	fold		ppm	fold	
0	8.4 ± 1.6	-	9.3 ± 2.1	-	-	9.2 ± 2.6	-	-
6	$76.8 \pm 11.6^*$	9	$74.2 \pm 12.9^*$	8	2.0	$20.1 \pm 5.2^*$	2	0.8
12	$45.3 \pm 13.2^*$	6	$34.1 \pm 8.8^*$	4	1.5	18.1 ± 7.3	2	0.7
24	$38.7 \pm 3.9^*$	5	34.2 ± 16.7	4	2.7	11.8 ± 8.0	1	1.0

*Significantly greater than 0 hour samples ($p<0.05$, $n=3$)

As expected, because of the cross-hybridization, similar amounts of nuclear RNA bound to the P450IIC1 and P450IIC2 probes (Table 2). In each case the amount of P450 RNA hybridized to cDNA increased from about 10 ppm to 80 ppm by 6 hours after phenobarbital treatment and then decreased at 12 and 24 hours but remained about 4 fold over the control levels. This increase corresponds well with the increases in mRNA of 3-fold for P450IIC2 when the increase of P450IIC1 mRNA is also considered. P450IIC1 mRNA was not detectable in control animals but increased to levels similar to that of P450IIC2 in treated aninmals by 24 hours after treatment (Leighton & Kemper, 1984). The combined relative increase of P450IIC1 and P450IIC2 should therefore be about twice that of P450IIC2 alone or about 6-fold. This value is consistent with the observed 4- to 8-fold increase in transcription. In contrast, the amount of P450IIC3 RNA synthesized was significantly, but only slightly, increased at 6 hours after phenobarbital treatment, and was not significantly increased at the later times. This agrees with the observation that P450IIC3 mRNA levels did not increase by 24 hours after phenobarbital treatment. The small, transient increase in P450IIC3 transcription observed may not have resulted in detectable differences in P450IIC3 mRNA levels.

These results indicate that increases in transcription of responsive P450IIC genes are largely responsible for the

Table 3. Comparison of intron splice sites of P-450IIC and P450IIB2[a] genes.

Intron	P450	Amino Acid[b]	Donor	Intron size bp	Acceptor
	IIC2	Asn56	ACCAATgtaagt	(1100)	ttccagCTGTCA
1	IIC3	Gln56		(>1500)	ttccagCTAGCA
	IIC4	Lys56	ACTAAGgtaagt	(2500)	tcccagTTCTCA
	IIB2	Gln57	ATGCAGgtgaga	(3200)	ttgcagCTTCGC
	IIC2	Gly111	GATTTGgtaggt	(250)	ctttagGAGTCA
2	IIC3	Gly111	GATTAGgtatgc	(4500)	tgatagGAATTG
	IIC4	Gly111	GCCTTGgtaagt	(168)	tggcagGAATTG
	IIB2	Gly112	AATATGgtaaga	(300)	tgtcagGTGTGA
	IIC2	Ala161	CCAATGgtgggt	(2750)	cttcagCCTCAC
3	IIC3	Ala161	CTAATGgtatat	(4000[c])	cctcagCATCTC
	IIC4	Ala161	CCAATGgtggtg	(3000)	cttcagCCTTAC
	IIB2	GLY162	CCCAGGgtgaat	(2300)	ctccagGAGCCC
	IIC2	Gln214	GTACAGgtgaga	(1750)	tttcagTTCTGC
4	IIC3	Gln214	ATACAGgtaaaa	(2500[c])	atacagCTCTAC
	IIC4	Gln214	TTGCAGgtgaga	(1500)	ttttagGTCTAC
	IIB2	Gln215	AGCCAGgtccgt	(800)	ttacagGTGTTT
	IIC2	Gln273	GAACAGgtaaag	(3800)	ttgtagGAAAAG
5	IIC3	Lys272	GAAAAGgtagaa	(1200[c])	tgctagGAAAAA
	IIC4	Lys274	GAGAAGgtaaat	(?)	
	IIB2	Lys274	GAGAAGgtgagt	(500)	aaccagGAGAAG
	IIC2	Ala321	TCATAGgtatgt	(>3500)	
6	IIC3	Ala320	TTACAGgtatga	(>3000)	
	IIC5			(>3500)	tatcagCTAGGG
	IIB2	Glu322	TCACAGgtatat	(1600)	acgcagAGAAAG
	IIC3	Lys382	CCAAAGgtgata		
7	IIC5	Lys383	CCTAAGgtaatc	(2000)	atccagGGTACG
	IIB2	Lys384	CCCAAGgtgagg	(300)	tcccagAACACT
	IIC3	Gly430			ttttagGAAAAA
8	IIC5	Gly431	CAGCAGgtgaaa	(2500)	tctcagGAAAAC
	IIB2	Gly432	CCACAGgtgagg	(3200)	ccacagGAAAGC

[a]IIB2 is the rat P450IIB2 sequence (Mizukami et al., 1983).
[b]The amino acid is the amino acid at the junction encoded by the underlined codon.
[c]The size has not been defined precisely.

increases in mRNA after phenobarbital treatment. For P450IIC3, and presumably other constituitive members of the P450IIC subfamily, the loss of response to phenobarbital probably results from mutations that inactivate the phenobarbital responsive elements of the gene. Thus, alterations in both the regulatory and structural regions of the gene are important in the evolution of these new P450s.

Characterization of P450IIC genes

We have characterized the 5' flanking region and exons 1-6 for P450IIC2, the 5' flanking region and exons 1-5 for P450IIC4, exons 2-9 for P450IIC3 and exons 7-9 for P450IIC5. In each case the introns contain the canonical gt...ag sequence. The locations of the introns for the P450IIC genes are at the same sites as the intron locations in the P450IIB2 gene although the sizes of the introns vary considerably (Table 3). The estimated sizes of the P450IIC genes are at least 25,000 base pairs. The identity of the intron locations within the P450IIC and P450IIB subfamilies contrasts with the lack of identity observed between gene families.

The locations of the RNA initiation sites have been determined for the P450IIC2 (Govind et al., 1986) and the P450IIC4 genes. The region around the TATA sequence is stongly conserved in the two genes, but less conservation is observed at the transcription start site (Figure 1). A second weaker start site was observed for P450IIC4 about 50 base pairs upstream of the major one and an appropriately placed second TATA sequence was also present. Clones of P450IIC4 (Johnson et al., 1987) and P450IIC5 (Tukey et al., 1985) cDNAs began near the major TATA sequence (arrow in Figure 1) and presumably resulted from RNA initiation at an upstream promoter. Comparison of the 5' flanking regions of these two genes indicated that significant homology extended only to -150 base pairs from the cap site suggesting that this sequence was functionally important in transcription of

Figure 1. Comparison of the promoter regions of P450IIC2 and P450IIC4 genes. The RNA initiation sites are indicated by the # symbols and the TATA sequence is underlined. The arrow indicates the 5' ends of cloned P450IIC5 (Tukey et al., 1985) and P450IIC4 cDNAs (Johnson et al., 1987).

```
                          ↓                      #
IIC2  ATTAGCAGGAGTGTTATAAAAGCCTG-GCTTGC-AGCTT-CAGTTGT-CAGG
      ******* ***** ************* **** * ****  ***     ****
IIC4  ATTAGCAAGAGTGGTATAAAAGCCTGTGCTTCCCAGCTCACAGGATCCCAGG
                    -------                      #
```

the gene. Comparison of the 5' flanking regions of the three phenobarbital responsive genes, rabbit P450IIC2 and P450IIC4 and rat P450IIB2, did not reveal any sequences longer than 10 nucleotides that were greater than 80% similar in all three genes.

There are several discrepancies between the amino acid sequence derived from the P450IIC3 cDNA (Leighton et al., 1984) and the amino acid sequence of P450IIC3 (Ozols et al., 1986). The determination of the gene sequence provides an additional independent derivation and except for position 257 confirms the cDNA sequence (Table 4). For positions 84, 90 and 339, the predicted amino acids are also present in the other four P450IIC proteins. At position 343 the protein sequence agrees with the other P450IIC sequences, but differs from the cDNA and gene sequences. It is interesting that the only difference between the cDNA and gene protein sequences at position 257 falls within a tryptic peptide that has altered chromatographic properties in the two known polymorphic forms of P450IIC3 (Ozols et al., 1985).

CONCLUSIONS AND SUMMARY

The rabbit P450IIC gene subfamily contains at least 7 genes of which at least 5 are expressed in the liver. The responses of the genes to phenobarbital are heterogeneous and for the responsive genes phenobarbital increases transcription of RNA. The P450IIB and P450IIC subfamilies clearly evolved from a single ancestral gene based on sequence homology, identical intron locations, and similar mechanisms of phenobarbital induction. This subfamily of

Table 4. Differences between derived and directly determined P450IIC3 protein sequences.

POSITION	PROTEIN	cDNA	GENE	OTHER IIC PROTEINS
82	GLY	-	THR	ALA(4)
84	ILE	LYS	LYS	LYS(4)
90	TYR	ASP	ASP	ASP(4)
223	GLY	GAP	GAP	LEU(2) PHE(1) ILE(1)
257	SER	SER	LEU	VAL(2) ILE(2)
339	SER	MET	MET	MET(4)
343	SER	THR	THR	SER(4)
430	THR	ALA	ALA	ALA(2) THR(2)

Protein sequence is from Ozols et al., 1985.

genes should be an excellent model system to study the mechanism and evolution of phenobarbital induction of P450s.

REFERENCES

Atchison, M. & Adesnik, M., 1983, Journal of Biological Chemistry, 258, 11280.

Nebert, D. W., Adesnik, M., Coon, M. J., Estabrook, R. W., Gonzalez, F. J., Guengerich, F. P., Gunsalus, I. C., Johnson, E. F., Kemper, B., Levin, W., Phillips, I. R., Sato, R. & Waterman, M. R. (1987) DNA, 6, 1-11.

Dieter, H. H., Muller-Eberhard, U., & Johnson, E. F., 1982, Biochemical and Biophysical Research Communications, 105, 515.

Govind, S., Bell, P. A. & Kemper, B., 1986, DNA, 5, 371.

Finlayson, M. J., Kemper, B., Browne, N. & Johnson, E. F., 1986, Biochemical and Biophysical Research Communications, 141, 728.

Francis, V. N. K., Swarki, V. I. J. & Padmanaban, G., 1986, Nucleic Acids Research, 14, 2497.

Hagen, D.A. & Coon, M.J., 1976, Journal of Biological Chemistry, 251, 7929.

Hardwick, J. P., Gonzalez, F. J. & Kasper, C. B., 1983, Journal of Biological Chemistry, 258, 8081.

Johnson, E. F., Barnes, H. J., Griffin, K. J., Okino, S., & Tukey, R. H., 1987, Journal of Biological Chemistry, 262, 5918.

Leighton, J. K. & Kemper, B., 1984, Journal Biological Chemistry, 259, 11165.

Leighton, J. K., Debrunner-Vossbrinck, B. A., & Kemper, B., 1984, Biochemistry 23, 204.

Mizukami, Y., Sogawa, D., Suwa, Y., Muramatsu, M. & Fujii-Kuriyama, Y., Proceedings of the National Academy of Sciences USA, 80, 3958.

Ozols, J., Heinemann, F. S. & Johnson, E. F., 1985, Journal of Biological Chemistry, 260, 5427.

Pike, S. G., Shephard, E. A., Rabin, B. R. & Phillips, I. R., 1985, Biochemical Pharmacology, 34, 2489.

Tukey, R. H., Okino, S., Barnes, H. J., Griffin, K. J., & Johnson, E. F., 1985, Journal of Biological Chemistry, 260, 13347.

Zhao, J., Leighton, J. K., & Kemper, B., 1987, Biochemical and Biophysical Research Communications, 146, 224.

Structure and Expression of the Chicken Phenobarbital-inducible P-450 Gene Family

B.K. MAY, A.J. HANSEN, L.A. MATTSCHLOSS, G. SRIVASTAVA, I.A. BORTHWICK, C.J. ELFERINK, M.J. BAWDEN and W.H. ELLIOTT

Commonwealth Special Research Centre for Gene Technology, Department of Biochemistry, University of Adelaide, Australia

INTRODUCTION

In our laboratory, we are investigating the molecular mechanism by which drugs increase both the levels of the phenobarbital-inducible cytochrome P450s (P450) and the mitochondrial enzyme, 5-aminolevulinate synthase (ALV-S). ALV-S is the first enzyme of the heme biosynthetic pathway. In the liver, ALV-S is rate-limiting and under negative feedback repression by heme (May *et al.*, 1986). The level of both P450 and ALV-S can be greatly elevated in the livers of experimental animals following treatment with a wide variety of drugs such as 2-allyl-2-isopropylacetamide (AIA), 3,5-diethoxycarbonyl-1,4-dihydrocollidine (DDC) and phenobarbital (PB). Since P450 requires heme, it is logical that both proteins should be co-ordinately induced. We have proposed that drug-induced P450 apoprotein conjugates heme, thereby removing the inhibitory effect of heme on ALV-S production (May *et al.*, 1986). However, the mechanism by which drugs elevate levels of the PB-inducible P450s is poorly understood.

In this paper, we report studies on the drug-induction of P450 mRNAs in the chick embryo and corresponding studies on the regulation of ALV-S mRNA levels in the rat.

1. Regulation of PB-Inducible Chicken P450s

We have investigated the regulation of the PB-inducible P450 mRNAs in the livers of chick embryos. P450 cDNA clones have been isolated from a library synthesized using mRNA extracted from the livers of AIA treated chick embryos (Hobbs *et al.*, 1986). Two P450 cDNA clones, pCHP3 and pCHP7, have been fully sequenced. A schematic representation of these clones is shown in Figure 1. Both pCHP3 and pCHP7 have identical 5' non-coding regions, but differ in the

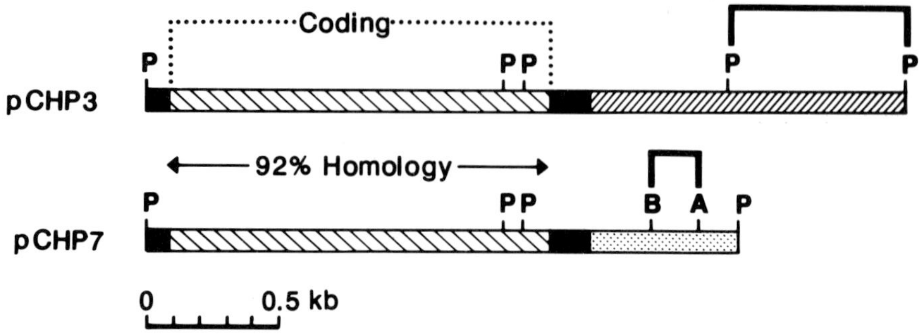

Figure 1. Schematic representation of the P450 cDNA clones, pCHP3 and pCHP7. The 5' non-coding and a portion of the 3' non-coding sequences of the clones are identical, as indicated by the darkened areas. The 3' non-coding restriction fragments used as specific probes for pCHP3 and pCHP7 are indicated (**P** = *Pst*1, **B** = *Bam*H1, **A** = *Ava*11).

length and sequence of their respective 3' non-coding regions. pCHP3 and pCHP7 each encode P450 proteins containing 491 amino acid residues, which share 92% homology in their primary sequences. Comparison of their derived protein sequences with those of the mammalian P450s reveals that the chicken P450s share the greatest extent of homology with members of the P450IIc subfamily. The chicken P450s share 56% homology with the poorly PB-inducible forms such as rabbit PBcI and 51% homology with a constitutive P450, rat P450f. Somewhat less homology (49%) was observed with the major rat PB-inducible form, P450b. When these comparisons were extended to members of other P450 families such as the TCDD-inducible rat isozyme P450c, only 30% homology was observed. Since the chicken P450s share <70% homology with other members of the P450IIc subfamily, they represent a unique P450 subfamily within the P450II family (Nebert *et al.*, 1987).

To characterize the chick embryo P450 mRNAs, Northern hybridization analysis was performed on total RNA isolated from the livers of chick embryos induced with AIA. Using either pCHP3 or pCHP7 as probes, three mRNA species, measuring 3.5, 2.8 and 2.2 Kb, were detected (Figure 2, lane 2). Although the three mRNAs were detectable in RNA isolated from the livers of non-induced chick embryos, they are not visible in the autoradiograph shown (Figure 2, lane 1). To determine the relationship between the three mRNAs and the two cDNA clones, pCHP3 and pCHP7, Northern hybridization analysis was performed using DNA probes from the 3' non-coding regions of the respective clones (refer to Figure 1). The 650 bp *Pst*1 DNA fragment at the extreme 3' end of pCHP3, hybridized only to the 3.5 Kb mRNA

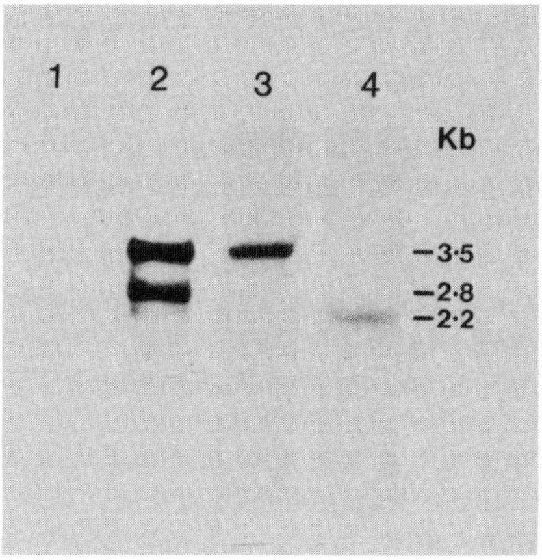

Figure 2. Characterization of chick embryo P450 mRNAs by Northern hybridization analysis. 20 μg of total RNA, isolated from non-induced (lane 1) and AIA-induced (lanes 2-4) chick embryo livers, was fractionated by electrophoresis on a 1% agarose gel containing 1.1 M formaldehyde. After transfer to nitrocellulose, the RNA was probed with nick-translated pCHP3 (lanes 1 & 2), or a nick-translated 3' non-coding fragment specific for either pCHP3 (lane 3) or pCHP7 (lane 4).

(Figure 2, lane 3), whereas the 228 bp *Bam*H1/*Ava*11 fragment, derived from the 3' non-coding region of pCHP7, hybridized specifically to the minor 2.2 Kb mRNA (Figure 2, lane 4). These results establish that the clones pCHP3 and pCHP7 are derived from the 3.5 Kb and 2.2 Kb mRNAs respectively.

The time course of AIA induction of the 3.5 and 2.2 Kb mRNAs by AIA was examined in the livers of chick embryos. As shown in Figure 3, AIA produces a rapid induction of both the 3.5 and 2.2 Kb mRNAs, with a 12-fold and 15-fold increase in the amount of these mRNAs respectively at 1 hour. The level of both mRNAs continued to

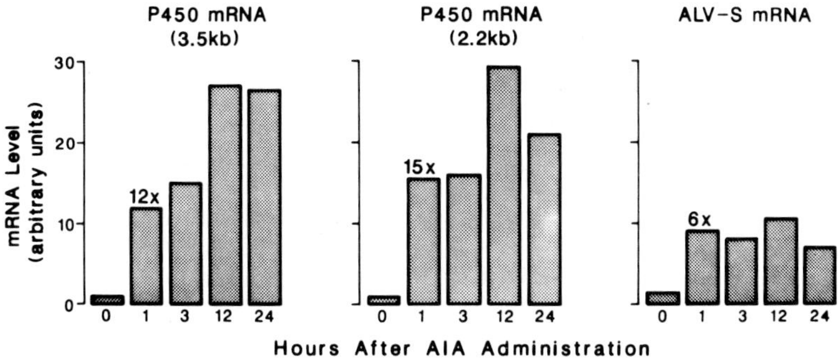

Figure 3. Time course of P450 and ALV-S mRNA induction by AIA. Total RNA (5 μg) from the livers of AIA-induced chick embryos, was applied to nitrocellulose and hybridized with specific [^{32}P]-labelled DNA probes. mRNA levels were quantitated by laser densitometry and are represented as arbitrary units in bar graphs.

increase until 12 hours, at which time, maximal induction of 25-30 fold was observed. In addition, ALV-S mRNA levels were increased 6-fold within 1 hour of AIA treatment and did not alter significantly for the duration of the experiment (Figure 3).

In order to determine whether increased levels of these mRNAs were due to gene activation, transcription "run-off" experiments were performed (Figure 4). Using nuclei isolated from the livers of chick embryos treated with AIA for 1 hour, it was established that transcripts for the 3.5 Kb mRNA were increased 2-fold. However, at 12 hours after AIA treatment, where the steady state level of the 3.5 Kb mRNA was maximal, the transcription rate of the gene encoding this P450 mRNA had declined to 50% of that observed in control nuclei. Since AIA did not affect the mRNA levels or the transcription of the genes encoding albumin and β-actin (results not shown), these results indicate that a significant post-transcriptional mechanism contributes to the increased level of the 3.5 Kb P450 mRNA by AIA. In contrast, the 6-fold increase in ALV-S mRNA measured at 1 hour following AIA administration, was due entirely to transcriptional activation of the gene (refer to Figures 3 & 4). The small size of the DNA probe specific for the 2.2 Kb mRNA made it difficult to accurately quantitate the level of this species.

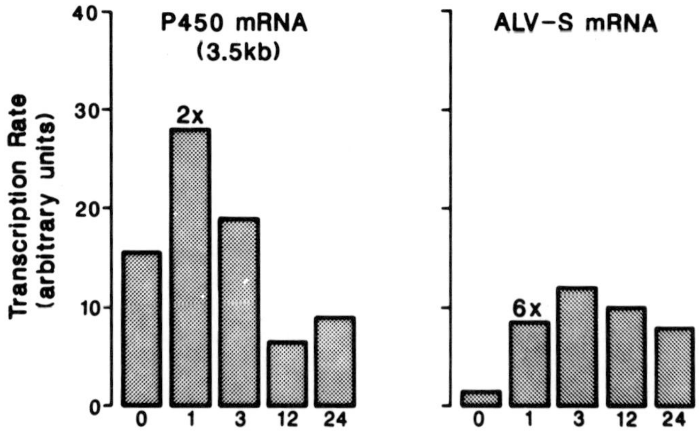

Figure 4. Time course of transcription rates for P450 and ALV-S genes. Nuclei were isolated from the livers of AIA-treated chick embryos and [^{32}P]-labelled RNA transcripts hybridized to filter-bound DNA. The level of RNA transcripts was quantitated by laser densitometry.

Studies were performed in adult hens to determine if these P450 mRNAs were expressed in a tissue-specific fashion. Total RNA was isolated from the liver, kidney, heart and spleen of both non-induced and AIA-induced hens. Northern hybridization analysis using pCHP3 as a probe revealed that all three mRNAs were expressed in non-induced liver and were elevated by AIA (Figure. 5). Using the specific probes, slot-blot hybridization analysis of the RNA indicated that the 3.5 Kb and 2.2 Kb mRNAs were induced 16 and 11-fold respectively (results not shown). Induction of all three mRNAs was also observed in the kidney (Figure 5), although in contrast to the liver, the basal levels were undetectable by Northern analysis. Neither the heart nor spleen expressed detectable levels of these P450 mRNAs. In parallel studies, it was shown that the AIA induction of ALV-S mRNA occurred only in the liver and kidney.

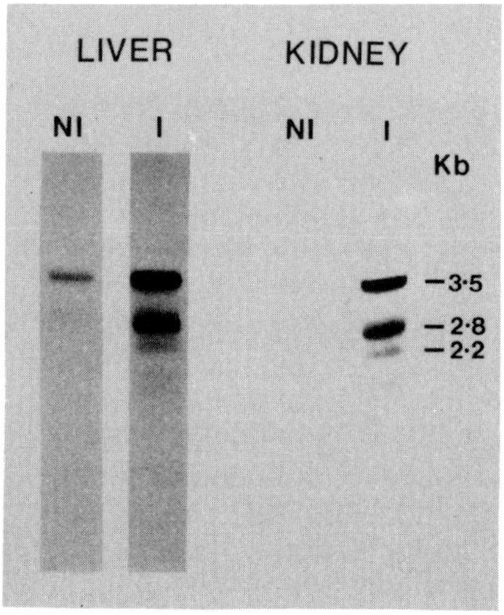

Figure 5. Tissue-specific expression of P450 mRNA in adult hens. Total RNA (50 µg), prepared from non-induced (NI) and induced (I) liver and kidney of adult hens, was fractionated by electrophoresis on a 1% agarose gel containing 1.1 M formaldehyde. RNA was transferred to nitrocellulose filters and probed with nick-translated pCHP3.

The cDNA clone, pCHP3, was used as a probe to screen a Charon 4A chicken genomic library and several λ clones were isolated. Restriction mapping and Southern hybridization analysis indicated that these clones represented two highly homologous genes, denoted A and B (Mattschoss *et al*., 1986). Preliminary results suggest that the chicken P450 A gene is at least 18 Kb in length, whereas the P450 B gene spans 14 Kb of DNA. Southern hybridization analysis of genomic clones using DNA probes specific for the 3.5 Kb and 2.2 Kb mRNAs, has established that both the chicken P450 A and B genes are transcriptionally active. Whereas the 3.5 Kb is derived from the A gene (Mattschoss *et al*., 1986), the 2.2 Kb mRNA is derived from the B gene.

2. Regulation of ALV-S Gene Expression

Parallel studies have determined how the synthesis of ALV-S mRNA is regulated by drugs. A full length cDNA clone for rat liver ALV-S has been used to characterize ALV-S mRNA in different tissues. A single mRNA species measuring 2.3 Kb is present in all tissues examined and RNase protection studies established that the ALV-S mRNA is identical in these tissues. Of particular interest is the finding that the erythroid and non-erythroid mRNAs are identical since it has been generally assumed that different isozymes exist in these tissues.

The drug induction of ALV-S mRNA is tissue-specific. Treatment of rats with AIA for 12 hours increased ALV-S mRNA levels 8-fold and 6-fold in the liver and kidney respectively (Figure 6). This increase in ALV-S mRNA is almost completely prevented by treatment with heme, or one of its precursors, 5-aminolevulinate, in the liver and kidney. Basal levels in these tissues are also repressed by heme treatment (Figure 6).

We have investigated whether AIA and heme affect ALV-S mRNA levels in the liver by regulating ALV-S gene transcription. As shown in Figure 7, treatment of rats for 12 hours with AIA, caused a 12-fold increase in the level of ALV-S gene transcription. In the presence of heme, ALV-S gene transcription was reduced to non-induced levels. Expression of the albumin gene was unaffected by AIA or heme (Figure 7). Since the rates of ALV-S gene transeription correlate well with alterations in ALV-S mRNA levels, it is concluded that both AIA and heme act predominantly at the transcriptional level to control ALV-S mRNA levels. As AIA induces ALV-S mRNA only in the liver and kidney of rats, it was therefore of interest to see if a similar tissue-specific expression occurred for the rat PB-inducible P450 b/e mRNAs. Although P450 b/e mRNA was detected in several tissues, induction occurred only in the liver and kidney (results not shown).

DISCUSSION

In the livers of chick embryos, three P450 mRNA species, measuring 3.5, 2.8 and 2.2 Kb, are elevated by drugs such as AIA. The 3.5 and 2.2 Kb mRNAs encode highly homologous P450 isozymes with the relative amounts of these mRNAs differing greatly in the induced liver. Interestingly, the AIA induced increase in the level of the 3.5 Kb mRNA is due to both gene activation and a significant post-transcriptional mechanism, possibly mRNA stabilization. The latter is not a general response to AIA-induction since drug-induced ALV-S mRNA levels were accounted for entirely by gene activation.

Previous studies (Mattschoss *et al.*, 1986) identified two homologous PB-inducible P450 genes in the chicken haploid genome. Although both genes are transcriptionally active, the relationship of the 2.8 Kb mRNA to these genes was previously unclear. However, recent analysis of cDNA clones corresponding to the 2.8 Kb mRNA indicates that it is derived from a separate gene.

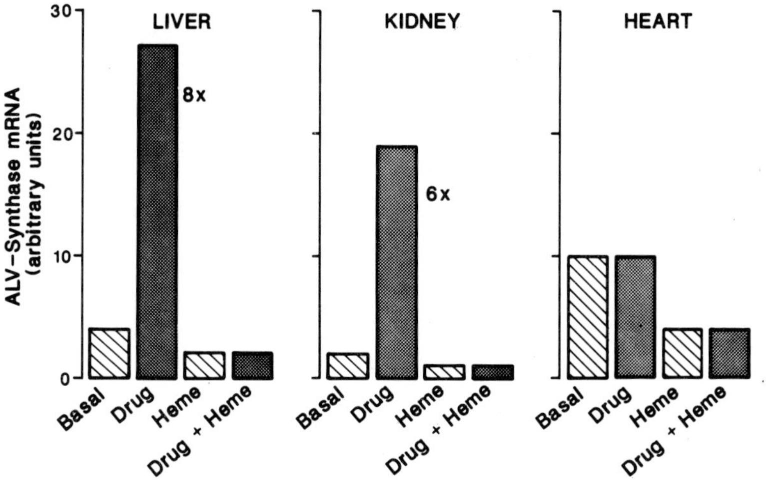

Figure 6. Effect of AIA and heme on ALV-S mRNA levels in different rat tissues. Rats were treated with AIA for 12 hours in the presence or absence of heme. Total RNA (5 μg) prepared from the liver, kidney and heart, was applied to nitrocellulose filters and hybridized with a nick-translated cDNA probe for rat ALV-S. mRNA levels were quantitated by laser densitometry and are represented as arbitrary units.

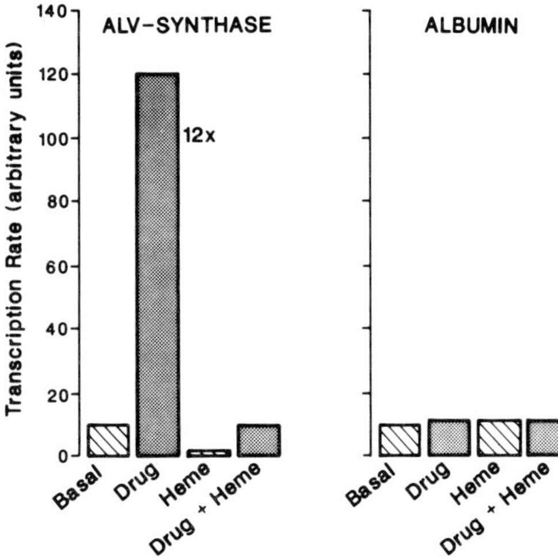

Figure 7. Effect of AIA and heme on rat ALV-S and albumin gene transcription. Nuclei were isolated from the livers of control rats and AIA-induced rat (12 hours), in the presence or absence of heme. [^{32}P]-labelled transcripts were hybridized to filter-bound DNA specific for ALV-S or albumin. RNA transcripts were quantitated by laser densitometry and are represented as arbitratry units by bar graphs.

We have shown that in both rats and adult hens, AIA-induction of ALV-S mRNA and the PB-inducible P450 mRNAs is limited to the liver and kidney, but heme repression occurs even in the heart, which is non-inducible. In addition, it has been established unequivocally, that heme inhibits transcription of the ALV-S gene. These results are there-fore consistent with the model proposed earlier, in which drug-induced P450 apoprotein removes heme, causing derepression of the ALV-S gene and a subsequent increase in the supply of heme for P450 apoprotein.

REFERENCES

Hobbs, A.A., Mattschoss, L.A., May, B.K., Williams, K.E., & Elliott, W.H., 1986, Journal of Biological Chemistry, 261, 9444.

Mattschoss, L.A., Hobbs, A.A., Steggles, A.W., May, B.K., & Elliott, W.H., 1986, Journal of Biological Chemistry, 261, 9438.

May, B.K., Borthwick, I.A., Srivastava, G., Pirola, B.A., & Elliott, W.H., 1986, Current Topics in Cellular Regulation, 28, 233.

Nebert, D.W., Adesnik, M., Coon, M.J., Estabrook, R.W., Gonzalez, F.J., Guengerich, F.P., Gunsalus, I.C., Johnson, E.F., Kemper, B., Levin, W., Phillips, I.R., Sato, R., & Waterman, M.R., 1987, DNA, 6, 1.

Characterization of the Human Homologues to Rat Liver Phenobarbital-inducible Cytochromes P-450

C.R.WOLF [1], R.R. MEEHAN [1,2], J.S. MILES [1], T.J.JOWETT [3], N.K. SPURR [4], K. STEVENSON [1], L.M. FORRESTER [1], J. GOSDEN [2], and N. HASTIE [2]

[1] Imperial Cancer Research Fund, Edinburgh, UK, [2] MRC Clinical and Population Cytogenetics Unit, Edinburgh, UK, [3] Department of Genetics, University of Newcastle, UK and [4] Imperial Cancer Research Fund, London, UK

INTRODUCTION

The characterization of human cytochrome P-450 genes and the extent to which their polymorphic expression relates to susceptibility to environmental diseases are outstanding problems in this research area. (1) In particular, the importance of constitutively expressed P-450 forms relative to those which are highly inducible is still unclear.

In the rat, the P450-2 family contains members which are either marginally (eg. PB_1*, locus P450-2C), or highly inducible by phenobarbital (eg. PB_3, locus P450-2B), as well as proteins which are not regulated by this compound (UT_1, locus P450-2A). Both PB_1 and UT_1 are constitutively expressed in high levels. The P450-3 family also contains phenobarbital-inducible forms.

In the work described below a molecular and genetic approach has been taken to study the chromosomal organization, structure, function and regulation of proteins within the human and rodent P450-2 and P450-3 families (2).

RESULTS AND DISCUSSION

Genomic Complexity of the P450-2 Families

Southern blots of genomic DNA probed with a cDNA coding for rat PB_1 or PB_3 demonstrate the complexity of the P450-2C and P450-2B families respectively (Figure 1). These blots also show the high degree of conservation of these genes through evolution. It is clear that in many species several genes must exist in each family. This

* Footnote: PB_1, PB_3 and UT_1 are equivalent to Levin forms k, b and a respectively (4).

Figure 1. <u>Southern Blot Analysis of a variety of genomic DNA samples probed with pTF-1 and pR17</u>

pTF1 (Fig. 1A cut with Bam H1) and pR17 (Fig. 1B) are cDNA probes for the P450-2C and P450-2B loci respectively. In Fig. 1A and 1B lanes 3-9 and 1-8 are DNA from different mouse strains.

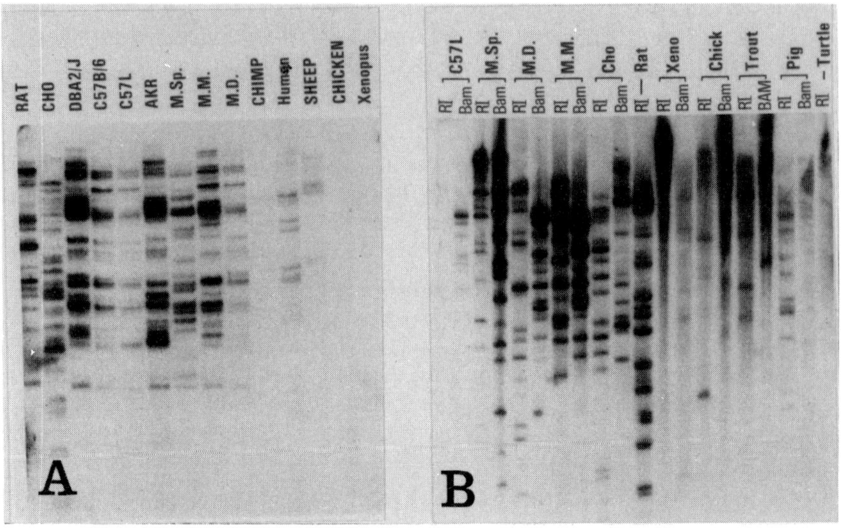

observation is confirmed for the PB_1 group by the isolation of at least six distinct proteins from rat liver (3,4). Only two PB_3 proteins have been isolated to date. However, one would predict that more must exist. The hybridization of the rat PB_3 cDNA probe to trout DNA is worthy of note in view of the lack of responsiveness of this species to phenobarbital (5). In small mammals the P450-2B and P450-2C families appear more complex than in human (6). There appear to be a minimum of 7-8 genes within the P450-2C family in rats and mice (Figure 2) and a minimum of 2-3 genes in this family in humans (6). The human P450-2B family also appears to be small and on Southern blots the bands cover approximately 20kb of genomic DNA. It would therefore be surprising if this family contained more than 2-3 genes. This is in marked contrast to the rat where at least 11 genes have been identified (7). In contrast the P450-2A family appears to be more complex in human than in the rat. These differences in gene number make it obviously difficult

Figure 2. <u>Gene Count for the Mouse Cytochrome P450-2C Family</u>.

DNA samples were cut with Eco R1 (A), Bam H1 (B) or Hind 111 (C) and probed with a 5' cDNA probe (A'B'C') or a 3' probe (A"B"C"). The dots represent restriction fragments in common.

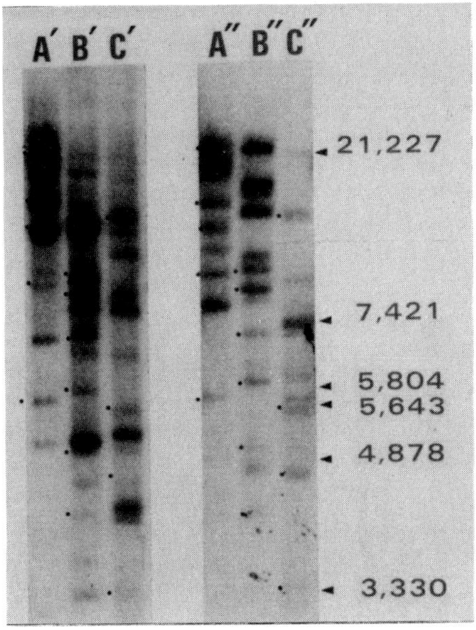

to equate specific P-450's between species. Table 1 compares the sequence homology of P-450's in rat and human. Three rat sequences within the P450-2C ie. PB_1, f and h, all show approximately the same homology to the human PB_1 sequence, exemplifying the problems of identifying orthologous genes.

Table 1. <u>Amino Acid Sequence Homology Between Human and Rat cDNA's Within the Families P450-2A, P450-2B, P450-2C and P450-3</u>.

	HUMAN				RAT					
LOCUS	C2C	C2B	C2A	C3	2C			2B	2A	3
NAME OF PROTEIN	PB_1	PB_3	UT_1	PCN	PB_1	f	h	PB_{3a}	UT_1	PCN
Human PB_1	100	52	44	21	75	71	75	50	48	18
Human PB_{3a}		100	51	25	53	53	54	75	54	24
Human UT_1			100	21	50	47	47	51	60	18
Human PCN				100	22	21	25	25	25	74

The PB_{3a} and UT_1 sequences used covered a region of 217 and 331 amino acids at the carboxyl terminal ends of the proteins.

Further comparison of the human cDNA sequences shows approximately 50% homology between the proteins from the P450-2A, P450-2B and P450-2C families. This extent of homology is also found between these families in the rat (8). Over the region studied the human PB_3 protein was highly homologous (75%) to the rat sequence. The UT_1 sequence (family 2A) was 60% homologous. In view of this level of homology it is at present not certain that the rat and human UT_1 clones are indeed from the same gene family. In general however, the high interspecies sequence homology (70-80%) for homologous P-450 families, and the low intraspecies homology (approx. 30-35%) between different families demonstrates that the families diverged before the divergence of mammals. This will explain why they are scattered through the human genome.

GENETIC ANALYSIS OF GENES WITHIN THE P450-2C FAMILY

The multiplicity of the mammalian P-450 system make it difficult to study. The use of molecular genetics provides a powerful tool in the characterization of the genomic organization of the genes and also in ascribing functions to specific gene loci. For such studies the mouse is the best characterized mammalian species. Recombinant inbred (RI) strains* were used to establish whether the apparently large number of genes are clustered together within the P450-2C locus and whether, through segregation with known genetic markers, the chromosomal location could be identified.

This type of analysis was possible because of restriction fragment length polymorphisms (RFLP's) in the P450-2C locus between mouse strains (9). Three clearly distinguishable DNA haplotypes were observed and the loci for which were termed $P450\text{-}2C^a$, $P450\text{-}2C^b$ and $P450\text{-}2C^c$. Out of 81 RI lines studied no recombination could be observed within the P450-2C locus, even when a variety of restriction enzymes were used (Figure 3). These data demonstrate that the mouse P450-2C locus is a family of clustered genes within 0.8cM of genomic DNA. Unfortunately, close linkage with

* Footnote: Recombinant inbred strains result from the systematic inbreeding of the F_2-cross between progenitor mouse strains. This essentially randomizes the genetic differences between the parent strains. Each line is typed for chromosomal markers and the strain distribution pattern (SDP) for each marker established (10). Unknown markers can be typed and their SDP compared to known markers, allowing linkage to be established.

Figure 3. Segregation pattern of the C57BL/6 vs DBA/2J haplotypes in recombinant inbred lines.
A) Southern blot analysis (Sst 1 digest) B) Summary of Data

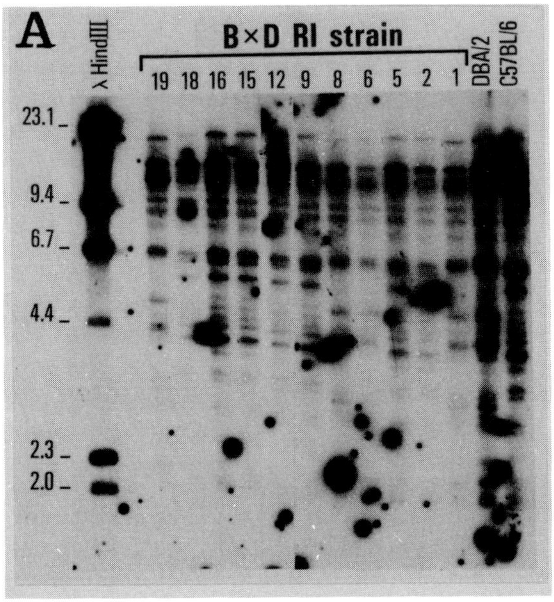

B DISTRIBUTION OF PB-I RFLP's in B x D RI strains.

RESTRICTION ENZYME	No. of fragments	Pattern for Strain No.																									
		1	2	5	6	8	9	11	12	13	14	15	16	18	19	20	21	22	23	24	25	27	28	29	30	31	32
Sst I	≥9	D	B	B	I	D	B	D	B	B	B	B	B	B	D	D	D	B	D	B	B	B	B	B	B	B	D
Hind III	3	D	B	B	D	D	B	D	B	B	B	B	B	B	D	D	D	B	D	B	B	B	B	B	B	B	D
Taq I	3	D	B	B	D	D	B	D	B	B	B	B	B	B	D	D	D	B	D	B	B	B	B	B	B	B	D

mapped genetic markers could not be established in these experiments. However loose linkage (20 out of 26 lines) with the Got-1 locus was observed. This locus maps to mouse chromosome 19. Somatic cell hybrids and *in situ* hybridization were therefore used to map the P450-2C. Using a series of mouse-chinese hamster hybrids there was 100% concordance between the presence of mouse specific P450-2C DNA fragments and Got-1. *In situ* hybridization confirmed the mapping of P450-2C to the teleomeric end of mouse chromosome 19.

In man the P450-2C locus maps to chromosome 10q 24.1 to 10q 24.3. (6) The Got-1 locus also maps to this chromosome, indicating this region is syntenic with the region on mouse chromosome 19.

In addition to the applications described above, RI analysis can be used to correlate P-450 functions with specific gene loci. Such a strategy has been used to correlate the coumarin hydroxylase locus with the major phenobarbital-inducible gene family (P450-2B) (11). Using this approach we found close linkage between the P450-2C and the constitutive metabolism of benzo(a)pyrene (AHH) (9). Concordance in 17 out of 18 lines was observed. These data provide strong evidence that constitutive (AHH) activity is mediated by a protein, or proteins, within this locus. This evidence is supported by the finding that antibodies to P450-PB_1 will inhibit this activity in hepatic microsomes from untreated or phenobarbital-treated rats (12,13).

POLYMORPHIC EXPRESSION OF CYTOCHROME P-450's

The probability that some human cytochrome P-450 families are less complex than in small mammals implies that there must be species differences in drug metabolizing capacity. This therefore complicates the extrapolation of rodent data to man. The expression of isolated P450 cDNA's in model systems will be important in evaluating where the differences lie. Cytochrome P-450-mediated drug metabolism exhibits significant variation in humans. This can be ascribed to a variety of causes including; differential exposure to environmental factors, differences in the levels of endogenous modulators such as steroid hormones and lymphokines, or genetic polymorphism. The latter could result from mutations in the structural-genes themselves or within genes which regulate their expression. In order to establish the individual variation in P-450 isozyme expression the levels of proteins within the P450-2A, P450-2B, P450-2C and P450-3 families was determined in a series of fifteen human liver samples by Western blot analysis

(Table 2). Antibodies to rat proteins UT_1, PB_{3a}, PB_1 and PB_{2c} were used to identify homologous proteins for these families respectively(3).

Table 2. <u>Fold Variation in Human Liver Cytochrome P-450 Isozymes</u>

LOCUS	P450-2A		P450-2B		P450-2C		P450-3
ANTIBODY USED	UT_1		PB_{3a}		PB_1		PB_{2c}
	A	B	A	B	A	B	A
FOLD VARIATION	340	6	53	5	21	5	300

In some cases two proteins were recognized by the antibody. A and B represent the lower and higher m.wt. bands respectively.

A subjective assessment of hepatic isozyme content indicated that in general proteins homologous to rat UT_1 and PB_1 were the most abundant (not shown). The levels of PB_3-like proteins being very low. However, significant differences between individuals in the expression of all the isozymes was observed (Table 2). The very high variation in some cases was primarily due to the extemely low levels of certain iszymes in some of the samples. One of the preparations studied was from an individual receiving the drug phenytoin. In this case the level of almost all of the isozymes was signficantly elevated (Burke <u>et</u> <u>al</u>. unpublished) indicating the modulation of proteins within these families by exogenous chemicals.

The extent to which genetic or environmental factors are responsible for the variability in the P-450 expression observed is unclear. In the case of PB_1-like proteins, which includes the protein involved in the polymorphic metabolism of the anticonvulsant mephenytoin (14,15), it has been proposed that the 'poor metabolizer' phenotype results from the synthesis of an aberrant protein rather than a reduced level of expression (Meyer <u>et</u> <u>al</u>. this issue). The variation observed here therefore would appear to be due to other factors. The above findings of Meyer <u>et</u> <u>al</u>. contrast to the apparent basis for the debrisoquine polymorphism (16) which has recently been ascribed to mRNA splicing defects (Gonzalez <u>et</u> <u>al</u>. this issue). This latter mechanism gains further interest as we have recently isolated a cDNA clone for a protein within the P450-2B family which is missing exon

8 and appears to still contain part of intron 5. Whether this is indicative of an additional genetic polymorphism remains to be established.

MECHANISM OF REGULATION OF P-450's WITHIN THE P450-2B AND P450-2C LOCI

It is probable that both environmental and genetic factors will be important in determining individual susceptibility to chemicals. The mechanisms involved in the regulation of cytochrome P-450 by environmental influences and the extent to which this occurs in man are still poorly understood. This is particularly the case for the regulation of P-450's in the families described above. In experimental animals some of these proteins are constitutively expressed in high levels and others are only found in the liver after the administration of drugs such as phenobarbital. In order to investigate factors governing P-450 regulation the inductive properties of an 'endogenous-type' inducer, the synthetic glucocorticoid dexamethasone, was investigated. It was highly surprising that this compound was as potent as phenobarbital in inducing P-450's in the mouse. This effect was species specific and did not occur to a significant extent in the rat. Measurement of mRNA levels indicated that the species difference was at the transcriptional level (Figure 4). Noshiro and Negishi (17) reported that Little mice, which have a growth hormone defficiency, have a high hepatic level of phenobarbital-inducible P-450's. It was therefore feasible that the induction by dexamethasone in the mouse was through effects on the pituitary. This possibility was strongly supported by the finding that hypophysectomy caused a large induction of PB_1 and PB_3 mRNA levels in the mouse liver but had only had marginal effects in rats. (Figure 4) From these experiments several potentially important conclusions can be reached. The pituitary produces factors which suppress the production of phenobarbital-inducible proteins in the mice and therefore plays a major role in the regulation of these proteins. The production of this suppressor may be inhibited by glucocorticoid-like molecules and thirdly, such compounds may prove to be phenobarbital-like inducing agents in man.

ACKNOWLEDGEMENTS

The authors would like to thank Dr. T. Friedberg and Dr. M. Adesnik for the use of the pTF-1 and pR17 cDNA probes and A. Ward for expert secretarial assistance.

Figure 4. <u>Quantitation of Hepatic mRNA Levels in Rats and Mice Following Dexamethasone-Treatment or Hypophysectomy</u>.

Figure 4a. A = Mouse, B = Rat. RNA samples (10 ug) were: 1, control; 2, phenobarbital and 3, dexamethasone-treated. Mouse blots were probed with pR17 and rat blots with a synthetic oligonucleotide for Family P450-2B.

Figure 4b. Mouse liver mRNA (10 ug) was probed with pTF_1, (P450-2C) or pR17 (P450-2B). C = untreated. Hypox = hypophysectomized, A + O = Adrenalectomized/ Ovarectomized. PB = phenobarbital-treated. Dex = dexamethasone-treated.

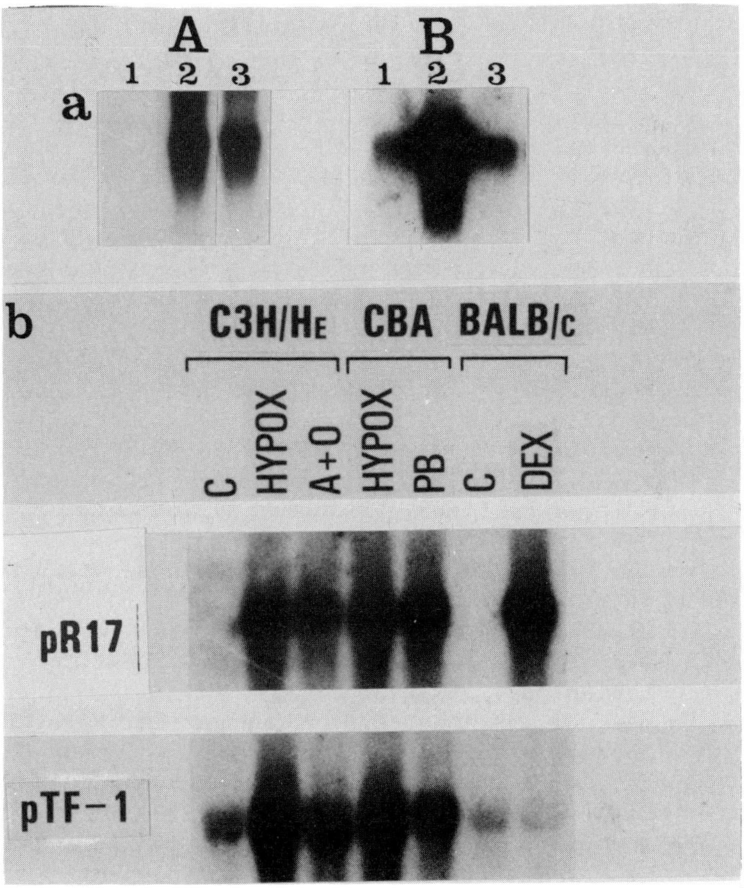

References

1. Wolf, C.R. (1986). Trends in Genetics, 8: 209-214.
2. Nebert, D.W., Adesnik, M., Coon, M.J., Estabrook, R.W., Gonzalez, F.J., Guengerich, F.P., Gunsalus, I.C., Johnson, E.F., Kemper, B., Levin, W., Phillips, I.R. Sato, R. & Waterman, M.R. (1987), DNA, 6: 1-11.
3. Wolf, C.R., Seilman, S., Oesch, F., Mayer, R.T. & Burke M.D. (1986), Biochemical Journal, 240: 27-33.
4. Ryan, D.E., Iida, S., Wood, A.W., Thomas, P.E., Lieber, C.S., & Levin, W. (1984), Journal of Biological Chemistry, 259: 1239-1250.
5. Elcombe, C.R. & Lech, J.J. (1979), Toxicology and Applied Pharmacology, 49: 437-450.
6. Meehan, R.R., Gosden, J.R., Rout, D., Hastie, N.D., Friedberg, T., Adesnik, M., Buckland, R. van Heyningen, V., Fletcher, J., Spurr, N. Sweeney, J. & Wolf, C.R. (1988), American Journal of Human Genetics (in press).
7. Adesnik, M. & Aitchison, M. (1986), C.R.C. Critical Review, Biochemistry, 19: 247-309.
8. Friedberg, T., Waxman, D.J., Aitchison, M., Kuman, A., Haaparanta, T., Raphael, C. & Adesnik, M. (1986) Biochemistry. 25: 7975-7983.
9. Meehan, R.R., Speed, R.M., Gosden, J.R., Rout, D. Hutton, J.J., Taylor, B.A., Hilkens, J., Kroezen, V., Hilgens, J., Adesnik, M., Friedberg, T., Hastie, N.D. & Wolf, C.R. (1987) (submitted)
10. Taylor, B.A. (1978), In "Origins of Inbred Mice", ed. Morse, C.H. (Academic Press, New York), 423-438.
11. Simmons, D.L., & Kasper, C.B. (1983), Journal of Biological Chemistry, 258: 9585-9588.
12. Adams, D.J., Seilman, S., Amelizad, Z., Oesch, F. & Wolf, C.R. (1985), Biochemical Journal, 232: 869-876.
13. Burke, M.D. & Wolf, C.R. (1987) 10th European Workshop on Microsomes and Drug Oxidations, In: "Drug Metabolism - from Molecules to Man, eds. Benford, D.J. Bridges, J.W. & Gibson, G.G. (Taylor & Francis, London) pp. 219-243, (1987).
14. Shimada, T., Misono, K.S. & Guengerich, F.P. (1986), Journal of Biological Chemistry. 261: 909-921.
15. Kupfer, A. & Preisig, R. (1984), European Journal of Clinical Pharmacology, 26: 753-759.
16. Mahgoub, A., Idle, J.R., Dring, L.G., Lancaster, R. & Smith, R.L. (1979), Lancet. 2: 584-586.
17. Noshiro, M. & Negishi, M. (1986), Journal of Biological Chemistry, 261: 15923-15927.

Structure and Expression of Genes Coding for Members of the Cytochrome P-450 II Family of Man

IAN R. PHILLIPS, ELIZABETH A. SHEPHARD [1], CHRISTOPHER GRIFFITH [1], INES SANTISTEBAN, BENJAMIN M. GAMMON, ARABELLA HUTTER, DAVID R. BELL and LESLEY A. FORREST [1]

Department of Biochemistry, St. Bartholomew's Hospital Medical College, University of London, Charterhouse Square, London EC1M 6BQ, UK
and [1] Department of Biochemistry, University College London, Gower Street, London WC1E 6BT, UK

INTRODUCTION

Cytochrome P-450-mediated mono-oxygenases play an important role in the synthesis of several essential endogenous compounds such as steroids, fatty acids and prostaglandins. They are also of central importance in the detoxification of a wide variety of foreign hydrophobic compounds, including many environmental pollutants, chemical carcinogens and therapeutic drugs. In effect, the role of cytochromes P-450 in protecting the body against foreign chemicals is as important as that of antibodies in dealing with invading organisms. Paradoxically, however, the mono-oxygenases can convert some compounds into active carcinogens. The ability of organisms to deal with such a tremendous range of naturally occurring and synthetic hydrophobic compounds is largely due to the versatility of the mono-oxygenase systems. This, in turn, can be ascribed to three major factors: (i) the multiplicity of cytochromes P-450, (ii) the ability of organisms to induce cytochromes P-450 in response to foreign chemicals; and (iii) the low substrate specificity of many cytochromes P-450.

Cytochromes P-450 are a superfamily of haemoproteins that range in amino acid sequence homology from < 20% to >97%. Amino acid sequence homology provides the basis for the recent classification of the superfamily into families and sub-families (Nebert et al., 1987). In the same species, cytochromes P-450 with ≪ 36% homology are assigned to different families (designated by Roman numerals). Within the same family, proteins between ~40 and ~65% homology are placed into different sub-families (designated by capital letters). Thus individual components of the same sub-family have homologies of ~70% or more and are designated by Arabic numerals. To date, as many as eight different mammalian cytochrome P-450 families have been identified. The largest of these (the P450II family) contains at least five sub-families (P450IIA-E).

Our group is investigating the structure and function of components of the mono-oxygenase system and the organization and expression of the corresponding genes. In the course of our work we have purified, and raised antibodies to, the major protein components of the mono-oxygenase of rat liver microsomal membranes, namely NADPH-dependent cytochrome P-450 reductase, cytochrome b_5, NADH-dependent cytochrome b_5 reductase and several different cytochromes P-450. We have isolated cloned DNAs coding for these proteins and for the corresponding proteins of man. The present communication will concentrate on some of our work on the structure and expression of the P450II gene family of man.

RESULTS AND DISCUSSION

Isolation of cDNA clones for human cytochromes P-450

Individual human variations are well known in the metabolism of certain therapeutic drugs, and it is likely that similar variations exist in the activation of chemical carcinogens. This metabolic variability may be due, in part, to genetically determined factors relating to the structure and expression of cytochromes P-450. As a first step towards investigating this, several groups, including our own, have isolated cloned DNAs coding for cytochromes P-450 of man. Our approach has been to use our rat cytochrome P-450 cDNA clones as molecular hybridization probes to screen human liver cDNA libraries. This has enabled us to isolate cloned cDNAs coding for several members of the P450I and II families of man. The inter-species sequence homology of components of the P450II family is such that we have been able, by varying the stringency of hybridization, to use a single rat cDNA clone coding for P450IIB2 (P-450e) (Phillips et al., 1983) to isolate cloned sequences coding for members of several different P450II sub-families of man. In this paper we will present data on the human P450IIA, B and C gene sub-families.

The P450IIA gene sub-family

Some of the cloned human cDNAs that we have isolated

code for a cytochrome P-450 whose amino acid sequence is 53% homologous to rat P450IIB2 and 65% homologous to rat P450IIA1 (P-450a) (Phillips et al., 1985a). At present it is thought that these cDNAs encode a previously uncharaterized member of the P450IIA sub-family which has been designated P450IIA2. Analysis of human genomic DNA by Southern blot hybridization revealed that the P450IIA2 cDNA hybridized to several DNA fragments whose total size was about 50,000 base pairs (Phillips et al., 1985a). No additional fragments were detected when the stringency of hybridization was decreased, whereas increasing the stringency caused no preferential loss of hybridization to any of the DNAs detected. Thus, in man, P450IIA is encoded by a small gene sub-family whose components are closely related. Analysis of genomic DNA from representatives of different types of primate showed that the size of this gene sub-family has been conserved throughout primate evolution. A small P450IIA gene sub-family was also found in non-primate mammals such as cow, horse and cat (E.A.S. and I.R.P., unpublished). The internal organization of the human P450IIA genes (A.H., I.R.P. and E.A.S. unpublished) is similar to that of the P450IIB gene of rat (Suwa et al.,1985).

Sequence analysis of cDNA and genomic DNA clones revealed that the P450IIA gene sub-family of man contains a minimum of three distinct genes, at least two of which are expressed in the liver. The amino acid sequence homologies of the proteins encoded by these genes are between 85 and 95% (Fig. 1). Cytochromes P-450 appear to have evolved at a rate of a 1% change in amino acid sequence about every 4 million years (Nebert et al., 1987). Thus it is likely that such closely related members of the P450IIA sub-family would have arisen via gene duplications that occurred some time after the radiation of mammals took place about 85 million years ago. If this were so, non-primate mammals would not be expected to possess the corresponding proteins, and this would have important implications for the extrapolation of drug metabolism data from experimental animals to man.

We have used Southern blot hybridization of DNA isolated from a panel of independent human-rodent somatic cell hybrids, together with in situ hybridization of human metaphase chromosome spreads, to map the P450IIA gene sub-family to the long arm of human chromosome 19 (Phillips et al., 1985b). Examination of stable reciprocal

Fig. 1. Nucleotide sequence comparison of three members of the human P450IIA sub-family in the region corresponding to exon 5. The top and bottom lines show the sequence of two cDNA clones isolated from a library prepared from a single individual. The middle line shows the sequence of a genomic DNA clone. Nucleotides in the middle and bottom lines are given only at positions where they differ from that in the top line.

```
CAG CTC TAT CAG ATG TTC TCT TCG GTG ATG AAA CAA CTG CCA GGA
                                              C
                                              C

CCA CAG CAA CAG GCC TTT CAG TTG CTG CAA GGG CTG GAG GAC TTC
                        A   GA


ATA GCC AAG AAG GTG GAG CAC AAC ACG CCA --- CTG GAT CCC AAT
 T            A                 CA  TGC ACG
                                CA   GC ACG

TCC CCA CGG GAC TTC ATT GAC TCC TTT CTC ATC CGC ATG CA
                              G
```

chromosomal translocations having break points on 19q enabled us to map the regional localization of the gene sub-family more precisely to 19q 13.1-13.3 (i.e. the central third of the long arm of chromosome 19) (Davis et al., 1986).

Analysis of genomic DNA from 30 European whites with 24 different restriction endonucleases identified only one restriction fragment length polymorphism (RFLP) associated with the P450IIA2 gene. This RFLP, which is identified by the enzyme SstI, has a frequency of 45% and a codominant segregation pattern of inheritance (Wainwright et al., 1985).

Northern blot hybridization of human liver RNA showed that the P450IIA2 cDNA hybridized to two mRNAs of 1850 and 2750 nucleotides. The lack of a suitable experimental system means that we are unable to demonstrate directly whether any members of the P450II family are inducible by foreign compounds in man. However, we have used marmoset as an experimental primate model to investigate cytochrome

P-450 induction. Northern blotting demonstrated that a P450IIA2 cDNA hybridizes, under relatively high stringency, to a phenobarbital-inducible marmoset liver mRNA of about 1800 nucleotides. A slot blot hybridization experiment showed that the mRNA was induced 8-fold. Under the conditions used, no cross-hybridization was obtained between the inducible mRNA species and rat or human P450IIB cDNAs or a human P450IIC cDNA (D.R.B., E.A.S. and I.R.P., unpublished). Thus it appears that a member of the P450IIA sub-family of marmosets is inducible by phenobarbital.Although P450IIA is not induced by phenobarbital in rats (Nagata et al., 1987), such species differences in the induction of cytochromes P-450 are not without precedent. Alternatively, it is possible that the human cDNA represents a previously uncharacterized P450II sub-family.

The P450IIB gene sub-family

By increasing the degree of stringency it is possible to retain hybridization of the rat P450IIB2 cDNA to human P450IIB cDNAs while eliminating cross-hybridization to cDNAs coding for members of other P450II sub-families. In this way we have been able to isolate cloned human cDNAs coding for proteins whose amino acid sequence is over 75% homologous to rat P450IIB, but less than 60% homologous to the other rat P450II sub-familes (Fig. 2). We have identified two closely related, but distinct members of the human P450IIB sub-family that differ in only a few amino acid residues. Interestingly, the locations of most of the differences correspond to positions of variability between rat P450IIB1 (P-450b) and P450IIB2 (P-450e).

Some of the human P450IIB cDNAs encode an mRNA from which the region corresponding to exons 5 and 6 of the gene is deleted (C.G., E.A.S. and I.R.P., unpublished). Thus the individual from whom the cDNA library was constructed possessed normal and defective P450IIB mRNAs. The deletion in the mRNA could be due to a genetic polymorphism or to an accidental mis-splicing that occurred during RNA processing. The mRNAs have a 3' non-coding region of about 1,000 nucleotides that contains an Alu-type repetitive sequence element.

Northern and slot blot analysis demonstrated that a human P450IIB cDNA hybridized, under high stringency, to P450IIB mRNA of rat, and to a phenobarbital-inducible marmoset liver mRNA of about 2,500 nucleotides (the same size as P450IIB mRNA in man) (B.M.G., I.R.P. and E.A.S.,

Fig. 2. Percent similarity of amino acid sequence between members of different P450II sub-families of rat and man.

Human \ Rat	IIA	IIB	IIC
IIA	65	53	54
IIB	56	76	54
IIC	48	50	66

unpublished). Under the conditions used no cross-hybridization was observed to the human P450IIA or IIC mRNAs or to the 1,800 nucleotide, phenobarbital-inducible P450IIA mRNA species of marmoset (see above).

The P450IIC gene sub-family

We have also isolated cloned cDNAs coding for members of the P450IIC sub-family of man. One of these codes for a previously unreported member of this sub-family whose amino acid sequence is ~85% homologous to a human P-450 mephenytoin hydroxylase (P-450mp) described by Umbenhauer et al. (1987) and ~80% homologous to rabbit liver P-450 steroid 21-hydroxylase (Tukey et al., 1985). Analysis of human genomic DNA by Southern blot hybridization showed that P450IIC is encoded by a multigene sub-family with a minimum size of 45,000 base pairs. Southern blot hybridization of DNA from human-rodent somatic cell hybrids demonstrated that the P450IIC gene sub-family of man is located on chromosome 10.

Northern blot hybridization of human liver RNA revealed the presence of a single P450IIC mRNA size class of about 1900 nucleotides. The human P450IIC cDNA hybridized to a similarly sized mRNA in marmoset liver. Slot blot hybridization showed that this mRNA was not induced by phenobarbital. No cross-hydridization was observed to human or marmoset P450IIA or IIB mRNAs.

CONCLUSION

The isolation and characterization of cDNA clones coding for members of the P450IIA, B and C sub-families of man will enable us to investigate the expression of the genes in different individuals and to determine the molecular genetic basis of polymorphic variations in drug metabolism that are associated with the corresponding proteins. Expression of the cDNAs in heterologous cells will provide a means of investigating the function of individual members of these human P-450 sub-families.

ACKNOWLEDGEMENTS

We thank Derek Woods and Savio Woo for providing human liver cDNA libraries. Our research is supported by grants from the Cancer Research Campaign, the Medical Research Council and the University of London Central Research Fund. D.R.B. is a recipient of a post-graduate research studentship from the Medical Research Council.

REFERENCES

Davis, M.B., West, L.F., Shephard, E.A. and Phillips, I.R. 1986, Annals of Human Genetics, 50, 237-240.

Nagata, K., Matsunaga, T., Gillette, J., Gelboin, H.V. and Gonzalez, F.J., 1987, Journal of Biological Chemistry, 262, 2787-2793.

Nebert, D.W., Adesnik, M., Coon, M.J., Estabrook, R.W., Gonzalez, F.J., Guengerich, F.P., Gunsalus, I.C., Johnson, E.F., Kemper, B., Levin, W., Phillips, I.R., Sato, R. and Waterman, M.R., 1987, DNA, 6, 1-11.

Phillips, I.R., Shephard, E.A., Ashworth, A. and Rabin, B.R., 1983, Gene, 24, 41-52.

Phillips, I.R., Shephard, E.A., Ashworth, A. and Rabin, B.R., 1985a, Proceedings of the National Academy of Sciences, U.S.A., 82, 983-987.

Phillips, I.R., Shephard, E.A., Povey, S., Davis, M.B., Kelsey, G., Monteiro, M., West, L.F. and Cowell, J.A., 1985b, Annals of Human Genetics, 49, 267-274.

Suwa, Y., Mizukami, Y., Sogawa, K. and Fujii-Kuriyama, Y., 1985, Journal of Biological Chemistry, 260, 7980-7984.

Tukey, R.H., Okino, S., Barnes, H., Griffin, K.J. and Johnson, E.F., 1985, Journal of Biological Chemistry, 260, 13347-13354.

Umbenhauer, D.R., Martin, M.V., Lloyd, R.S. and Guengerich, F.P., 1987, Biochemistry, 26, 1094-1099.

Wainwright, B.J., Watson, E.K., Shephard, E.A. and Phillips, I.R., 1985, Nucleic Acids Research, 13, 4610.

CYTOCHROMES P-450 FOR ENDOGENOUS SUBSTRATES

Regulation of the Biosynthesis and Catalytic Activity of Steroid Hydroxylases in Rabbit Liver

E.F. JOHNSON, G.E. SCHWAB, K.J. GRIFFIN, J. RAUCY, R. TUKEY [1] and L. VICKERY [2]

Research Institute of Scripps Clinic, La Jolla, CA 92037, USA, [1] Cancer Center, University of California, La Jolla, CA 92093, USA and [2] University of California at Irvine, CA 92717, USA

INTRODUCTION

The hepatic steroid hydroxylases are a subset of the microsomal cytochrome P-450 system. These enzymes are remarkable for the wide range of structurally distinct substrates they metabolize, for the capacity of different P-450s to generate distinct products from the same substrate, and for the many ways in which the expression of this metabolic capacity is regulated. The basis for this diversity is rooted in the multiplicity of P-450 proteins. In the rabbit there are three principal hydroxylative metabolites derived from progesterone, 6β-, 16α- and 21-hydroxyprogesterone, whereas 17β-estradiol is principally hydroxylated at the 2 position. The relative activities of these pathways are governed by genetic and ontogenetic restrictions on expression, by the induction of specific P-450 enzymes by foreign compounds, and by the modulation of catalytic activity by positive effectors.

The 6β-hydroxylation of progesterone is catalyzed in the rabbit by both P-450s 3b (Dieter et al., 1982a) and 3c (Lange et al., 1985), IIC3 and IIIA3 respectively in the systematic nomenclature (Nebert et al., 1987). P-450 3c is induced by exposure to macrolide antibiotics and steroids (Wrighton et al., 1985), whereas inducers of P-450 3b have not been identified. P-450 3c is also expressed at earlier ages than P-450 3b. Yet in most rabbits, P-450 3b accounts for a major portion of the constitutive progesterone 6β-hydroxylase activity of adult hepatic microsomes (Reubi et al., 1984). The exception occurs in animals exhibiting genetically determined differences in the capacity of P-450 3b to catalyze this reaction without compromising its capacity to metabolize other substrates (Dieter & Johnson, 1982). On the other hand, heritable differences in elevated biosynthesis of P-450 1 in some animals increasing the rates of 2-hydroxylation of estradiol (Schwab & Johnson, 1985b) and 21-hydroxylation of progesterone (Dieter et al., 1982b) catalyzed by liver microsomes.

ACTIVATION OF P-450 3b

The full metabolic potential of two of these proteins may require modulatory interactions with substrates and allosteric effectors. Work in our laboratory has shown that rabbit microsomal cytochrome P-450 3b isolated from rabbit strain IIIVO/J exhibits a progesterone 16α-hydroxylase activity that is stimulated by other steroids (Johnson et al., 1983). A clear structure activity relation was evident for the compounds which stimulated this activity. In general, 5β-pregnanes are more stimulatory than 5α-pregnanes, and 3β-hydroxyl compounds are more active than 3-keto or 3α-hydroxyl compounds. The most potent of these effector steroids is 5β-pregnane-3β,20α-diol. The stimulatory action of the diol was saturable with half-maximum stimulation achieved at about 3 μM. The reconstituted enzyme exhibits a higher efficiency, Vmax/Km, in the presence of the diol than in its absence. As is shown in Figure 1, the enzyme appears to be activated by the substrate, progesterone, in such a manner that at high substrate concentrations, the kinetic properties of the enzyme approach that determined when the diol is present, whereas the efficiency of the enzyme declines rapidly at lower substrate concentrations. This activation did not appear to be an artifact of enzyme reconstitution because a modest stimulation of microsomal progesterone 16α-hydroxylation by the diol is apparent which can be inhibited by a monoclonal antibody directed to P-450 3b, thereby indicating that P-450 3b exhibits these properties *in situ* (Schwab & Johnson, 1985a).

Our results indicate that the products of the reductive metabolism of progesterone could serve as allosteric effectors that activate the 16α-hydroxylation of progesterone by P-450 3b. Thus, an increased flux of progesterone through the reductive pathway could stimulate a second, oxidative pathway for catabolism in response to high concentrations of progesterone. Despite this apparent point of interaction between the two pathways for the catabolism of progesterone, a convincing rationale for this mode of regulation in the metabolism of progesterone seems unnecessary because 16α-hydroxylation does not contribute greatly to products observed for the catabolism of progesterone *in vivo*. Moreover, there was no evidence for conservation of this mode of regulation among other species. P-450 3b is the product of a class IIC P-450 gene (Nebert et al., 1987), and the activation of P-450 3b may be another of the idiotypic properties associated with members of the P-450 IIC gene family. The class IIC P-450s are an extensive family of P-450 genes which exhibit poor conservation of function and of regulation between species and which display a high degree of diversity within a species (Nebert et al., 1987).

Figure 1. Double reciprocal plot for 16α-hydroxylation of progesterone catalyzed by P-450 3b reconstituted with reductase. The rate was determined in the presence (□) and absence (○) of 5 μM 5α-pregnane-3β,20α-diol.

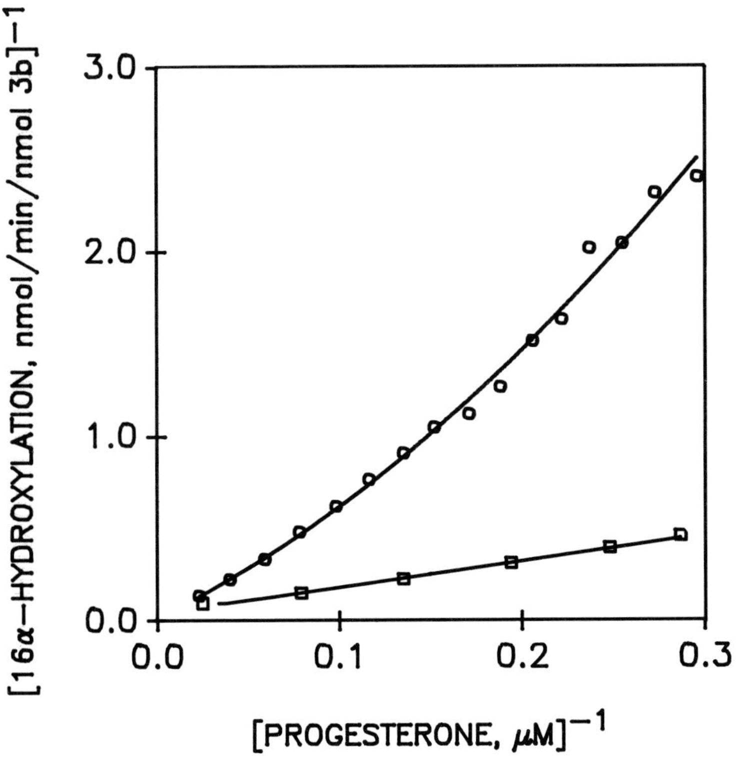

ACTIVATION OF P-450 3c

The stimulation of the P-450 mediated, microsomal metabolism of other substrates by heterologous activators such as acetone (Anders, 1971), metyrapone (Leibman, 1969), flavonoids (Lasker et al., 1984; Wiebel et al., 1971) and steroids (Benford & Bridges, 1980) had been noted in earlier work suggesting that the activation kinetics noted for P-450 3b might also apply to other forms of P-450. In this regard, Waxman and Walsh (Waxman & Walsh, 1983) reported that rat P-450 Pb-1, P-450 IIC6, is activated by metyrapone when reconstituted with reductase. Their results indicate that metyrapone increases the apparent Vmax for this enzyme leaving Km unaffected. In addition, Huang et al., (1981) reported that the hydroxylation of benzo(a)pyrene by reconstituted rabbit P-450 3c is stimulated by flavone and α-naphthoflavone. Further support for the participation of P-450 3c in the activities stimulated by α-naphthoflavone was provided by the observation that rifampicin which induces P-450 3c (Lange et al., 1985) leads to an elevation of the benzo(a)pyrene activity stimulated by α-naphthoflavone (Raucy & Johnson, 1985). Rifampicin also induces progesterone 6β-hydroxylation (Lange et al., 1985) and, estradiol 2-hydroxylation, and we have found that both of these steroid hydroxylations are stimulated by flavone and α-napthoflavone.

The kinetic properties of these two enzymes were analyzed for microsomes prepared from rifampicin treated B/J rabbits. This inbred strain does not express the form of P-450 3b that catalyzes progesterone 6β-hydroxylation and expresses low concentrations of P-450 1, a high efficiency estradiol 2-hydroxylation. In the case of estradiol 2-hydroxylation, the apparent Km for the substrate decreases and the apparent efficiency of the reaction, Vmax/Km, increases when α-napthoflavone is present. This increase of efficiency is also apparent for progesterone 6β-hydroxylation but a clear determination of the effect on the Km for the substrate is confounded by apparent substrate activation observed in the absence of the effector. The substrate activation seen for 6β-hydroxylation is similar to that observed for the 16α-hydroxylation catalyzed by the variant form of P-450 3b which is stimulated by the diol. Moreover, human microsomal catalyzed reactions were found to exhibit a similar response to α-napthoflavone suggesting that this mode of regulation is conserved between man and rabbit.

The decrease of the apparent Km seen for 17β-estradiol suggests that α-napthoflavone may alter the affinity of the enzyme for the substrate. We, therefore, examined whether this effector would alter the affinity of P-450 3c for an amino compound which can serve as a ligand to the heme iron. The alterations in the spectral properties of the P-450 provided a means to monitor the extent of binding of the ligand to the P-450 by following "type II" spectral changes (Schenkman et al., 1967). The amino steroid, 22-amino-23,24-bisnor-5-cholen-3β-ol, 22-ABC, a potent inhibitor of several steroid hydroxylases (Sheets & Vickery, 1982; Sheets et al., 1985; Johnson et al., 1986), proved to be suitable for this purpose. 22-ABC is a competitive inhibitor of the activated progesterone 6β-hydroxylase with an apparent Ki of 0.3 μM. The compound elicits a type II difference spectrum with microsomes from both rifampicin-treated and untreated rabbits. The magnitude of the type II difference spectrum is much greater for the rifampicin-treated *versus* untreated microsomes, and these results suggest that 22-ABC binds to the major protein induced by rifampicin, P-450 3c. This was confirmed with the purified protein. The apparent binding constant as determined by the dependence of the type II spectral change on the concentration of 22-ABC was about 5 μM for both P-450 3c and microsomes from rifampicin-treated B/J rabbits. In the presence of the effector, α-napthoflavone, the apparent binding constant decreases to less than 1 μM for both the purified protein and microsomes from rifampicin treated B/J rabbits indicating that the effector alters the affinity of the cytochrome for the ligand in a manner similar to the change in the apparent Km for the substrate. Based on the extinction coefficient estimated for the binding of 22-ABC to P-450 3c in the absence of α-napthoflavone, the stoichiometry of binding in the presence of the effector appears to be 1:1. Thus, it is likely that the effector molecule elicits a change in the conformation of the P-450 in a manner that alters its catalytic properties.

EXPRESSION OF A P-450 3c cDNA IN MAMMALIAN CELLS

The activity of the reconstituted P-450 3c was very low compared to the magnitude of the microsomal activity and the activities of other benzo(a)pyrene hydroxylases such as P-450 6 or P-450 1 (Raucy & Johnson, 1985), a progesterone 6β-hydroxylase such as P-450 3b (Dieter et al., 1982a) or an estradiol 2-hydroxylase such as P-450 1 (Schwab & Johnson, 1985b). The orthologous enzymes from rat and man are also resistant to reconstitution (Elshourbagy & Guzelian, 1980; Waxman et al., 1985; Guengerich et al., 1986). In the case of P-450 3c, we observe this apparent loss of activity after the first purification step, and the activity is not associated with other fractions obtained from the procedure.

This lead us to clone a cDNA encoding P-450 3c and to express the enzyme by transfection of cells with the cDNA. The cDNA that we isolated codes for a protein that exhibits the N-terminal amino acid sequence of P-450 3c and which is immunoreactive with monoclonal antibodies developed to P-450 3c. The sequence of P-450 3c deduced from the cDNA exhibits 75% similarity to the human nifedipine oxidase (Beaune et al., 1986) and 70% to rat P-450 PCN1 (Gonzalez et al., 1985). Based on the data reported by Dalet et al. (1986), this cDNA may be closely related to the one that they described. Both cDNAs code for the same N-terminal amino acid sequence and hybridize to two mRNA species of about 1700 and 1900 bp on Northern blots. The latter corresponds closely to the size of the cDNA we have characterized which is 1881bp. The smaller polyadenylated RNA may arise through the use of an alternative polyadenylation signal found in the sequence.

We have used the SV40 late promoter to drive transcription of the cDNA in COS1 cells. The transfected cells secrete elevated amounts of 6β-hydroxyprogesterone when the media was supplemented with 10 μM [^{14}C]progesterone. A two fold stimulation of this activity was observed when α-naphthoflavone was also added to the media. This provides additional evidence that P-450 3c is the locus of the stimulatory action of α-napthoflavone on these activities and indicates that the enzyme can be activated in a cellular environment.

SUMMARY

Rabbit hepatic microsomes catalyze the formation of three major metabolites of progesterone, 6β-, 16α- and 21-hydroxyprogesterone. The first two reactions are principally catalyzed by P-450 3b in untreated rabbits, whereas formation of 21-hydroxyprogesterone is catalyzed predominantly by P-450 1. These two forms of P-450 are members of a relatively large family of P-450 genes for which extensive genetic diversity is evident. Genetic polymorphisms in rabbits lead to a deficiency in the expression of one nonallelic variant of P-450 3b that leads to diminished 6β-hydroxylase activity in some animals and on the other hand, to elevated expression of P-450 1 in others. Rifampicin induces a second progesterone 6β-hydroxylase, P-450 3c, in rabbits. Both the activities catalyzed by this enzyme and the 16α-hydroxylation catalyzed by a nonallelic variant of P-450 3b exhibit substrate and allosteric effector activation. The progesterone 6β-hydroxylase, estradiol 2-hydroxylase and benzo(a)pyrene hydroxylase activities induced by rifampicin are all stimulated by α-napthoflavone. Studies with P-450 3c indicate that the effector α-napthoflavone increases the affinity of P-450 3c for the amino steroid, 22-ABC. The binding stoichiometry of the resulting complex appears to be 1:1 suggesting that the effect observed does not arise by cooperative interaction between P-450 3c monomers and half-the-sites binding for ligand and effector.

ACKNOWLEDGMENT

The work reviewed in this reported was supported by USPHS grants HD04445, CA37139, AM30109, and AM01005, and by the Clayton Foundation for Research-California Division.

REFERENCES

Anders, M.W., 1971, Annual Reviews in Pharmacology, 11, 37.

Beaune, P.H., Umbenhauer, D.R., Bork, R.W., Lloyd, R.S., & Guengerich, F.P., 1986, Proceedings of the National Academy of Sciences, USA, 83, 8064.

Benford, D.J. & Bridges, J.W., 1980, Biochemical Pharmacology, 29, 2345.

Dalet, C., Blanchard, J.M., Guzelian, P., Barwick, J., Hartle, H. & Maurel, P., 1986, Nucleic Acids Research, 14, 5999.

Dieter, H.H. & Johnson, E.F., 1982, Journal of Biological Chemistry, 257, 9315.

Dieter, H.H., Muller-Eberhard, U. & Johnson, E.F., 1982a, Biochemical and Biophysical Research Communications, 105, 515.

Dieter, H.H., Muller-Eberhard, U. & Johnson, E.F., 1982b, Science, 217, 741.

Elshourbagy, N.A. & Guzelian, P.S., 1980, Journal of Biological Chemistry, 255, 1279.

Gonzalez, F.J., Nebert, D.W., Hardwick, J.P. & Kasper, C.B., 1985, Journal of Biological Chemistry, 260, 7435.

Guengerich, F.P., Martin, M.V., Beaune, P.H., Kremers, P., Wolff, T. & Waxman, D.J., 1986, Journal of Biological Chemistry, 261, 5051.

Huang, M-T., Johnson, E.F., Muller-Eberhard, U., Koop, D.R., Coon, M.J. & Conney, A.H., 1981, Journal of Biological Chemistry, 256, 10897.

Johnson, E.F., Schwab, G.E. & Dieter, H.H., 1983, Journal of Biological Chemistry, 258, 2785.

Johnson, E.F., Schwab, G.E., Singh, J. & Vickery, L.E., 1986, Journal of Biological Chemistry, 261, 10204.

Lange, R., Larroque, C., Balny, C. & Maurel, P., 1985, Biochemical and Biophysical Research Communications, 126, 833.

Lasker, J.M., Huang, M-T. & Conney, A.H., 1984, Journal of Pharmacology and Experimental Therapeutics, 299, 162.

Leibman, K.C., 1969, Molecular Pharmacology, 5, 1.

Nebert, D.W., Adesnik, M., Coon, M.J., Estabrook, R.W., Gonzalez, F.J., Guengerich, F.P., Gunsalus, I.C., Johnson, E.F., Kemper, B., Levin, W., Phillips, I.R., Sato, R. & Waterman, M.R., 1987, DNA, 6, 1.

Raucy, J.L. & Johnson, E.F., 1985, Molecular Pharmacology, 27, 296.

Reubi, I., Griffin, K.J., Raucy, J. & Johnson, E.F., 1984, Biochemistry, 23, 4598.

Schenkman, J.B., Remmer, H. & Estabrook, R.W., 1967, Molecular Pharmacology, 3, 113.

Schwab, G.E. & Johnson, E.F., 1985a, Biochemistry, 24, 7222.

Schwab, G.E. & Johnson, E.F., 1985b, Archives of Biochemistry and Biophysics, 237, 17.

Sheets, J.J. & Vickery, L.E., 1982, Proceedings of the National Academy of Sciences, USA, 79, 5773.

Sheets, J.J., Zuber, M.X., McCarthy, J.L., Vickery, L.E. & Waterman, M.R., 1985, Archives of Biochemistry and Biophysics, 242, 297.

Waxman, D.J., Dannan, G.A. & Guengerich, F.P., 1985, Biochemistry, 24, 4409.

Waxman, D.J. & Walsh, C., 1983, Biochemistry, 22, 4846.

Wiebel, F.J., Leutz, J.C., Diamond, L. & Gelboin, H.V., 1971, Archives of Biochemistry and Biophysics, 144, 78.

Wrighton, S.A., Schuetz, E.G., Watkins, P.B., Maurel, P., Barwick, J., Bailey, B.S., Hartle, H.T., Young, B. & Guzelian, P., 1985, Molecular Pharmacology, 28, 312.

Multi-Factorial Regulation of Steroid Hydroxylase Gene Expression

MICHAEL R. WATERMAN, J. IAN MASON and EVAN R. SIMPSON

Departments of Biochemistry and Obstetrics and Gynecology,
University of Texas Health Science Center, Dallas, TX 75235, USA

Steroid hydroxylases localized in steroidogenic tissues catalyze the biosynthesis of steroid hormones including glucocorticoids and mineralocorticoids (adrenal cortex) and sex hormones (testis and ovary).

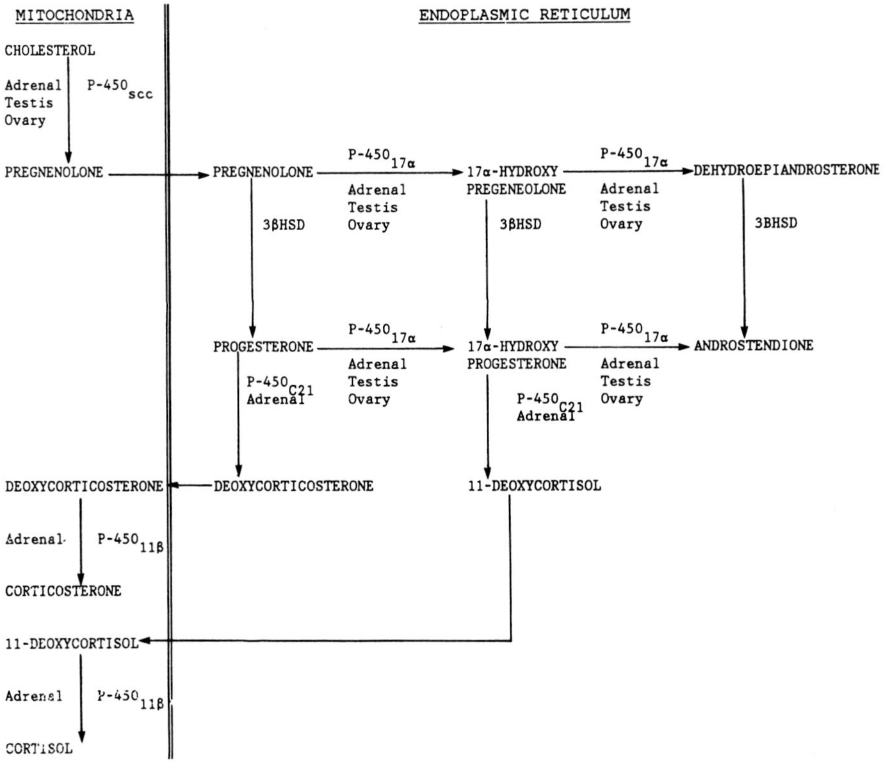

Fig. 1. Tissue localization and subcellular localization of steps in the steroidogenic pathways. 3βHSD = 3β-hydroxysteroid dehydrogenase.

As noted in Figure 1, these pathways include both mitochondrial (P-450scc and P-45011β) and microsomal (P-45017α and P-450C21) forms of cytochrome P-450. The activities of the microsomal components of these pathways are supported by the ubiquitous microsomal flavoprotein, NADPH-cytochrome P-450 reductase, while the mitochondrial components are supplied reducing equivalents via a mitochondrial electron transport system involving a flavoprotein (adrenodoxin reductase) and an iron-sulfur protein (adrenodoxin) (Fig. 2).

NADPH → Adrenodoxin Reductase → Adrenodoxin → P-450scc or P-45011β

Fig. 2. Steroidogenic mitochondrial electron transport pathway

As reported herein, we have found that expression of genes encoding these four steroid hydroxylases as well as that encoding adrenodoxin are regulated by at least three distinct mechanisms.

1. <u>cAMP-Dependent Regulation</u>

The best understood of these three regulatory mechanisms is that mediated by cAMP. Peptide hormones derived from the anterior pituitary bind to specific cell surface receptors on steroidogenic cells and as a consequence activate adenylate cyclase leading to elevated levels of intracellular cAMP. These hormones include ACTH (adrenal cortex), FSH (ovary) and LH (testis and ovary). Cyclic AMP has two distinct actions on steroidogenesis which can be separated from one another temporally. The acute response occurs within seconds or minutes following elevation of cAMP levels and involves mobilization of cholesterol (the steroidogenic substrate) into the inner mitochondrial membrane in the vicinity of P-450scc which catalyzes the conversion of this cholesterol to pregnenolone, the initial step of all steroidogenic pathways. The chronic response to cAMP occurs over a much longer time frame (hours) and involves enhanced synthesis of steroid hydroxylases and related enzymes. This action of cAMP is required for maintenance of optimal steroidogenic capacity in steroidogenic cells.

Utilizing primary cultures of bovine adrenocortical cells (Gospodarowicz et al., 1977) and cDNA probes specific for bovine P-450scc (John et al., 1984), P-45017α (Zuber et al., 1986), P-450C21 (John et al., 1986a), P-45011β (John et al., 1985) and adrenodoxin (Okamura et al., 1985) we have shown that treatment of monolayer cell cultures with ACTH or dibutyryl cAMP leads to increased levels of mRNA encoding each of these enzymes. The elevated levels of mRNA could arise from enhanced transcription, increased mRNA half-lives or both. Of the enzymes listed above, only the P-450scc mRNA showed an increased half-life in response to ACTH (V. Boggaram, unpublished). However, in all cases increased transcription was observed by nuclear run-on experiments (John et al., 1986b). Thus we conclude that a major result of the chronic action of cAMP in steroidogenic cells is enhanced transcription of steroid hydroxylase and adrenodoxin genes.

From studies of the mechanism by which cAMP enhances the transcription of these genes we have found that cycloheximide treatment in conjunction with dibutyryl cAMP inhibits steroid hydroxylase mRNA accumulation. This indicates that labile protein factors are required for transcriptional activation of these genes and that the levels of these putative trans-acting factors are regulated by cAMP. As presented at the 6th International Symposium on Microsomes and Drug Oxidations, these trans-acting factors have been designated SHIP (steroid hydroxylase inducing proteins) (Fig. 3).

Genes encoding bovine P-45017α (B. Adler, unpublished), adrenodoxin (M. Kagimoto, unpublished) and P-450C21 (Chung et al., 1986), human P-45017α (M. Kagimoto, unpublished) P-450scc (Y. Fujii-Kuriyama, unpublished), P-450C21 (White et al., 1986; Higashi et al., 1986) and mouse P-450C21 (White et al., 1984) have now been isolated and characterized and are being utilized to identify cis-acting regions to which SHIP proteins bind. Once such regions are known it will be possible to utilize them to isolate the SHIP proteins and determine what degree of homology exists between the cAMP-mediated transcriptional activators of steroid hydroxylase genes.

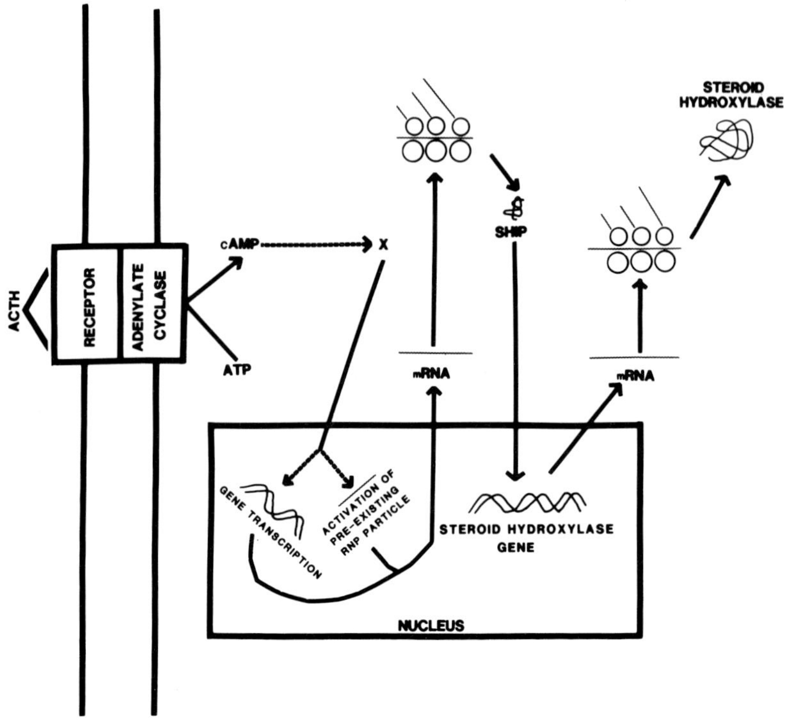

Fig. 3. Hypothetical scheme for cAMP-mediated transcriptional regulation of steroid hydroxylase gene expression (Waterman & Simpson, 1985)

2. Tissue-Specific Regulation

It is apparent from Fig. 1 that tissue specificity plays an important role in steroid hydroxylase gene expression. P-45011β and P-450C21 are expressed solely in the adrenal cortex while P-450scc, P-45017α and adrenodoxin are expressed in all steroidogenic tissues. Furthermore none of these enzymes are expressed in non-steroidogenic tissues, other than the brain (LeCoascogne et al., 1987). This expression of the steroid hydroxylases demonstrates an all-or-none phenotype and good evidence exists for localization of such a tissue-specific regulatory region in the 5'-end of the mouse P-450C21 gene (Parker et al., 1986).

In addition to the all-or-none type of expression of steroid hydroxylase genes an additional type of cell-specific expression of steroid hydroxylase genes involving physiological regulation has been identified. Two steroidogenic cell types exist in the ovarian follicle, the granulosa and the theca interna. Both cell types contain P-450scc and the cells of the theca interna also contain P-45017α. Following ovulation these two cell types differentiate into luteal cells comprising the corpus luteum, a very active steroidogenic tissue. P-450scc mRNA levels increase dramatically during this differentiation to meet the demand of the corpus luteum for progesterone production. At the same time P-45017α mRNA disappears from the cells of the bovine theca interna. Thus in contrast to the adrenal where P-450scc and P-45017α mRNA levels increase and decrease concordinately, during the differentiation of the bovine theca interna into the corpus luteum, P-450scc mRNA content increases and P-45017α mRNA disappears (Rodgers et al., 1987). Thus physiological conditions can regulate the cell-specific expression of these two steroid hydroxylase genes.

3. cAMP-Independent Regulation (Developmental)

A third type of steroid hydroxylase gene expression is developmental in nature and appears to involve a cAMP-independent mechanism. This type of regulation became apparent during studies of steroid hydroxylase mRNA levels in human fetal adrenal. We had the occasion in investigate adrenals from anencephalic human fetuses which we presumed would be a good control for our studies on cAMP-dependent regulation. Anencephalic fetuses are lacking a functional hypothalamic-pituitary axis and consequently contain low or undetectable levels of ACTH in their blood. Also there is essentially no detectable adenylate cyclase activity in these adrenal membranes. Thus we predicted the absence of detectable steroid hydroxylase mRNA in such adrenals. This hypothesis proved to be incorrect, the mRNA levels for P-450scc, P-45011β, P-45017α, P-450C21 and adrenodoxin being indistinguishable from those of normal human fetal adrenals (John et al., 1987). Furthermore these mRNAs are functional since normal levels of steroid hydroxylase proteins were detected by Western blot analysis. Thus we conclude that the imprinting of expression of these genes during fetal development occurs prior to the appearance of the cAMP-dependent mechanism and presumably this fetal imprinting occurs via a cAMP-independent mechanism. This same pattern has been observed in fetal sheep (John et al., 1987) and fetal cows and the mechanism of this developmental regulation is currently being investigated in these systems.

Conclusion

We conclude that there are at least three groups of transcription factors which are required to establish optimal regulation of steroid hydroxylases and related enzymes and consequently steroidogenic capacity in steroidogenic cells. The 3 different classes of trans-acting factors are summarized in Table 1.

Table 1. Three different classes of trans-acting factors postulated to regulate steroid hydroxylase gene expression.

1.	cAMP-dependent - SHIP protein(s).
2.	Tissue-specific - all-or-none and physiological.
3.	Developmental - fetal-imprinting.

It may turn out that this multi-factorial system of regulation outlined in Table 1 is too simplistic and that additional levels of regulation are also necessary for maintenance of optimal steroidogenic activity. However, as complex as this system appears, the availability of genes encoding these enzymes makes it potentially possible to isolate and characterize these different transcriptional factors and to understand in detail how they function to support steroidogenesis.

ACKNOWLEDGMENTS

The authors wish to acknowledge the intellectual and experimental contributions of their colleagues who contributed so heavily to this research program. Also the support of USPHS Grants DK28350, HD11144, HD13234 and Grant I-624 from The Robert A. Welch Foundation is appreciated.

REFERENCES

Chung, B.C., Matteson, K.J. & Miller, W.L., 1986, Proceedings of the National Academy of Sciences, 83, 4243.

Gospodarowicz, D., Ill, C.R., Hornsby, P.J. & Gill, G., 1977, Endocrinology, 100, 1080.

Higashi, Y., Yoshioka, H., Yamane, M., Gotoh, O., & Fujii-Kuriyama, Y., 1986, Proceedings of the National Academy of Sciences, 83, 2841.

John, M.E., John, M.C., Ashley, P., MacDonald, R.J., Simpson, E.R. & Waterman, M.R., 1984, Proceedings of the National Academy of Sciences, 81, 5628.

John, M.E., John, M.C., Simpson, E.R. and Waterman, M.R., 1985, The Journal of Biological Chemistry, 260, 5760.

John, M.E., Okamura, T., Dee, A., Adler, M.C., John, M.C., White, P.C., Simpson, E.R., & Waterman, M.R., 1986a, Biochemistry, 25, 2846.

John, M.E., John, M.C., Boggaram, V., Simpson, E.R., & Waterman, M.R., 1986b, Proceedings of the National Academy of Sciences, 83, 4715.

John, M.E., Simpson, E.R., Carr, B, R., Magness, R.R., Rosenfeld, C.R., Waterman, M.R., & Mason, J.I., 1987, Molecular and Cellular Endocrinology, 50, 263.

LeCoascogne, C., Robel, P., Gauezou, M., Sananes, N., Baulieu, E.-E., & Waterman, M.R., 1987, Science, (In Press).

Okamura, T., John, M.E., Zuber, M.X., Simpson, E.R., & Waterman, M.R., 1985, Proceedings of the National Academy of Sciences, 82, 5705.

Parker, K.L., Schimmer, B.P., Chaplin, D.D., & Seidman, J.G., 1986, The Journal of Biological Chemistry, 261, 5353.

Rodgers, R.J., Waterman, M.R., & Simpson, E.R., 1987, Molecular Endocrinology, 1, 274.

Waterman, M.R. & Simpson, E.R., 1984, Mechanisms of induction of endogenous substrate oxidation. In Microsomes and Drug Oxidations, Edited by A.R. Boobis, J. Caldwell, F. deMatteis & C.R. Elcombe (Taylor & Francis, London), p. 136.

White, P.C., Chaplin, D.D., Weis, J.H., DuPont, B., New, M.I., & Seidman, J.G., 1984, Nature, 312, 465.

White, P.C., New, M.I., & DuPont, P., 1986, Proceedings of the National Academy of Sciences, 83, 5111.

Zuber, M.X., John, M.E., Okamura, T., Simpson, E.R., & Waterman, M.R., 1986, The Journal of Biological Chemistry, 261, 2475.

Biogenesis of Mitochondrial Cytochrome P-450

TSUNEO OMURA

Department of Molecular Biology, Graduate School of Medical Science,
Kyushu University, Fukuoka 812, Japan

INTRODUCTION

Steroidogenic organs contain high concentrations of cytochrome P-450 in mitochondria, which participate in the biosynthesis of steroid hormones in concert with their microsomal counterparts. Mitochondrial forms of cytochrome P-450 are also found in various non-steroidogenic organs including liver and kidney, and they catalyze hydroxylations of various steroid substrates. Mitochondrial P-450s are firmly bound to the inner mitochondrial membrane, and receive electrons from NADPH for monooxygenase reactions via two soluble matrix proteins, NADPH-adrenodoxin reductase and adrenodoxin, whereas microsomal P-450s receive electrons from NADPH via membrane-bound NADPH-cytochrome P-450 reductase. Mitochondrial P-450s are thus clearly distinct from their microsomal counterparts not only in their topographical location in the cell but also in the supply of reducing equivalents for monooxygenase reactions.

The steroid hydroxylation systems of adrenal cortex have been most intensively studied among various steroidogenic organs. Adrenal cortex cells contain two forms of cytochrome P-450, P-450(SCC) and P-450(11β), in mitochondria, in addition to a few other forms of cytochrome P-450 in endoplasmic reticulum membrane. In common with most mitochondrial proteins, these two mitochondrial P-450s are encoded by nuclear genes and synthesized by cytoplasmic polysomes as precursor peptides larger than the mature forms. NADPH-adrenodoxin reductase and adrenodoxin are also synthesized as larger precursors by cytoplasmic polysomes. Concerted expression of P-450s with NADPH-adrenodoxin reductase and adrenodoxin, and selective incorporation of their newly synthesized precursor peptides into mitochondria are important problems peculiar to mitochondrial P-450s.

RESULTS AND DISCUSSION

Primary Structures of P-450(SCC) and P-450(11β)

The primary structures of bovine P-450(SCC) (Morohashi, et al., 1984) and P-450(11β) (Morohashi, et al., 1987) have been determined from the nucleotide sequences of their cloned cDNAs. P-450(SCC) peptide consists of 520 amino acids including an extension peptide of 39 amino acids at the amino terminus. P-450(11β) peptide consists of 503 amino acids, and its extension peptide is 24 amino acid long. As can be clearly seen from Table I, the two mitochondrial P-450s are significantly more related with each other than with microsomal P-450s, P-450(C21) (Yoshioka, et al., 1986) and P-450(17α) (Zuber, et al., 1986), in the same bovine organ. The sequence relatedness between P-450(SCC) or P-450(11β) with various liver microsomal P-450s is also around 15%. P-450cam, a bacterial P-450, is also not much related to the mitochondrial P-450s, the sequence relatedness being about 14%.

Distinction between mitochondrial and microsomal P-450s is more clearly seen when the amino acid sequences around the heme-binding cysteine residue are compared (Fig. 1). The two mitochondrial P-450s apparently form a unique group of cytochrome P-450 distinct from their microsomal counterparts.

Table I. Sequence relatedness among mitochondrial and microsomal P-450s of bovine adrenal cortex. Percentages of matched residues are shown in the table.

	P-450(SCC)	P-450(11β)
	%	
P-450(SCC)	-	35.6
P-450(11β)	35.6	-
P-450(C21)	16.5	15.4
P-450(17α)	15.9	12.7

P-450(SCC)	F G W G V R Q C V G R R I A E L E M T L F
P-450(11β)	F G F G V R Q C L G R R V A E V E M L L L
P-450(C21)	F G C G A R V C L G E S L A R L E L F V V
P-450(17α)	F G A G P R S C V G E M L A R Q E L F L F
P-450(M-1)	F S A G K R I C A G E A L A R T E L F L F
P-450b	F S T G K R I C L G E G I A R N E L F L F
P-450d	F G L G K R R C I G E I P A K W E V F L F
P-450PCN	F G N G P R N C I G M R F A L M N M K L F

Figure 1. Comparison of the amino acid sequences around the heme-binding cysteine residue of mitochondrial and microsomal P-450s. The amino acids conserved in only mitochondrial or microsomal type of cytochrome P-450 are shaded (Morohashi, et al., 1987b).

Structure of P-450(SCC) gene

P-450(SCC) gene was cloned from human genomic libraries using bovine P-450(SCC) cDNA, and characterized (Morohashi, et al., 1987a). It has 9 exons, the same number of exons as the genes of microsomal phenobarbital-inducible P-450s, but the insertion sites of the introns are different between P-450(SCC) gene and phenobarbital-inducible P-450 gene subfamily except for the 7th intron. Judging from the gene structure, this mitochondrial P-450 does not belong to any of the known families of microsomal P-450s.

Role of extension peptide in the import of P-450(SCC) precursor into mitochondria

P-450(SCC) (DuBois, et al., 1981) and P-450(11β) (Nabi, et al., 1980) are synthesized by cytoplasmic ribosomes as larger precursor peptides, and post-translationally imported into mitochondria to be bound to the inner membrane as mature forms. NADPH-adrenodoxin reductase (Kramer, et al., 1982) and adrenodoxin (Nabi & Omura, 1980) are also synthesized as larger precursors by cytoplasmic ribosomes and selectively imported into the matrix compartment of mitochondria. The extension peptide at the amino terminal portion of the precursor peptides is essential for the selective incorporation of the precursors into mitochondria. Figure 2 shows the amino acid sequences of the extension peptides of P-450(SCC) and P-450(11β) precursors.

```
              1        10        20        30        40
                  +  +    +             -  +++    -     +
P-450(SCC)   MLARGLPLRSALVKACPPILSTVGEGWGHHRVGTGEGAGISTKT
                                                     ▲
                 + + +          +- +    +
P-450(11β)   MALWAKARVRMAGPWLSLHEARLLGTRGA
                                    ▲
```

Figure 2. Extension peptides of bovine P-450(SCC) and P-450(11β) precursors. Positively or negatively charged amino acids are marked with symbols. The processing point of each precursor is shown by an arrowhead.

In order to determine which portion of the extension peptide is functional in the import of the precursor peptides into mitochondria, various portions of the extension peptide of P-450(SCC) precursor were chemically synthesized, and their effects on the import of P-450(SCC) precursor were examined (Furuya, et al., 1987). The synthetic peptides were named SEP's, and the amino acid sequence contained in a particular SEP is shown by the numbers of the amino acids in the primary sequence of P-450(SCC) precursor (Fig. 2). SEP1-15 and SEP1-20 strongly inhibited the import of P-450(SCC) precursor into mitochondria in a cell-free system, whereas SEP25-39, which corresponded to the carboxy terminal portion of the extension peptide, showed no inhibitory effect on the import. These observations indicated the importance of the amino terminal portion of the extension peptide in the import of the precursor. Since SEP1-11, SEP6-11, and SEP6-15 were only weakly inhibitory to the import, a certain length of the amino terminal portion, about 15 amino acids long, seems to be required for the function of the extension peptide.

SEP1-15 and SEP1-20 were also inhibitory to the import of P-450(11β) and adrenodoxin precursors. Conversely, a synthetic peptide corresponding to the amino terminal portion of adrenodoxin precursor, AEP1-14, inhibited not only the import of adrenodoxin precursor into mitochondria but also the import of P-450(SCC) and P-450(11β) precursors (Furuya, et al., 1987). Same import machinery seems to be utilized by these three precursors in the import into mitochondria.

The extension peptides of mitochondrial protein precursors are generally rich in basic amino acids. The importance of the positively charged amino acids in the amino terminal portion of the extension peptide of P-450(SCC) precursor was confirmed by replacing them with uncharged amino acids (Kumamoto, et al., 1987).

```
  1       4              9              14
MetLeuAlaArgGlyLeuProLeuArgSerAlaLeuValLysAlaCysProPro

ATGCTAGCAAGGGGGCTTCCCCTCCGCTCAGCCCTGGTCAAAGCCTGCCCACCC
        ---            ---            ---
         ↓              ↓              ↓
        AGC            AGC            ACA

        Ser            Ser            Thr
```

Figure 3. Replacement of positively charged amino acids with un-charged amino acids in the extension peptide of P-450(SCC) precursor. Replacements of the amino acids by single base changes of cDNA are shown by arrows.

As shown in Fig. 3, Arg(4), Arg(9), and Lys(14) in the extension peptide could be replaced by Ser, Ser, and Thr, respectively, by single base changes of the cloned cDNA. Single base substitutions were introduced into the extension peptide region of the cDNA by site-directed mutagenesis using synthetic oligonucleotides. Mutated P-450(SCC) precursors were then synthesized in vitro by the translation of the mRNAs produced from the mutated cDNAs using a SP6 in vitro transcription system.

When any one of Arg(4), Arg(9), and Lys(14) in the extension peptide was replaced with an uncharged amino acid, serine or threonine, the import of the mutated P-450(SCC) precursors into mitochondria was significantly slower than the original precursor. Mutated P-450(SCC) precursors with two or all of Arg(4), Arg(9), and Lys(14) substituted with serine or threonine were no longer imported into mitochondria (Kumamoto, et al., 1987). The basic amino acid residues in the amino terminal portion of the extension peptide have decisive importance in the import of the precursor peptide into mitochondria.

The amino terminal portion of the peptide also plays an essential role in the integration of newly synthesized molecules of microsomal forms of cytochrome P-450 into the membrane of endoplasmic reticulum (Sakaguchi, et al., 1987). Intracellular sorting of both mitochondrial and microsomal P-450s depends on the characteristic short amino acid sequences consisting of 15-20 amino acids at the amino terminal portion of the peptides.

Mechanism of import, processing, and membrane integration of P-450(SCC) precursor

The import of P-450(SCC) precursor into mitochondria

is accompanied by its processing to the mature form, and the mature form is integrated into the inner membrane to become a membrane-bound protein. The processing of P-450(SCC) precursor is catalyzed by a metal protease bound to the matrix-side surface of the inner membrane (Ogishima, et al., 1985), whereas adrenodoxin precursor is processed by a soluble metal protease in the materix, which is inactive to P-450(SCC) and P-450(11β) precursors (Kumamoto, et al., 1986).

The internalization of P-450(SCC) precursor into mitochondria is, however, not tightly coupled with its processing to the mature form. When the processing protease was inhibited by incubating the mitochondria with o-phenanthroline, P-450(SCC) precursor was imported but not processed to the mature form (Ogishima, et al., 1985, Ou, et al., 1986). This observation is in accord with the report by Zwitzinski and Neupert (1983) concerning the import of unprocessed ATPase subunits in the presence of the metal chelator. When the incubation of P-450(SCC) precursor with mitochondria in a cell-free system was carried out at relatively low temperature, unprocessed precursor was first accumulated in the inside of the mitochondria, and it was gradually converted to the mature form (Ou, et al., 1986). The import of enzyme precursors into mitochondria and their processing to the mature forms are not tightly coupled.

The unprocessed P-450(SCC) precursor internalized into mitochondria was apparently rather loosely associated with the matrix-side surface of the inner membrane. The precursor was recovered in the inner membrane vesicle fraction when the mitochondria were disrupted by sonication, but it was released from the membrane vesicles in a high ionic strength medium, whereas mature P-450(SCC) was tightly bound to the membrane.

The correct integration of mature P-450(SCC) molecules into the inner membrane must be accompanied by the incorporation of heme to the molecules. However, we have no information at present about the step of heme incorporation in the whole process of the import and membrane integration of P-450(SCC).

REFERENCES

DuBois,R.N., Simpson,E.R., Kramer,R.E., & Waterman,M.R., 1981, Journal of Biological Chemistry, 256, 7000-7005.

Furuya,S., Ito,A., Aoyagi,H., Kanmera,T., Kato,T., Sagara,Y., Horiuchi,T., & Omura,T., 1987, Journal of Biochemistry, 102, in press.

Kramer,R.E., DuBois,R.N., Simpson,E.R., Anderson,C.M., Kashiwagi,K., Lambeth,J.D., Jefcoate,C.R., & Waterman, M.R., 1982, Archives of Biochemistry and Biophysics, 215, 478-485.

Kumamoto,T., Ito,A., & Omura,T., 1986, Journal of Biochemistry, 100, 247-254.

Kumamoto,T., Morohashi,K., Ito,A., & Omura,T., 1987, Journal of Biochemistry, 102, in press.

Morohashi,K., Fujii-Kuriyama,Y., Okada,Y., Sogawa,K., Hirose,T., Inayama,S., & Omura,T., 1984, Proceedings of National Academy of Science USA., 81, 4647-4651.

Morohashi,K., Sogawa,K., Omura,T., & Fujii-Kuriyama,Y., 1987a, Journal of Biochemistry, 101, 879-887.

Morohashi,K., Yoshioka,H., Gotoh,O., Okada,Y., Yamamoto, K., Miyata,T., Sogawa,K., Fujii-Kuriyama,Y., & Omura,T., 1987b, Journal of Biochemistry, 102, 545-554.

Nabi,N. & Omura,T., 1980, Biochemical and Biophysical Research Communications, 97, 680-686.

Nabi,N., Kominami,S., Takemori,S., & Omura,T., 1980, Biochemical and Biophysical Research Communications, 97, 687-693.

Ogishima,T., Okada,Y., & Omura,T., 1985, Journal of Biochemistry, 98, 781-791.

Ou,W., Ito,A., Morohashi,K., Fujii-Kuriyama,Y., & Omura, T., 1986, Journal of Biochemistry, 100, 1287-1296.

Sakaguchi,M., Mihara,K., & Sato,R., 1987, EMBO Journal, 6, in press.

Yoshioka,H., Morohashi,K., Sogawa,K., Yamane,M., Kominami, S., Takemori,S., Okada,Y., Omura,T., & Fujii-Kuriyama, Y., 1986, Journal of Biological Chemistry, 261, 4106-4109.

Zuber,M.X., John,M.E., Okamura,T., Simpson,E.R., & Waterman,M.R., 1986, Journal of Biological Chemistry, 261, 2475-2482.

Zwitzinski,C. & Neupert,W., 1983, Journal of Biological Chemistry, 258, 13340 13346.

Regulation of the Cytochrome P-450 ω-Hydroxylase from Rabbit Lung *

BETTIE SUE SILER MASTERS, RICHARD T. OKITA, A. SCOTT MUERHOFF, MARIE T. LEITHAUSER, DAVID L. ROERIG [1] and JOAN E. CLARK

Departments of Biochemistry, Anesthesiology [1] and Pharmacology/Toxicology [1], Medical College of Wisconsin and VA Medical Center [1], Milwaukee, WI 53226, USA

* supported by USPHS GM 31296 to BSSM and RTO

INTRODUCTION

As more endogenous compounds are discovered to be hydroxylated at the terminal, or omega (ω) carbon, interest has increased in ascertaining the function of these metabolites. It has been suggested recently that the oxygenation of the ultimate and penultimate carbons of eicosanoids formed via the cyclooxygenase and lipoxygenase pathways could have physiological significance aside from that of preparation for excretion. Beginning with the studies of Ichihara, et al. (1969) on the liver microsomal oxidation of medium chain fatty acids and of Orrenius and his colleagues (Ellin, et al., 1972) on kidney microsomal oxidation of lauric acid, the involvement of cytochrome P-450 was demonstrated. Samuelsson, et al. (1971) showed that the first step in the oxidative sequence leading to the formation of dicarboxylic acid products of prostaglandins was the introduction of a hydroxyl group at the 20(ω)-carbon. This laboratory became interested in kidney microsomal hydroxylation reactions with lauric acid as a substrate (Fan, et al., 1976) which led us later to examine ω- and (ω-1)-hydroxylation of fatty acids and prostaglandins in hepatic and extrahepatic tissues (Okita & Masters, 1980; Williams, et al, 1984). Recent reports of the ω-hydroxylation of 12-hydroxyeicosatetraenoic acid (12-HETE) by Marcus, et al. (1984) and of leukotriene B_4 (LTB_4) by Jubiz, et al. (1982) in neutrophils extend the list of extrahepatic tissues catalyzing these reactions. Soberman, et al. (1987) have shown that the LTB_4 ω-hydroxylase exhibits unique catalytic and immunochemical properties.

Following the reports by Powell's laboratory (1978 a,b) of the ω-hydroxylation of prostaglandins E_1 and $F_{2\alpha}$ by lung microsomes from pregnant or progesterone-treated rabbits, this laboratory (Williams, et al., 1984) and that of Kusunose (Yamamoto, et al., 1984) have purified the cytochrome P-450 isozyme responsible for this activity. Powell (1978b) examined the gestational age dependence of prostaglandin ω-hydroxylation activity. Masters, et al. (1985) confirmed these data and, in addition, demonstrated by immunoblotting techniques that the protein levels followed a similar pattern. The regulation of this highly induced enzymatic activity has since become a subject of intense interest leading this laboratory into experimental approaches to determine the level of regulation. Masters, et al. (1987) and Muerhoff, et al. (1987) have now shown that prostaglandin ω-hydroxylase translatable mRNA levels are concomitantly increased with enzyme activity and protein levels as a function of gestational age in the lungs of pregnant rabbits. Kusunose, et al. (1987) have purified the progesterone-induced cytochrome P-450 from rabbit lung and prepared a lung cDNA library from which they were able to isolate a clone expressing the enzyme and allowing the determination of its amino acid sequence.

This report will present experiments designed to characterize the cytochrome P-450 prostaglandin ω-hydroxylase more fully in order to understand its physiological function. A variety of techniques has been utilized to examine the enzyme, including isolated lung perfusion and reconstitution studies using purified enzymes.

EXPERIMENTAL PROCEDURES

Pregnant New Zealand white rabbits (4-5 kg) were obtained from Hickory Hill, Inc., Flint, TX and sacrificed at 25-28 days of gestation. Three forms of cytochrome P-450 were purified according to Williams, et al. (1984). NADPH-cytochrome P-450 reductase was purified by the method of Yasukochi & Masters (1976), as described by Masters & Okita (1980), and cytochrome b_5 was purified as described by Strittmatter, et al. (1978) from porcine liver. Prostaglandin dehydrogenase was purified according to the method of Bergholte and Okita (1986) from the lungs of pregnant rabbits. Protein sequencing was performed on an Applied Biosystems Model 477AV liquid phase sequenator.

Purified cytochrome $P\text{-}450_{PG\text{-}\omega}$ (0.05 nmol) was reconstituted with 0.1 nmol of porcine liver NADPH-cytochrome P-450 reductase and cytochrome b_5 (Williams, et al., 1984). All analyses for ω-hydroxylation of PGE_1, 15-hydroxyeicosatetraenoic acid (15-HETE), arachidonate, and myristate, were performed by separation on reverse phase high performance liquid chromatography (RP-HPLC) on a Bio-Sil ODS-5S (150 x 4 mm) column using radioactive substrates. A Varian 5000 liquid chromatograph on line with a Flo-One Model HP (Radiomatic) was used to monitor elution of the metabolites.

Experiments with isolated, perfused lungs were performed with an artificial thorax in which physiological respiration, blood gases, temperature, and perfusion flow rate were controlled. All perfusions were performed with 4.5% albumin, 5 mM glucose, Krebs-Ringer bicarbonate buffer, pH 7.4, 37°C, at 200 ml/min. A 0.5 ml bolus of perfusate containing 2.5 μCi $[^3H]\text{-}PGE_1$ (60 pmol) was infused through an injection port connected to the arterial cannula with the lung arrested at maximum inspiratory capacity. Fractions were collected from the venous cannula at one per sec for the entire 45 sec of the single-pass experiments.

EXPERIMENTAL RESULTS

Purification of three isozymes of cytochrome P-450 from the lungs of pregnant rabbits has been achieved (Williams, et al., 1984). Two of these forms, $P\text{-}450_{PG\text{-}\omega}$ and $P\text{-}450_{II}$ (isozyme 5), catalyze ω-hydroxylation reactions, the former of prostaglandins and 15-HETE and the latter of medium-chain fatty acids.

Sequences of the N-termini of the lung cytochromes P-450 prostaglandin ω-hydroxylase ($P\text{-}450_{PG\text{-}\omega}$) and the fatty acid hydroxylase (isozyme 5-type) have been obtained. There is little homology between the N-terminal sequences of these two cytochromes P-450.

Interestingly, the N-terminal sequence of our isozyme 5 is nearly identical with the isozyme 5 reported by Parandoosh, et al. (1987). Whether or not these are actually identical isozymes awaits comparison of catalytic activities and immunochemical characteristics, as well as further sequencing information.

Several compounds were tested as potential substrates for the P-$450_{PG-\omega}$. This work has been prompted by the findings of Soberman, et al., 1987, who showed that the leukotriene B_4 (LTB_4) ω-hydroxylase is a distinct enzyme in neutrophils which will not utilize prostaglandins as substrates. Subsequently, Okita, et al., 1987, have shown that the P-$450_{PG-\omega}$ will not catalyze the ω-hydroxylation of LTB_4 (Okita, et al., 1987), further substantiating the differences between the ω-hydroxylases from these two distinct tissue sources. The P-$450_{PG-\omega}$ is very active with substrates having an ω-6 hydroxyl, such as PGE_1, $PGF_{2\alpha}$, and PGA_1. We have also found that it catalyzes the ω-hydroxylation of 15-HETE and, additionally, the 15-keto-PGE_1 and 15-ketoeicosatetraenoic acid (15-KETE), the latter of which have keto groups at the ω-6 position. The P-$450_{PG-\omega}$ will not utilize 5-HETE, 12-HETE, LTB_4, or lauric, myristic, or arachidonic acids as substrates under the assay conditions utilized by Williams, et al. (1984).

The rabbit lung P-$450_{PG-\omega}$ is minimally expressed in normal female rabbits but is induced over 100-fold during pregnancy. This induction of P-$450_{PG-\omega}$ has been demonstrated by an increase in PGE_1, $PGF_{2\alpha}$, and PGA_1 ω-hydroxylase activities, in immunoreactive protein, and in translatable P-$450_{PG-\omega}$ mRNA levels (Muerhoff, et al., 1987). Increased activity is clearly substantiated when PGE_1 metabolism is studied in isolated perfused lungs of pregnant and nonpregnant rabbits. The results are shown in Figure 1.

It can be seen that a nonpolar metabolite of PGE_1 was the major product from the nonpregnant lung with very low quantities of polar metabolites being formed. The lungs from pregnant rabbits revealed a remarkably different profile with polar metabolites of PGE_1 being the predominant species. The products have been tentatively identified by co-elution HPLC chromatography with standard compounds generated by the isolated P-$450_{PG-\omega}$, PGDH, and $^{13}\Delta$ reductase. The nonpolar metabolite was identified as 13,14-dihydro-15-keto-PGE_1 and the polar metabolites identified as the 20-OH-PGE_1 and 13,14-dihydro-15-keto-20-OH-PGE_1 (See Figure 2).

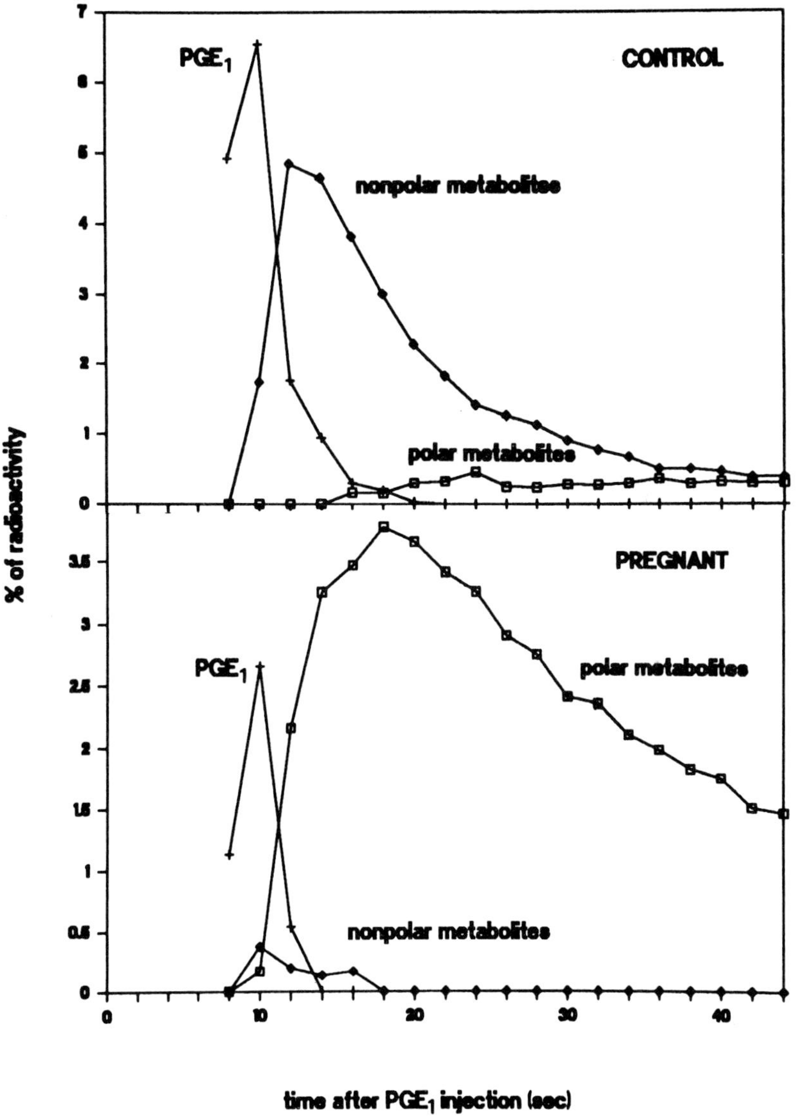

FIGURE 1. PGE_1 metabolites in perfused lung.

$[^3H]$-PGE_1 (60 pmol, 2.5 μCi) was injected at time zero. The % radioactivity refers to that amount in the sample relative to the total radioactivity collected during the 45 sec period.

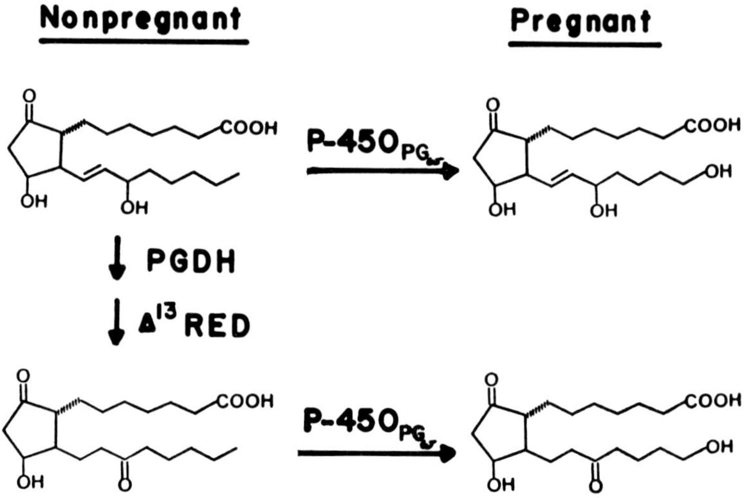

Figure 2 illustrates the alteration in metabolic fate of PGE_1 observed in the lung perfusion studies of pregnant rabbits. As the scheme shows, the pregnant animal produces more ω-hydroxylated products (polar) either directly from PGE_1 or from the 13,14-dihydro-15-keto compound. Conversely, the relatively nonpolar 13,14-dihydro-15-keto derivative is found in the perfusate from nonpregnant rabbits.

DISCUSSION

The data presented support the original conclusion of Williams, et al. (1984) that the cytochrome P-450 prostaglandin ω-hydroxylase is, in fact, a unique isozyme. Its immunochemical distinctness from other known rabbit liver cytochromes P-450 and from rabbit lung isozymes purified during its isolation had already strongly suggested its unique nature. Although our laboratory has not performed a total sequence analysis of the $P\text{-}450_{PG\text{-}\omega}$, the sequence obtained of the first twenty amino acids reveals its lack of homology with the P-450 isozyme 5, also purified from the lungs of pregnant rabbits. Kusunose, et al. (1987), have recently reported, from cDNA sequence data, a 74% homology between their $P\text{-}450_{P\text{-}2}$ (progesterone-induced) and the $P\text{-}450_{LA\omega}$ (clofibrate-induced) from rat liver cloned and sequenced by Hardwick, et al. (1987).

The regio- and stereospecificities of the ω-hydroxylases have only recently come to light with the studies of Soberman, et al. (1987) on neutrophil LTB_4 ω-hydroxylase being noteworthy. Our continuing studies of lung $P\text{-}450_{PG\text{-}\omega}$ have now demonstrated a substrate specificity which suggests its physiological importance to the female rabbit during pregnancy. The isolated, perfused lung experiments in which more polar, ω-hydroxylated metabolites were produced by lungs from pregnant rabbits than from non-pregnant controls reinforce this notion.

The gestational age dependence of the prostaglandin ω-hydroxylase activities, protein concentration (determined by immunoblotting), and translatable mRNA levels presents a compelling argument for some as yet undetermined role for this enzyme during pregnancy.

REFERENCES

Bergholte, J.M. & Okita, R.T., 1986, Archives of Biochemistry and Biophysics, 245, 308.

Ellin, A., Jakobsson, S.V., Schenkman, J.B. & Orrenius, S., 1972, Archives of Biochemistry and Biophysics, 150, 64.

Fan, L.L., Masters, B.S.S. & Prough, R.A., 1976, Analytical Biochemistry, 1976, 265.

Hardwick, J.P., Song, B.-J., Huberman, E. & Gonzalez, F.J., 1987, Journal of Biological Chemistry, 262, 801.

Ichihara, K., Kusunose, E. & Kusunose, M., 1969, Biochimica et Biophysica Acta, 176, 704.

Jubiz, W., Radmark, O., Malmsten, C., Hansson, G., Lindgren, J.A., Palmblad, J., Uden, A.-M. & Samuelsson, B., 1982, Journal of Biological Chemistry, 257, 6106.

Kusunose, M., Sogawa, K., Yamamoto, S., Matsubara, S., Yokotani, N., Kusunose, E. & Fujii-Kuriyama, Y., 1987, Fatty acid and prostaglandin ω-hydroxylase cytochrome P-450. In Yamada Conference: Cytochrome P-450, New Trends, Nara, Japan (in the press).

Marcus, A., Safier, L., Ullman, H., Broekman, M., Islam, N., Oglesby, T.D. & Gorman, R.R., 1984, Proceedings of the National Academy of Sciences (USA), 81, 903.

Masters, B.S.S. & Okita, R.T., 1980, The history, properties, and function of NADPH-cytochrome P-450 reductase. In Pharmacology and Therapeutics, 11, edited by J.B. Schenkman & D. Kupfer (Pergamon Press) p. 227.

Masters, B.S.S., Okita, R.T., Williams, D.E., Hale, S.E., Muerhoff, A.S. & Hevner, R.F., 1985, Prostaglandin and fatty acid ω- and (ω-1)-hydroxylation by cytochromes P-450 in extrahepatic tissues. In Microsomes and Drug Oxidations, edited by A. Boobis, J. Caldwell, F. DeMatteis, & C. Elcombe (Taylor and Francis) p. 74.

Masters, B.S.S., Okita, R.T., Muerhoff, A.S., Bergholte, J.M., and Leithauser, M.T., 1987, Studies on the metabolism, specificity, and regulation of a pregnancy-induced rabbit lung microsomal cytochrome P-450 prostaglandin ω-hydroxylase. In Yamada Conference: Cytochrome P-450, New Trends, Nara, Japan, in the press.

Muerhoff, A.S., Williams, D.E., Leithauser, M.T., Jackson, V.E., Waterman, M.R. & Masters, B.S.S., 1987, Proceedings of the National Academy of Sciences (USA), in the press.

Okita, R.T. & Masters, B.S.S., 1980, Drug Metabolism and Disposition, 8, 147.

Okita, R.T., Soberman, R.J., Bergholte, J.M., Masters, B.S.S., Hayes, R. & Murphy, R.C., 1987, Submitted.

Parandoosh, Z., Fujita, V.S., Coon, M.J. & Philpot, R.M., 1987, Drug Metabolism and Disposition, 15, 59.

Powell, W.S., 1978b, Journal of Biological Chemistry, 253, 6711.

Powell, W.S. & Solomon, S., 1978a, Journal of Biological Chemistry, 253, 4609.

Samuelsson, B., Granstrom, E., Green, K. Hamberg, M., 1971 Annals of the New York Academy of Sciences, 180, 138.

Soberman, R.J., Okita, R.T., Fitzsimmons, B., Rokach, J., Spur, B. & Austen, K.F., 1987, Journal of Biological Chemistry, 262, in the press.

Strittmatter, P., Fleming, P., Connors, M. & Corcoran, D., 1978, Purification of cytochrome b_5. In Methods in Enzymology, Vol. LII, Biomembranes, Part C., edited by S. Fleischer & L. Packer (Academic Press) p. 97.

Williams, D.E., Hale, S.E., Okita, R.T. & Masters, B.S.S., 1984, Journal of Biological Chemistry, 251, 14600.

Yamamoto, S., Kusunose, E., Ogita, K., Kaku, M., Ichihara, K. & Kusunose, M., 1984, Journal of Biochemistry (Japan), 96, 593.

Yasukochi, Y. & Masters, B.S.S., 1976, Journal of Biological Chemistry, 251, 5337.

HORMONAL REGULATION OF CYTOCHROMES P-450

Complete Sequence and Hormonal Regulation of Cytochrome P-450 Isozymes 15β and 16α

AGNETA MODE, PETER ZAPHIROPOULOS, ANDERS STRÖM, CATRIONA MacGEOCH, CHRISTER MÖLLER, EDWARD MORGAN and JAN-ÅKE GUSTAFSSON

Department of Medical Nutrition, Karolinska Institute, Huddinge University Hospital F69, S-141 86 Huddinge, Sweden

SUMMARY

Full length cDNA clones for the sexually differentiated rat liver cytochrome P-450 isozymes 15β and 16α have been isolated and completely sequenced. The deduced amino acid sequence of $P\text{-}450_{15\beta}$ is 66% homologous to that of $P\text{-}450_{16\alpha}$. Furthermore, either of these two cytochromes has a homology of about 50% with the major phenobarbital inducible form, cytochrome P-450b, while the homology with the major carcinogen inducible form, cytochrome P-450c, is less than 30%. Interestingly, the homology of these two sex-dependent cytochromes with the developmentally regulated cytochrome P-450 isozymes f and PB-1 is in the range of 70%. Therefore, it is concluded that P-450 isozymes 15β, 16α, f, and PB-1 represent a subfamily within the phenobarbital related cytochrome P-450 gene family.

The secretory pattern of growth hormone has been known to regulate the expression of these cytochromes at the apoprotein level. Northern analysis now indicates that growth hormone regulates the expression of $P\text{-}450_{15\beta}$ and $P\text{-}450_{16\alpha}$ at a pretranslational level.

INTRODUCTION

The microsomal 15β-hydroxylase of rat liver is a sexually differentiated cytochrome P-450 isozyme ($P\text{-}450_{15\beta}$) which catalyzes the specific conversion of water soluble steroid sulfates to 15β-hydroxylated products (Gustafsson & Ingelman-Sundberg 1975, MacGeoch et al. 1984). The expression of 15β-hydroxylation is female predominant in adult rats and represents a major physiological pathway in this gender (Gustafsson & Ingelman-Sundberg 1975). The sexually dimorphic plasma pattern of growth hormone (GH) (Edén 1979) has been shown to be a major determinant for sexually differentiated hepatic steroid metabolism in the rat (Mode et al. 1982). In males, the plasma GH levels peak every 3 to 4 h with very low levels between the peaks, whereas in females, the GH levels are less variable (Edén 1979).

It has also been established that the male-specific testosterone 16α-and 2α-hydroxylase and androstenedione 16α-hydroxylase activities of rat liver (Morgan et al. 1985a, Waxman 1984) as well as vitamin D_3 25-hydroxylase activity (Andersson & Jörnvall 1986) are due to the presence of a male-specific isozyme of cytochrome P-450, which we have named P-$450_{16\alpha}$ (Morgan et al. 1985b). The major regulatory factor for the sexually differentiated expression of the P-$450_{15\beta}$ and P-$450_{16\alpha}$, was found to be the secretory pattern of GH secretion (Morgan et al. 1985b, MacGeoch et al. 1985). Thus, continuous infusion of human GH (hGH) completely feminized the levels of P-$450_{15\beta}$ in the livers of male rats, i.e. the levels of P-$450_{15\beta}$ increased (MacGeoch et al. 1985), whereas the same daily dose of hGH administered by injection at 12-hourly intervals had no effect (MacGeoch et al. 1985). On the contrary, continuous infusion of human GH reduces the level of P-$450_{16\alpha}$ expression, while intermittent injections of the same daily dose induce the enzyme (Morgan et al. 1985b).

To further investigate the mechanism of growth hormone regulation of expression of P-$450_{15\beta}$ and P-$450_{16\alpha}$ in the rat, specific antibodies have been developed and used to isolate the corresponding cDNA clones from rat liver cDNA libraries (MacGeoch et al. 1987, Ström et al. 1987). In this report we describe the isolation and complete sequence of full length cDNA clones for these two cytochromes as well as evidence for pretranslational regulation of expression by growth hormone.

MATERIALS AND METHODS

Molecular cloning and sequencing.

A female rat liver cDNA library was purchased from Clontech-Intermedica while a male rat liver cDNA library was constructed in this laboratory (Ström et al. 1987) by the method of Gubler and Hoffman (12). Antibody against P-$450_{15\beta}$ (MacGeoch et al. 1985) and P-$450_{16\alpha}$ (Morgan et al. 1985a) were used for screenings according to the method of Helfman et al. (Helfman et al. 1983). cDNA's were characterized by restriction mapping and cloned either in pGEM blue or M13 mp18 and mp19. Sequencing was done by the dideoxy chain termination method (Sanger et al. 1980). In certain cases custom synthesized primers were used.

Hormone treatments.

Sprague-Dawley rats, 8-9 weeks of age were used. Each experimental and control group consisted of four rats. Hx rats were obtained from Møllegaards Avelslaboratorium (Skensved, Denmark). Where indicated, bovine GH (bGH) or hGH were administered by continuous infusion from osmotic minipumps for 7 days at rates of 10 μg/h and 5 μg/h, respectively, as described previously (MacGeoch et al. 1984). bGH was also administered by 12-hourly s.c. injections of 120 μg for 7 days.

Northern analysis of RNA.

Total liver RNA isolated by the method of Chirgwin et al. (Chirgwin et al. 1979) (20 μg per lane) was electrophoresed in 1% (w/v) agarose gels in the presence of formaldehyde and blotted on to nitrocellulose sheets. Prehybridization was performed in 50% (v/v) formamide and 5 x SSC containing 50 mM sodium phosphate buffer, pH 6.5, 1 x Denhardts solution and 0.25 mg/ml salmon sperm DNA at 42^{0} for 20 h (Andersson & Jörnvall 1986). Hybridization was performed in the same buffer with 2 x 10^{6} cpm/ml of the nick-translated P-$450_{15\beta}$ cDNA probe or the multiprime labeled P-$450_{16\alpha}$ probe for 20 h at 42^{0}. Filters were washed 4 times for 5 min in 2 x SSC, 0.1% SDS (w/v) at room temperature, and twice for 20 min at 65^{0} (P-$450_{15\beta}$) or 50^{0} (P-$450_{16\alpha}$) in 0.1 x SSC, 0.1% SDS (w/v).

Materials.

[α^{32}P]dCTP, [α^{32}P]dATP, and multiprime kit was supplied by Amersham, England. Nick-translation kits were from Bethesda Research Laboratories, Bethesda, MD. Oligo(dT) was from Collaborative Research (Lexington, MA). Human GH (Crescormon, 2 IU/mg) was kindly provided from KabiVitrum, Stockholm, Sweden. Bovine GH (USDA-B1, 1.4 IU/mg) was from the NIADKK.

RESULTS AND DISCUSSION

Initial screenings with antibody probes for cytochrome P-$450_{15\beta}$ and P-$450_{16\alpha}$ resulted in the isolation of non-full length cDNA clones. These clones were used as probes for further screenings which resulted in the isolation of full length cDNA probes. DNA sequencing revealed a 71% homology of the coding sequences (Fig. 1) while the deduced amino acid sequences of the two cytochromes had a 66% homology to each other (Fig. 2). Furthermore, the developmentally regulated cytochromes P-450 f and P-450 PB-1 (Gonzalez et al. 1986) were both 68% homologous with P-$450_{15\beta}$ while with P-$450_{16\alpha}$ the homology was 69% and 73%, respectively.

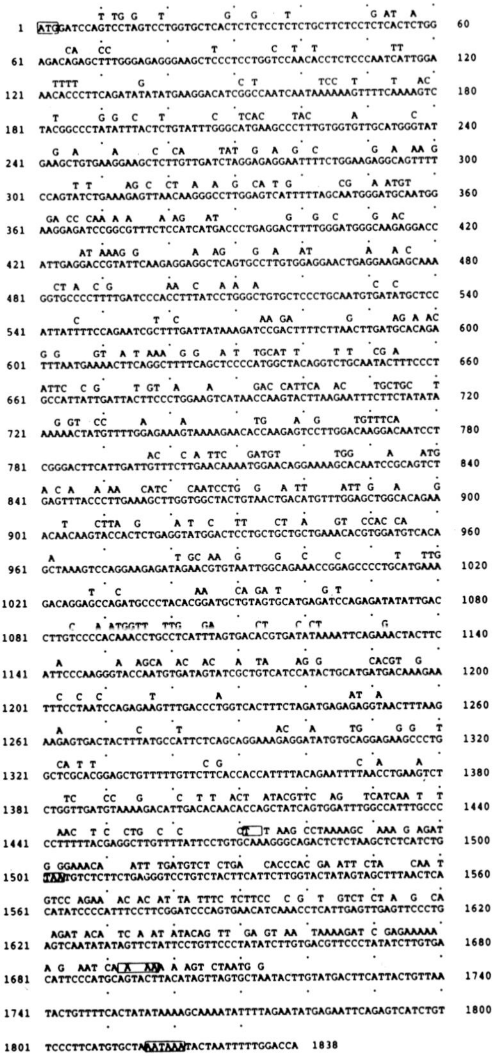

Fig. 1. Nucleotide sequence comparison of P-$450_{15\beta}$ (line 1) and P-$450_{16\alpha}$ (line 2). Mismatches are presented in the P-$450_{15\beta}$ sequence (line 1) while identical nucleotides are omitted. The initiator ATG codon, the termination codons and the polyadenylation signals are boxed.

```
             .         .         .         .         .
    MDPFVVLVLSLSFLLLLYLWRPSPGRGKLPPGPTPLPIFGNFLQIDMKDI
    |||  |||| || |||| ||| | |||||||||||||| || ||| ||||
  1 MDPVLVLVLTLSSLLLLSLWRQSFGRGKLPPGPTPLPIIGNTLQIYMKDI   50
             .         .         .         .         .
    RQSISNFSKTYGPVFTLYFGSQPTVVLHGYEAVKEALIDYGEEFSGRGRM
     |||  ||| ||| |||| |  | ||||||||||||| | ||||||||
 51 GQSIKKFSKVYGPIFTLYLGMKPFVVLHGYEAVKEALVDLGEEFSGRGSF  100
             .         .         .         .         .
    PVFEKATKGLGISFSRGNVWRATRHFTVNTLRSLGMGKRTIEIKVQEEAE
    || |   |||| || |  |   | |   |||  ||||||||   ||||
101 PVSERVNKGLGVIFSNGMQWKEIRRFSIMTLRTFGMGKRTIEDRIQEEAQ  150
             .         .         .         .         .
    WLVMELKKTKGSPCDPKFIIGCAPCNVICSIIFQNRFDYKDKDFLSLIEN
     || || | || | || || |||||||||||||||||||||  || |
151 CLVEELRKSKGAPFDPTFILGCAPCNVICSIIFQNRFDYKDPTFLNLMHR  200
             .         .         .         .         .
    VNEYIKIVSTPAFQVFNAFPILLDYCPGNHKTHSKHFAAIKSYLLKKIKE
     ||     | |  || | ||   || || |    | |  || | | | ||
201 FNENFRLFSSPWLQVCNTFPAIIDYFPGSHNQVLKNFFYIKNYVLEKVKE  250
             .         .         .         .         .
    HEESLDVSNPRDFIDYFLIQRCQENGNQQMNYTQEHLAILVTNLFIGGTE
    | ||||  ||||||| ||    ||  | |   | | |   ||  |  |||
251 HQESLDKDNPRDFIDCFLNKMEQEKHNPQSEFTLESLVATVTDMFGAGTE  300
             .         .         .         .         .
    TSSLTLRFALLLLMKYPHITDKVQEEIGQVIGRHRSPCMLDRIHMPYTNA
    | | |||  |||| |    | ||||||  |||| ||||| ||  |||| |
301 TTSTTLRYGLLLLLKHVDVTAKVQEEIERVIGRNRSPCMKDRSQMPYTDA  350
             .         .         .         .         .
    MIHEVQRYIDLAPNGLLHEVTCDTKFRDYFIPKGTAVLTSLTSVLHARKE
      ||  |||||| |  | | || | ||| ||||||| |  || | ||  ||
351 VVHEIQRYIDLVPTNLPHLVTRDIKFRNYFIPKGTNVIVSLSSILHDDKE  400
             .         .         .         .         .
    FPNPEMFDPGHFLDENGNFKKSDYFMPFSAGKRKCVGEGLASMELFLFLT
    ||||| ||||||||| ||||||||||||||||| | || ||  ||||| |
401 FPNPEKFDPGHFLDERGNFKKSDYFMPFSAGKRICAGEALARTELFLFFT  450
             .         .         .         .         .
    TILQNFKLKSLSDPKDIDINSIRSEFSSIPPTFQLCFIPV
    |||||| |||| | ||||     | |   ||    |||||
451 TILQNFNLKSLVDVKDIDTTPAISGFGHLPPFYEACFIPVQRADSLSSHL  500
```

Fig. 2. Amino acid sequence comparison of P-$450_{15\beta}$ (line 1) and P-$450_{16\alpha}$ (line 2).

Therefore, it was concluded that these four cytochromes represent a subfamily within the phenobarbital related P-450 gene family.

Interestingly, the regions of homology between these two sexually differentiated cytochromes do not seem to be evenly spread out. There are long stretches of complete identity, while other segments show a significantly lower homology. It is tempting to speculate that the regions of weaker homology play a role in substrate specificity. Furthermore, P-$450_{15\beta}$ which is known to accept steroid sulfates as substrates with much higher affinity than nonsulfated ones has a positively charged lysine residue (lys 434) next to cysteine 435 which is presumed to provide the sulfur for the heme fifth ligand. P-$450_{16\alpha}$ has instead an isoleucine residue at position 434. Thus, it is possible that the presence of this positively charged residue next to the heme moiety may account for the preference of $P\text{-}450_{15\beta}$ for negatively charged steroid sulfates.

It has been suggested that the region between position 199-219 corresponds to the putative substrate binding site (Yoshioka et al. 1987). Interestingly, no predominance of positively charged residues in $P\text{-}450_{15\beta}$ relative to $P\text{-}450_{16\alpha}$ are present in that region. On the contrary, the reverse seems to be the case. However, another variable region between positions 230-240 reveals such a predominance. Clearly, site directed mutagenesis experiments will be necessary to clarify all these important issues.

Northern blots of total rat liver RNA from animals subjected to different growth hormone regimens were performed using the cDNA clones for either $P\text{-}450_{15\beta}$ or $P\text{-}450_{16\alpha}$. The results, as indicated in Fig. 3. and Fig. 4., demonstrate that GH infusion caused an increase in $P\text{-}450_{15\beta}$ mRNA levels, both in normal and hypophysectomized males, but had no effect on $P\text{-}450_{15\beta}$ mRNA levels when given by 12-hourly injections. Furthermore, levels of $P\text{-}450_{16\alpha}$ mRNA were significantly depressed in Hx male rats, and almost abolished in males infused with GH. However, in this experiment intermittent administration of the same dose of GH gave essentially no response. This lack of induction of $P\text{-}450_{16\alpha}$ mRNA by GH injected every 12 h may be an indication of the obvious difficulty in precisely mimicking a male-type GH secretory pattern by subcutaneous injections.

These data combined with what is known from studies at the protein level demonstrate that the secretory pattern of GH is critical in expression of $P\text{-}450_{15\beta}$ and $P\text{-}450_{16\alpha}$.

CONCLUDING REMARKS

In this manuscript we report the complete sequence of the sexually differentiated cytochromes $P\text{-}450_{15\beta}$ and $P\text{-}450_{16\alpha}$ as deduced from the corresponding cDNA's. These two cytochromes are 66% homologous to each other and together with the developmentally regulated cytochromes P-450 f and PB-1 belong to the same subfamily within the phenobarbital regulated P-450 gene family.

The secretory pattern of growth hormone has been shown to regulate the expression of these cytochromes at the apoprotein level. Here we report that regulation occurs also at the mRNA level. Experiments to determine whether this is due to transcriptional regulation or not are under way in this laboratory.

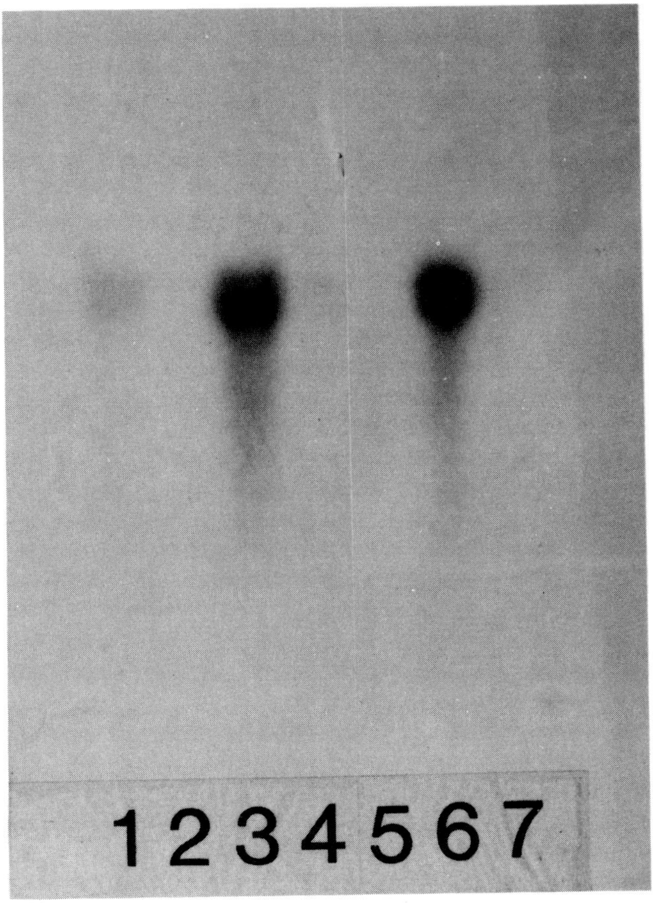

Fig. 3. Northern blot of rat liver RNA probed with P-$450_{15\beta}$ cDNA. 20 μg per well of total rat liver RNA was electrophoresed in 1% agarose and blotted onto nitrocellulose. The blot was probed with a P-$450_{15\beta}$ cDNA clone containing half of the coding sequence as well as the 3'-end nontranslatable sequence. The RNA samples applied to the wells were from rats treated as follows: 1, Hx male infused with bovine GH; 2, Hx male; 3, male infused with human GH; 4, control male; 5, Hx female; 6, control female; 7, hypophysectomized male injected with bovine GH.

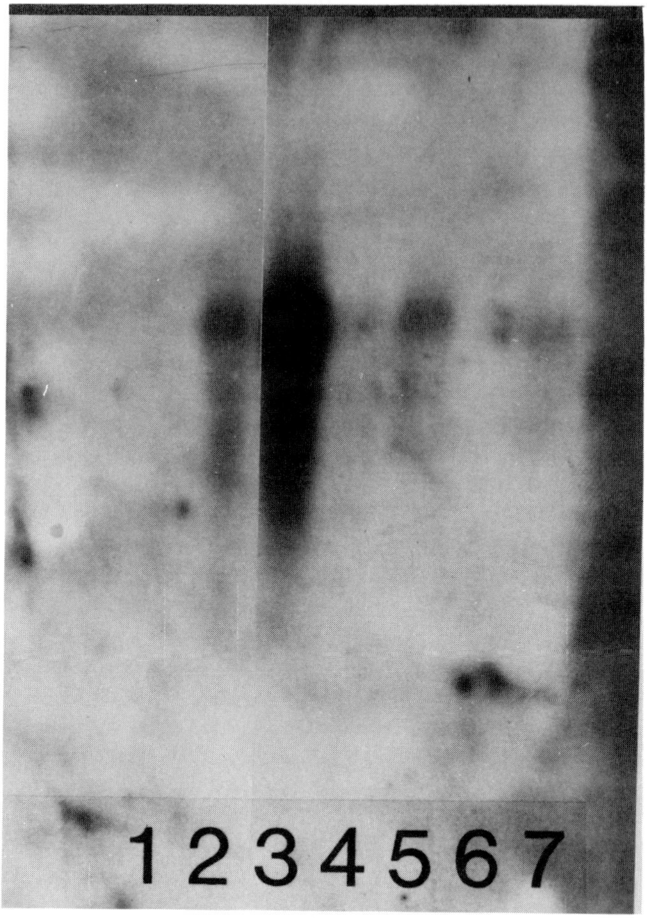

Fig. 4. Northern blot of rat liver RNA probed with $P\text{-}450_{16\alpha}$ cDNA RNA was electrophoresed and blotted as described in Fig. 3 and probed with the full length $P\text{-}450_{16\alpha}$ cDNA. The RNA source for each lane was liver from: 1. Control female; 2. Hx female; 3. Control male; 4. Male infused with hGH; 5. Hx male; 6. Hx male infused with bGH; 7. Hx male injected with bGH. Hx, Hypophysectomised. hGH, Human growth hormone. bGH Bovine growth hormone.

ACKNOWLEDGEMENTS

We wish to thank Anne-Gerd Nilsson, Stefan Westin and Elisabeth Wiersma for their skilful technical assistance. This study was supported by a grant from the Swedish

Medical Research Council (No. 13X-2819). P.Z. is a visiting scientist sponsored by the Swedish Medical Research Council.

REFERENCES

Andersson, S. & Jörnvall, H., 1986, Journal of Biological Chemistry, 261, 16932-16936.

Chirgwin, J.M., Przybyla, A.E., MacDonald, R.J. & Rutter, W.J., 1979, Biochemistry, 18, 5294-5299.

Edén, S., 1979, Endocrinology, 105, 555-560.

Gonzalez, F.J., Kimura, S., Song, B.J., Pastewka, J., Gelboin, H.V. & Hardwick, J.P., 1986, Journal of Biological Chemistry, 261, 10667-10672.

Gubler, U. & Hoffman, B.J., 1983, Gene, 25, 263-269.

Gustafsson, J.-Å. & Ingelman-Sundberg, M., 1975, Journal of Biological Chemistry, 250, 3451-3458.

Helfman, D.M., Feramisco, J.R., Fiddes, J.C., Thomas, G.P. & Hughes, S.H., 1983, Proceedings of the National Academy of Sciences U.S.A., 80, 31-35.

MacGeoch, C., Morgan, E.T., Cordell, B. & Gustafsson, J.-Å., 1987, Biochemical and Biophysical Research Communications, 143, 782-788.

MacGeoch, C., Morgan, E.T. & Gustafsson, J.-Å., 1985, Endocrinology, 117, 2085-2092.

MacGeoch, C., Morgan, E.T., Halpert, J. & Gustafsson, J.-Å., 1984, Journal of Biological Chemistry, 259, 15433-15439.

Mode, A., Gustafsson, J.-Å., Jansson, J.-O., Edén, S. & Isaksson, O., 1982, Endocrinology, 111, 1692-1696.

Morgan, E.T., MacGeoch, C. & Gustafsson, J.-Å., 1985a, Molecular Pharmacology, 27, 471-479.

Morgan, E.T., MacGeoch, C. & Gustafsson, J.-Å., 1985b, Journal of Biological Chemistry, 260, 11895-11898.

Sanger, F., Coulson, A.R., Barrell, B.-G., Smith, A.J.H. & Roe, B.A., 1980, Journal of Molecular Biology, 143, 161-178.

Ström, A., Mode, A., Zaphiropoulos, P.G., Morgan, E.T. & Gustafsson, J.-Å., 1987, Manuscript submitted for publication.

Waxman, D.J., 1984, Journal of Biological Chemistry, 259, 15481-15490.

Yoshioka, H., Morohashi, K.-I., Sogawa, K., Miyata, T., Kawajiri, K., Hirose, T., Inayama, S., Fujii-Kuriyama, Y. & Omura, T., 1987, Journal of Biological Chemistry, 262, 1706-1711.

Hormonal Regulation of Sex-Specific Forms of Cytochrome P-450 Involved in the Hydroxylation of Steroids and Drugs

RYUICHI KATO and YASUSHI YAMAZOE

Department of Pharmacology, School of Medicine, Keio University, Shinjuku, Tokyo 160, Japan

Sex-related difference in drug and steroid metabolism is a well-known phenomenon in rats (Kato & Yamazoe, 1987). Recent progress on the purification of cytochrome P-450 indicates that sex-specific forms of cytochrome P-450 are responsible for the occurrence (Kamataki et al., 1983b). From rat livers, more than five forms of sex-specific or sex-dependent cytochrome P-450 have been isolated and shown to have distinct properties catalytic on drugs and steroids, although the endogenous role of these cytochrome P-450 is not yet well understood (Bandiera et al., 1986; Jansson et al., 1985; Kamataki et al., 1983b; Waxman et al., 1985). On the regulation of these sex-related forms, we showed that P-450-male and P-450-female, major sex-specific and constitutive forms in the respective male and female rat livers, are under the regulation of sex-steroids and pituitary growth hormone (Kamataki et al., 1983b; Morgan et al., 1985; Kato et al., 1986; Yamazoe et al., 1986). However, the extent of the sex-difference in drug metabolism is known to be affected by the administrtion of drugs and by changes in physiological state. Thus we have studied the mechanism by which drugs and physiological states modify the levels of sex-specific forms of cytochrome P-450. In addition, we also discuss the mechanism of interaction of steroids on the growth hormone-dependent regulation.

Changes in the contents of P-450-male and P-450-female in livers of rats treated with various drugs

In our previous paper (Kamataki et al., 1986),

contents of hepatic P-450-male and P-450-female were not increased, but rather decreased, by the treatment of male and female rats, respectively, with phenobarbital, PCB or 3-methylcholanthrene. To understand the mechanism of the modulation, effects of chemicals were further examined as described in Table 1.

Among the 14 chemicals examined, a low dose (1 mg/kg) of dexamethasone showed the most effective suppression on the total cytochrome P-450 contents. A large dose of dexamethasone rather increased the total cytochrome P-450 contents, while the treatment showed the most obvious inhibitory effect on the contents of P-450-male and P-450-female. Treatment of male rats with alloxan, reserpine, morphine or sodium selenite also resulted in the clear repression of P-450-male contents. The amount of female-specific hemoprotein, P-450-female, was less affected by the treatment with various chemicals, except for dexamethasone. Hepatic levels of P-450-male and P-450-female are regulated by steroids and pituitary growth hormone. Thus, the chemicals can possibly modify the hepatic levels of these cytochrome P-450s both directly in livers and indirectly through the hypothalamo-pituitary system. To verify the possibilities, effects of typical chemicals were also examined with hypophysectomized rats. As a result, treatment of hypophysectomized rats with dexamethasone (100 mg/kg) strikingly reduced the level of P-450-male, but 3-methylcholanthrene did not decrease the contents. Phenobarbital and morphine also showed a repressive effect, but the effects were less in hypophysectomized rats than in normal rats. These results suggest that dexamethasone, phenobarbital and morphine repress the appearance of P-450-male acting in both the liver and hypothalamo-pituitary system, and 3-methylcholanthrene modulates the level of P-450-male mainly through the hypothalamo-pituitary system.

Effects of changes in physiological states on levels of hepatic cytochrome P-450

Hepatic drug metabolizing activities, which show a clear sex-related difference, are decreased in male, but not in female, diabetic rats (Kato & Gillette, 1965). Thus, the changes in the amounts of sex-specific cytochrome P-450 were investigated with alloxan-induced diabetic rats. Associated with the decrease in benzo[a]-pyrene hydroxylation and testosterone 2α- and 16α-

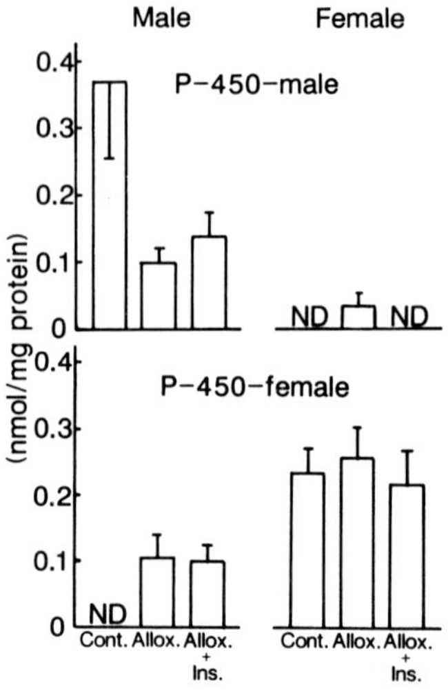

Figure 1. Effect of alloxan-induced diabetic state and insulin on the hepatic contents of P-450-male and P-450-female.
Cont., Allox. and Ins. indicate control, alloxan-treated (170 mg/kg) and insulin treatment (25 IU/day for 6 days).
N.D.: not detectable

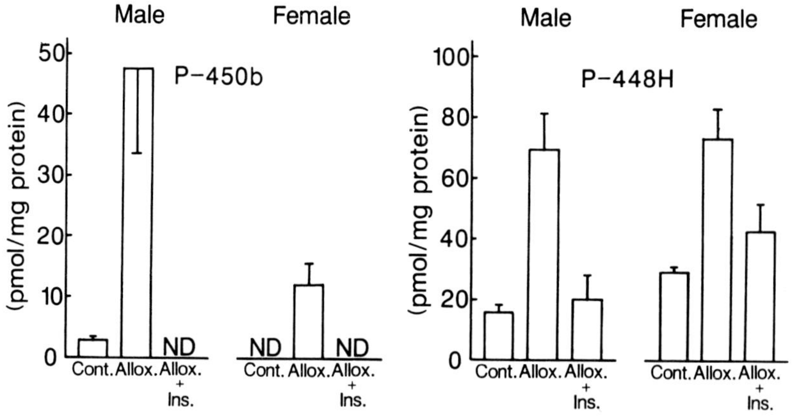

Figure 2. Effect of alloxan-induced diabetic state and insulin on the hepatic contents of phenobarbital-inducible P-450b and 3-methylcholanthrene-inducible P-448H. Experimental details are the same as described in Fig. 1.

hydroxylations, the content of P-450-male was reduced in livers of male rats after alloxan-treatment (Fig. 1). In addition, low levels of P-450-female also appeared in alloxan-treated male rats, while insulin-treatment (25 IU/kg for 6 days) had no significant effect on the contents of P-450-male and P-450-female. In female rats, the hepatic level of P-450-female was not affected by alloxan treatment, but low levels of P-450-male were detected in the hepatic microsomes. The result was further supported with inhibition by anti-P-450-male IgG of benzo[a]pyrene and testosterone hydroxylations. Using specific antibodies, we also examined the levels of phenobarbital-inducible P-450b, a high-spin form P-448H (Kamataki et al., 1983a) and $P\text{-}450_{6\beta}$, which is a male-specific testosterone 6β-hydroxylase in rat livers (Fig. 2). Amounts of P-450b and P-448H were increased in alloxan-treated male and female rats, and their levels were restored by insulin administration to the diabetic rats. The hepatic level of $P\text{-}450_{6\beta}$ was low in diabetic rats and did not recover by treatment with insulin (data not shown).

Adult rats treated with mono sodium glutamate (MSG) during the neonatal period showed disorders characterized by reduced anterior pituitary, stunted growth and abnormal reproductive function (Millard et al., 1982). Thus the hepatic levels of 5 specific forms of cytochrome P-450 in MSG-treated rats were also examined to understand the growth hormone-dependent regulation of these cytochrome P-450s. Neonatal MSG-treatment, which caused the disruption of the growth hormone secretory pattern in the adult rats, decreased the contents of P-450-male and $P\text{-}450_{6\beta}$ in the male and slightly decreased the content of P-450-female in the female adult rat (Table 2). Although the hepatic level of phenobarbital-inducible forms, P-450b and P-450e, were non-detectable in normal and MSG-treated female rats, higher induction of phenobarbital-inducible cytochrome P-450 was observed in MSG-treated female rats (Table 3). Hepatic levels of P-450b, P-450e and the dealkylation of ethylmorphine and O-pentylresorufin were 2-3 fold higher in MSG-treated than control female rats after phenobarbital treatment. These results are consistent with our recent finding that growth hormone suppressively regulates the hepatic level of P-450b and P-450e in rat livers (Yamazoe et al., 1987).

Table 1. Effect of various drugs on the hepatic contents of P-450-male, P-450-female and total cytochrome P-450.

Treatment	(mg/kgxd)	Male		Female	
		Total	P-450-male	Total	P-450-female
None		100(%)	100(%)	100(%)	100(%)
Phenobarbital	(80x3)	177	41	130	98
Dexamethasone	(100x4)	144	28	199	21
Dexamethasone	(1x4)	52	60	49	46
3-Methylcholanthrene	(40x3)	135	41	159	64
Isosafrole	(150x3)	85	44	104	55
PCB	(500x1)	338	80	319	56
Pregnenolone-16α-carbonitrile	(75x3)	68	55	111	132
Clofibrate	(100x3)	77	53	ND	
Isoniazid	(0.1%)[a]	66	102	ND	
Alloxan	(170x1)	108	29	159	108
Streptozotocin	(65x1)	ND		127	97
Reserpine	(5x4)	68	39	ND	
Haloperidol	(1x2x4)[b]	83	85	ND	
Morphine	(15→35 x4)[c]	57	35	ND	
Sodium Selenite	(2.5x1)	69	38	ND	

a Drinking water for 10 days.
b 1 mg/kg twice a day for 4 days.
c Twice a day for 4 days with the following dosing schedule. (15,20,25,25,30,30,35, and 35 mg/kg)
ND: not determined.

Table 2. Effect of neonatal MSG-treatment on the hepatic levels of sex-specific cytochrome P-450s.

P-450	Male		Female	
	Control	MSG-treated	Control	MSG-treated
P-450-male	100	48	N.D.	N.D.
P-450-female	N.D.	N.D.	100	82
P-450$_{6\beta}$	100	41	N.D.	6

Amounts are represented as the relative amounts corresponding to the respective controls.
N.D.: not detectable (less than 3% of the respective control)

Mechanism of growth hormone-dependent regulation of hepatic cytochrome P-450

Hepatic contents of P-450-male are affected by the secretory pattern of growth hormone. Intermittent injection (twice a day) of hGH stimulates the amounts of P-450-male in hypophysectomized rats, while continuous infusion of hGH decreases the amounts of P-450-male (Fig. 3). Another male specific form, $P\text{-}450_{6\beta}$, is also regulated by serum level of growth hormone in rats (Yamazoe et al., 1986). Hypophysectomy of female rats resulted in the expressions of $P\text{-}450_{6\beta}$, and the level was higher than that in normal male rats. Administration of hGH by either intermittent injection or continuous infusion decreased the hepatic amount of $P\text{-}450_{6\beta}$. However, treatment of hypophysectomized rats with insulin had no appreciable effect on the contents of P-450-male and $P\text{-}450_{6\beta}$. Similar results were also obtained by RNA blots using specific cDNA probes for P-450-male, $P\text{-}450_{6\beta}$ and P-450b. These results indicate that growth hormone regulates, stimulatively and suppressively, hepatic forms of cytochrome P-450, mainly at pretranslational steps in rat livers. Our recent results indicate that other forms of cytochrome P-450, P-450-female, P-450b, P-450e and probably P-450g are also under the regulation of growth hormone in rats. To investigate the mechanism of growth hormone-dependent regulation of these cytochrome P-450s, time-dependent changes in the levels of microsomal testosterone hydroxylation and specific cytochrome P-450 were quantified with hGH-infused hypophysectomized rats. Microsomal testosterone 2α-, 6β-, 16α- and 16β-hydroxylations were decreased by nearly half the levels of non-infused hypophysectomized rats after 1 day infusion of hGH and reached plateau levels after 3 days infusion (Fig. 4). The rate of testosterone 7α-hydroxylation was also slightly reduced after the 1 day infusion, but the rate was increased during the prolonged infusion. Consistent with testosterone hydroxylations, hepatic contents of P-450-male, $P\text{-}450_{6\beta}$ and P-450b were decreased similarly to one another and attained low basal levels within 5 days infusion of hGH. The amounts of P-450-female were rapidly increased from non-detectable to 65% of the maximal level after 1 day infusion of hGH and reached the maximal level after 3 days infusion. These results suggest that stimulative and suppressive actions of growth hormone on hepatic cytochrome P-450 are mediated through a mutual receptor mechanism.

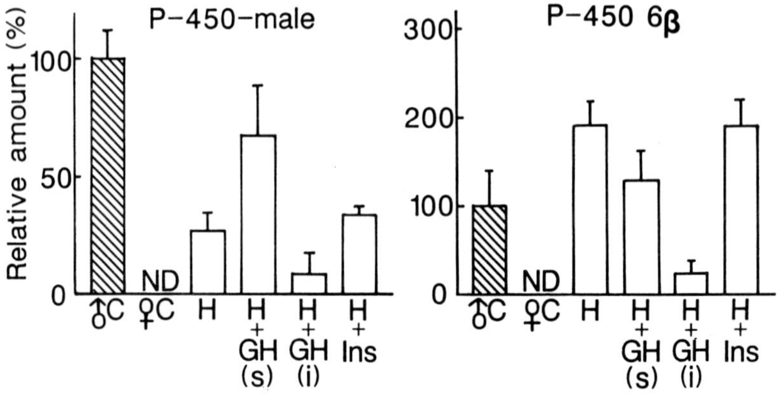

Figure 3. Effects of growth hormone and insulin on the hepatic contents of P-450-male and P-450$_{6\beta}$ in hypophysectomized female rats.
H, GH(s), GH(i) and Ins indicate the treatments of hypophysectomy, intermittent injection of hGH, continuous infusion of hGH and insulin-treatment.

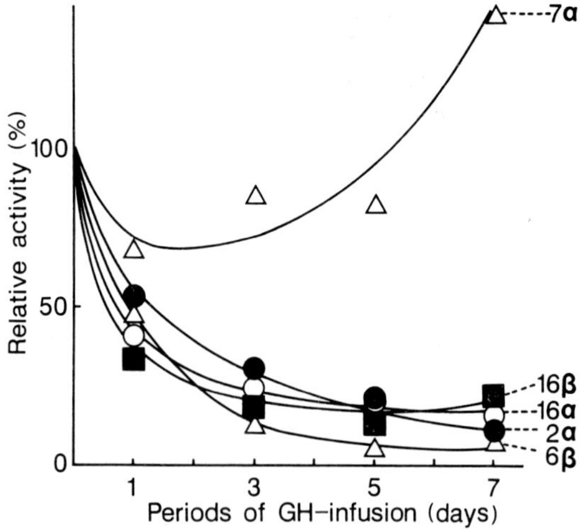

Figure 4. Changes in the rates of testosterone hydroxylation during hGH infusion in hypophysectomized female rats.
Microsomal testosterone hydroxylations were measured by HPLC.

REFERENCES

Bandiera, S., Ryan, D.E., Levin, W. & Thomas, P., 1986, Archives of Biochemistry and Biophysics, 248, 658.

Jansson, I., Mole, J. & Schenkman, J.B., 1985, Journal of Biological Chemistry, 260, 7084.

Kamataki, T., Maeda, K., Shimada, M. & Kato, R., 1986, Journal of Biochemistry, 99, 841.

Kamataki, T., Maeda, K., Yamazoe, Y., Matsuda, N., Ishii, K. & Kato, R., 1983a, Molecular Pharmacology, 24, 146.

Kamataki, T., Maeda, K., Yamazoe, Y., Nagai, T. & Kato, R., 1983b, Archives of Biochemistry and Biophysics, 225, 758.

Kato, R. & Gillette, J.R., 1965, Journal of Pharmacology and Experimental Therapeutics, 150, 279.

Kato, R., Yamazoe, Y., Shimada, M., Murayama, N. & Kamataki, T., 1986, Journal of Biochemistry, 100, 895.

Kato, R. & Yamazoe, Y., 1987, In Frontiers in Biotransformation, edited by K. Ruckpaul, (Akademie-Verlang, Berlin), in press.

Millard, W.J., Martin, J.B., Jr., Sagar, A.S.M. & Martin, J.B., 1982, Endocrinology, 110, 540.

Morgan, E.T., MacGeoch, C. & Gustafsson, J.-Å., 1985, Journal of Biological Chemistry, 260, 11895.

Waxman, D.J., Dannan, G.A. & Guengerich, F.P., 1985, Biochemistry, 99, 841.

Yamazoe, Y., Shimada, M., Kamataki, T. & Kato, R., 1986, Japanese Journal of Pharmacology, 42, 371.

Yamazoe, Y., Shimada, M., Murayama, N. & Kato, R., 1987, Journal of Biological Chemistry, 262, 7423.

Yamazoe, Y., Shimada, M., Murayama, N., Kawano, S. & Kato, R., 1986, Journal of Biochemistry, 100, 1095.

Regulation of the Glucocorticoid-inducible Cytochromes P-450

PHILIP S. GUZELIAN

Division of Clinical Toxicology and Environmental Medicine, Medical College of Virginia, Box 267, MCV Station, Richmond, VA 23298, USA

Nearly 20 years ago Hans Selye identified pregnenolone-16α-carbonitrile (PCN) as a potent "catatoxic" steroid capable of affording protection from toxic substances by accelerating their metabolic elimination from animals (Selye 1971). It was recognized also that PCN induced liver microsomal drug metabolizing enzymes that catalyzed 6β hydroxylation of progesterone and, thus, differed from enzymes induced by either the barbiturate and pesticide class of compounds or the polycyclic aromatic hydrocarbons (Lu, et al., 1972). Definitive evidence that PCN was an inducer of a unique microsomal enzyme was the purification from rat liver of a cytochrome P-450 (now termed P-450p) which could be distinguished biochemically and immunochemically from the forms of cytochrome P-450 responsive to other classes of inducers (Elshourbagy & Guzelian, 1980). With the availability of anti-P-450p antibodies and of cloned cDNAs it has become evident that P-450p is one of a family of related forms of hepatic (Wrighton, et al., 1985) and extra-hepatic cytochrome P-450 proteins found in many species.

Table I presents the P-450p proteins for which NH_2-terminal amino acid sequences are available. Although errors in sequencing led to some initial confusion (Hostetler, et al., 1987) it is now clear that P-450p is the same as P-450 PCN1 (Gonzalez, et al., 1985). P-450PCN2 is a cloned cDNA encoding a protein similar but not identical to P-450PCN1 (Gonzalez et al., 1986). P-450PCN2 may be one of the two proteins, P-450PCNb and P-450PCNc, whose sequences through 28 residues are identical to P-450PCN2. P-450PCNb and P-450 PCNc appear to be different proteins based on their chromatographic characteristics and reactivity to monoclonal antibodies directed against P-450p (J. Halpert, personal communication). P-450PCNa and $P\text{-}450PB_{2b}$ (Wolf, et al., 1986) raise to five the apparent number of "P-450p-like" proteins in rat liver. Given the multiplicity of closely related P-450p-like proteins, it is unclear which of these forms (or their combination) is represented in the preparations termed P-450PB-2a/PCN-E (Guengerich, et al., 1982). In rabbit liver, the single form of cytochrome P-450 homologous to P-450p is LM3c (Dalet, et al., 1986) although evidence for additional "P-450p-like" forms in the rabbit has been adduced (Molowa, et al., 1986). The orthologous form in humans is HLp, which has been shown to be related to P-450p and to LM3c in its structure, function, and regulatory characteristics (Molowa, et al., 1986; Watkins, et al., 1985). This same protein has also been identified based on its ability to catalyze nifedipine oxidation (Guengerich, et al., 1982). There are at least three genes encoding HLp and related forms (unpublished observations). The predicted sequences of two proteins that encode HLp and HLp2 are listed in Table I.

Substrates for the P-450p Family

Analysis of the substrates for P-450p has been complicated by the peculiar lack of activity when P-450p, LM3c, or HLp are tested as purified proteins in a reconstituted system. The loss of enzyme activity appears to occur early in the purification process. Therefore, evidence for the involvement of P-450p proteins in various oxidase reactions has come, by necessity, from indirect studies using P-450p inducers and anti-P-450p antibodies to inhibit the induced enzyme activity (see Table II). Steroids themselves are substrates. P-450p and HLp catalyze 6β-hydroxylation of testosterone, progesterone, and cortisol. These enzymes may also be involved in formation of catechols from estradiol. Macrolide antibiotics are a second group of substrates that also are inducers of the P-450p family. All species tested (Wrighton, et al., 1985) exhibit a liver microsomal erythromycin demethylase activity and the ability to convert triacetyloleandomycin (TAO) to a metabolite that forms a spectral complex with P-450p. The P-450p family of proteins preferentially hydroxylates warfarin at the R-10 position. LM3c catalyzes oxidation of cyclosporin A (Bertault-Peres, et al., 1987) In rats P-450p is a major 4-hydroxylase of mephenytoin (Shimada & Guengerich, 1985). In humans, however, HLp is not a mephenytoin hydroxylase. This important example illustrates that conservation of protein structure across species lines need not be accompanied by conservation of function.

Inducers of the P-450p Family

As listed in Table III, there are four main classes of compounds that induce P-450p or related proteins. In rats,

Table I. NH_2-Terminal Amino Acid Sequence of Glucocorticoid-Inducible Cytochromes P-450

	1				5					10					15					20					25		
RAT																											
P-450p (PCN1)	M	D	L	L	S	A	L	T	L	E	T	W	V	L	L	A	V	V	L	V	L	L	Y	G	F	G	T
P-450PCN2	M	D	L	L	S	A	L	T	L	E	T	W	V	L	L	A	V	I	L	V	L	L	Y	R	L	G	T
P-450 PCNa	M	D	L	A	S	A	L	V	L	E	Y	W	V	L	L												
P-450 PCNb	M	D	L	L	S	A	L	T	L	E	T	W	V	L	L	A	V	I	L	V	L	L	Y	R	L	G	T
P-450 PCNc	M	D	L	L	S	A	L	T	L	E	T	W	V	L	L	A	V	I	L	V	L	L	Y	R	L	G	T
P-450 PB_{2b}	M	K	L	L	V	A	L	P	L	E	T	W	V	L	-	A	V										
Rabbit																											
LM3c	M	D	L	I	F	S	L	E	T	W	V	L	L	A	A	S	L	V	L	L	Y	L	Y	G	G		
Human																											
HLp	M	A	L	I	P	D	L	A	M	Q	T	W	L	L	L	A	V	S	L	V	L	L	Y	L	Y		
HLp2	M	D	L	I	P	N	L	A	V	E	T	W	L	L	L	A	V	S	L	V	L	L	Y	L	Y		

Table II. Substrates for the P-450p Family

Steroids	testosterone β 6-hydroxylase	rat, human
	progesterone β 6-hydroxylase	rat
	cortisol β 6-hydroxylase	human
	estrogen catechol formation	rat, human
Macrolides	erythromycin demethylase	human, rat, rabbit
	TAO complex	
Warfarin	R-10 hydroxylase	human, rabbit, rat
Mephenytoin	hydroxylase	rat
Nifedipine	hydroxylase	human
Quinidine	hydroxylase	human
Aldrin	hydroxylase	rat, human
Cyclosporin A	hydroxylase, demethylase	rabbit

Table III. Inducers of the P-450p Family

Steroids	Species
Glucocortocoids (Synthetic & Endogenous)	rat, human
Antiglucocortoids (PCN, Dex-Mes)	rat
Antimineralocoid (Spironolactone)	rat
"Phenobarbital-like" Inducers	
Anticonvulsants (Phenobarbital, Dilantin)	rat
Organochlorine Pesticides	rat
Polychlorinated Biphenyls	rat
Macrolide Antibiotics	
Triacetyloleandomycin (TAO)	rat, rabbit, human
Erythromycin (estolate, base)	rat
Rifampicin	rabbit, human
Antifungal (Imidazoles)	
Clotrimazole	rat
Ketoconazole	rat

glucocorticoids, but not estrogen, testosterone, or progesterone, are inducers of P-450p (Heuman, et al., 1982). Because P-450p can be induced in primary monolayer cultures of adult rat hepatocytes (Elshourbagy, et al., 1981) it has been possible to carry out detailed studies of the dose-response relationships, agonist-antagonist interactions and time course of induction of P-450p (Schuetz & Guzelian, 1984). It became evident that PCN actually is an anti-glucocorticoid, and thus, the mechanism of induction of P-450p by steroids differs from the classical pathway whereby glucocorticoids bind to their receptor protein and interact with specific DNA enhancer elements (Yamamoto, 1985). Indeed, PCN and dexamethasone differ in the extent and time course of stimulated transcription of the P-450p gene in rats (Simmons et al., 1987).

A second class of inducers of the P-450p family are the barbiturate and pesticide compounds that are well-known to induce P-450b/e in the liver. Tests of organochlorine pesticides and polychlorinated biphenyls both in rats and in cultured rat hepatocytes (Schuetz, 1986a) disclosed different structure activity relationships between induction of P-450p versus that of P-450b/e. Therefore, separate recognition mechanisms may be involved in the co-induction of P-450p and P-450b/e by phenobarbital. A further complication is that the regulatory control is not identical for all members of the P-450p family. P-450p is induced by steroids and by phenobarbital whereas the mRNA for P-450 PCN2 is induced only by phenobarbital (Gonzales, et al., 1985). In some instances, therefore, it will be necessary to utilize form-specific probes to establish definitively the function and regulation of individual members of the P-450p family.

Macrolide antibiotics represent a group of P-450p inducers which exhibit species differences. For example, rifampicin is the most potent inducer of LM3c in rabbits and yet does not at all induce P-450p in rats. In contrast, PCN, an inducer of P-450p in the rat, does not induce, and, may actually decrease, LM3c in rabbits. It is not clear whether this striking qualitative interspecies difference in expression of homologous genes is attributable to dissimilarities in their structure or to extragenic cellular factors. Nevertheless, this example illustrates that conservation of regulatory control need not accompany the homology in gene structure across species lines.

Because TAO failed to induce de novo synthesis of P-450p in primary cultures of adult rat hepatocytes (Watkins, et al., 1986) we proposed that TAO acts indirectly in rats by augmenting endogenous glucocorticoid inducers. However, when hepatocytes are cultured on a reconstituted basement membrane matrix TAO induces P-450p mRNA directly in hepatocyte cultures (unpublished observations). The availability of an improved culture system should greatly facilitate future investigations of this interesting class of clinically useful compounds (Schuetz, et al., 1986b).

A final group of recently identified inducers of P-450p are the anti-fungal agents exemplified by clotrimazole and ketaconazole. Like TAO, these compounds appear to induce by increasing the amount of P-450 mRNA and by preventing the turnover of P-450p protein (unpublished observations).

P-450p is found only in male adult rat liver although monoclonal antibodies to P-450p proteins have identified an apparent female specific form of this family as well (Hostetler, et al., 1987). Two studies, each using antibody against P-450PB2a/PCN-E, have reached completely contradictory results as to whether immunoreactive P-450p rises selectively to adult levels in males or is selectively repressed in developing females (Cresteil, et al., 1986; Waxman, et al., 1985). Northern blot analysis suggests that P-450PCN2 represents the ascending male-specific form in developing rats whereas P-450PCN1 mRNA is not represented in either male or female rats in the basal steady state (Gonzales, et al., 1985). Control of the male-specific steady state level of the P-450p-like proteins appears to require growth hormone (Yamazoe, et al., 1986). In contrast to the absence of P-450p in rat fetal liver, human fetal liver contains HFLa, a testosterone 6β-hydroxylase activity (Kitada, et al., 1987) which likely represents HLp. Thus the presence of HLp as a major form in human fetal liver and the lack of sex differentiation of HLp among adult humans (Watkins, et al., 1985) reinforces the importance of evaluating the regulatory control of P-450 genes in humans directly.

References

Beaune, P.H., Umbenhauer, D.R., Bork, R.W., Lloyd, R.S. & Guengerich, F.P., 1986, Proceedings of the National Academy of Sciences, USA, 83, 8064.

Bertault-Peres, P., Bonfils, C., Fabre, G., Just, S., Cano, J.P. & Maurel, P., 1987, Drug Metabolism & Disposition, 15, 391.

Cresteil, T., Beaune, P., Celier, C., LeRoux, J.P. & Guengerich, F.P., 1986, The Journal of Pharmacology and Experimental Therapeutics, 236, 269.

Dalet, C., Blanchard, J.M., Guzelian, P., Barwick, J., Hartle, H. & Maurel, P., 1986, Nucleic Acids Research, 14, 5999.

Elshourbagy, N.A. & Guzelian, P.S., 1980, The Journal of Biological Chemistry, 25, 1279.

Elshourbagy, N.A., Barwick, J.L. & Guzelian, P.S., 1981,

The Journal of Biological Chemistry, 256, 6060.
Gonzalez, F.J., Nebert, D.W., Hardwick, J.P. & Kasper, C.B. 1985, The Journal of Biological Chemistry, 260, 7435.
Gonzalez, F.J., Song, B.-J. & Hardwick, J.P., 1986, Molecular and Cellular Biology, 6, 2969.
Guengerich, F.P., Dannan, G.A., Wright, S.T., Martin, M.V. & Kaminsky, L.S., 1982, Biochemistry, 21, 6019.
Heuman, D.M., Gallagher, E.O., Barwick, J.L., Elshourbagy, N.A. & Guzelian, P.S., 1982, Molecular Pharmacology, 21, 753.
Hostetler, K.A., Wrighton, S.A., Kremers, P. & Guzelian, P.S., 1987, Biochemical Journal, in press.
Kitada, M., Kamataki, T., Itahashi, K., Rikihisa, T. & Kanakubo, Y., 1987, Biochemical Pharmacology, 36, 453.
Lu, A.Y.H., Somogyi, A., West, S., Kuntzman, R. & Conney, A.H., 1972, Archives of Biochemistry and Biophysics, 152, 457.
Molowa, D.T., Schuetz, E.G., Wrighton, S.A., Watkins, P.B., Kremers, P., Mendez-Picon, G., Parker, G.A. & Guzelian, P.S., 1986, Proceedings of the National Academy of Sciences, USA, 83, 5311.
Schuetz, E.G. & Guzelian, P.S., 1984, The Journal of Biological Chemistry, 259, 2007.
Schuetz, E.G., Wrighton, S.A., Safe, S.H. & Guzelian, P.S., 1986a, Biochemistry, 25, 1124.
Schuetz, E.G., Elswick, B., Li, D., Wrighton, S.A., & Guzelian, P.S., 1986b, Hepatology, 6, 1188.
Selye, H., 1971, Journal of Pharmaceutical Sciences, 60, 1.
Shimada, T. & Guengerich, F.P., 1985, Molecular Pharmacology, 28, 215.
Simmons, D.L., McQuiddy, P. & Kasper, C.B., 1987, The Journal of Biological Chemistry, 262, 326.
Watkins, P.B., Wrighton, S.A., Maurel, P., Schuetz, E.G., Mendez-Picon, G., Parker, G.A. & Guzelian, P.S., 1985, Proceedings of the National Academy of Sciences, USA, 82, 6210.
Watkins, P.B., Wrighton, S.A., Schuetz, E.G., Maurel, P. & Guzelian, P.S., 1986, The Journal of Biological Chemistry, 261, 6264.
Waxman, D.J., Dannan, G.A. & Guengerich, F.P., 1985, Biochemistry, 24, 4409.
Wolf, C.R., Seilman, S., Oesch, F., Mayer, R.T. & Burke, M.D., 1986, Biochemical Journal, 240, 27.
Wrighton, S.A., Schuetz, E.G., Watkins, P.B., Maurel, P., Barwick, J., Bailey, B.S., Hartle, H.T., Young, B. & Guzelian, P.S., 1985, Molecular Pharmacology, 28, 312.
Yamamoto, K.R., 1985, Annual Review of Genetics, 19, 209.
Yamazoe, Y., Shimada, M., Murayama, N.,, Kawano, S. & Kato, R., 1986, Journal of Biochemistry, 100, 1095.

CHEMISTRY AND PHYSICS OF CYTOCHROME P-450

Crystallographic Studies on Cytochrome P-450cam

THOMAS L. POULOS

Center for Advanced Research in Biotechnology, 9600 Gudelsky Drive,
Rockville, MD 20850, USA

Introduction

Through the years, cytochrome P450cam has served as the single best P450 system for detailed structure/function studies and was the obvious choice for a high resolution crystal structure analysis. A collaborative effort between our group and and that of Prof. Gunsalus lead to the crystallization of P450cam in a form suitable for high resolution structural studies (Poulos et al., 1982). The initial structural work at 2.6Å (Poulos et al., 1985) now has been extended to 1.63Å (Poulos et al., 1987). Theses studies focused on the structure of P450cam with its substrate, camphor, bound at the active site. We also have solved the crystal structure of camphor-free P450cam (Poulos et al., 1986) and have carried out a detailed comparison of the camphor-free and bound structures. In what follows we provide a summary of this work with special emphasis placed on how the crystal structure work relates to the wealth of biophysical data available on P450cam.

Another question of central importance is how good of a model the P450cam system is for eukaryotic P450s. While a wealth of sequence data have been generated for eukaryotic P450s (Black & Coon, 1986), only the crystal structure of P450cam is known. Thus, it is tempting to use the P450cam structure as a guide to eukaryotic P450s especially with respect to which regions of the structure control substrate specificity. We examine this question as well and offer a working hypothesis and prediction on what we might expect to find when a eukaryotic P450 structure becomes available.

Summary of Structural Work

The camphor-bound structure has been refined at 1.63Å using data obtained from only two crystals and the newly developed Nicolet-Xentronics imaging proportional counter or area detector. The crystallogrpahic R factor=0.19 and the geometry of the final model has been tightly constrained (deviation of bond lengths=0.022Å from ideality). The camphor-free structure was refined to R=0.17 at 2.2Å using only a single crystal.

Fig. 1 depicts the overall P450cam molecule. The molecule approximates a triangular prism in shape with about half the structure in an alpha helical configuration. Notice that most of the helical structure clusters together on the right half of the molecule while a majority of the beta structure occurs in the left half. P450cam therefore closely resembles both cytochrome c peroxidase and the globins in that helices form the heme pocket and contribute residues to the active site (Poulos & Finzel, 1984).

Figure 1. Schematic diagram of P450cam. Helices are represented as rods and beta pairs as flat arrows.

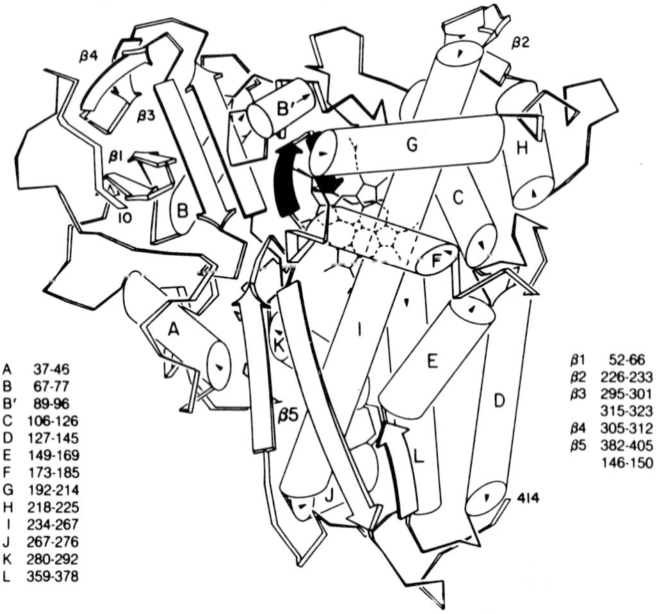

The camphor lies deeply embedded in the molecule and sits just above the heme adjacent to the oxygen binding site (Fig. 2). In Fig. 1 the camphor is hidden from view by helix F. Complementary hydrophobic contacts hold the camphor in place while the only polar contact is a hydrogen bond between the camphor carbonyl oxygen atom and the hydroxyl group of Tyr 96. Interestingly, Fisher & Sligar (1985) correctly predicted the presence of a single tyrosine residue in the P450cam active site using spectroscopic methods.

Figure 2. Edge-on view of the heme and its immediate environment.

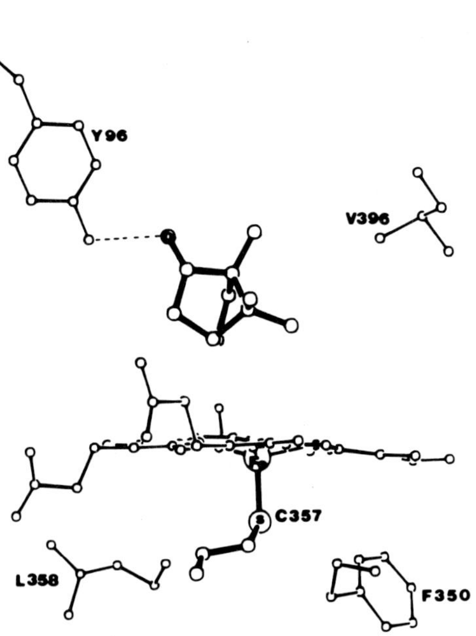

Fig. 2 also shows the expected Cys ligand. Refined electron density maps gave no evidence of a second axial ligand so in the camphor-bound state, P450cam is pentacoordinate. As might be expected for a high spin heme system, the iron atom is displaced about 0.43Å from the mean plane formed by the four pyrrole ring nitrogens.

A question of central interest is the structure of the enzyme without substrate bound. A low to high spin transition occurs when substrate binds and the redox equilibrium shifts toward higher values (Sligar & Murray, 1986; Sligar & Gunsalus, 1976).

$$\underset{-300mV}{S\text{-}Fe^{+3}\text{-}X} + R \rightleftharpoons \underset{-173mV}{S\text{-}Fe^{+3}\,R} \quad R=\text{camphor}$$

Some ligand, X, which spectroscopic probes suggested was an oxygen ligand (Dawson et al.,1982; Philson et al., 1979; Griffin & Peterson, 1975), is displaced when substrate binds. Moreover, the camphor-bound enzyme is significantly more stable than the substrate free enzyme (Jung et al., 1985). Clearly it is of interest to determine how these two forms of P450cam differ structurally. Fortunately, it has been possible to prepare crystals of substrate-free P450cam that are isomorphous with substrate-bound allowing for a direct comparison between the two structures.

A detailed comparison between the two structures revealed very few differences. All changes were confined to the active site where solvent replaces the substrate and a water molecule or hydroxide ion coordinates with the iron atom (Fig. 4). With the exception of the iron-linked aqua ligand, none of the active site waters were resolved as isolated lobes of electron-density. This very likely is due to microheterogeneity since none of the waters are fixed in place by hydrogen bonds with neighboring protein groups. In other words, the active site waters are free to "swim" within the confines of the camphor pocket while only the aqua ligand is fixed in place. The release of the these waters in addition to desolvation of the substrate upon entering the active site provides an explanation for why substrate binding is entropically driven (Griffin & Peterson, 1975).

Figures 3. Edge-on view of the heme without substrate bound. A cluster of hydrogen bonded water molecules occupies the substrate pocket one of which coordinates with the iron atom.

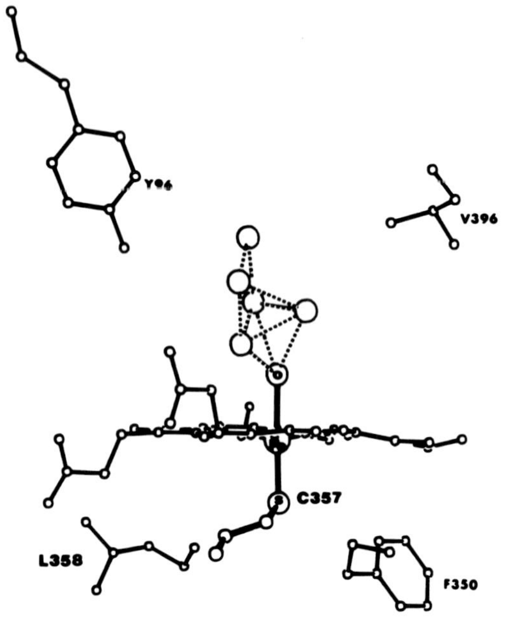

Only a slight repositioning in active site residues was observed and nothing that could be construed as a significant conformational change takes place. Thus, the various spectroscopic diffferences are due to the change in coordination environment of the iron atom and dielectric environment of the heme pocket. For example, the shift in redox potential from -300mV to -173mV is most simply accounted for by the change in dielectric environment. Without camphor bound, the active site waters provide a polar envirnonment that favors Fe^{3+} relative to Fe^{2+} while when camphor binds, the polarity decreases and the Fe^{3+}/Fe^{2+} shifts toward Fe^{2+}

Conserved regions

The one region of P450 that is expected to be highly conserved is the heme linked Cys segment. There is very little doubt now that the Cys corresponding to Cys 357 in P450cam provides the axial thiolate ligand in all P450s. Key residues are highly conserved or invariant. Invariant Phe 350, for example, flanks the Cys just beneath the heme and helps to provide a pocket on the molecular surface that houses the S-Fe interaction. The polypeptide conformation in this region also is likely to be conserved. The Cys ligand lies at the C terminal end of a short antiparallel beta region and beta bulge just beneath the heme. An invariant Gly at position 353 apparently is required for the correct sharp reversal of the polypeptide chain. It seems reasonable to conclude that this same structure is maintained in all P450s.

Since all P450s have similar requirements for O_2 binding and cleavage of the O-O bond, one might expect a similar structure here as well. Sequence comparisons suggest that this may be true. As shown in Table I, the central region of helix I that forms the O_2 pocket in P450cam exhibits good homology.

Table I. Alignment of distal helices in various P450s in the vicinity of the highly converved Thr 252.

P-450cam	$GLLLVGGLDT_{252}VVNFLSF$
P-450d (rat)	$NDIFGAGFET_{319}VTTAIFW$
P-450c (rat)	$FDLFGAGFDT_{325}ITTAISW$
P-450LM2 (rabbit)	$LSLFFAGTET_{302}TTSTTLR$
P-450b (rat)	$LSLFFAGTET_{302}TSSTTLR$
P-450e (rat)	$LSLFFAGTET_{302}TGSTTLR$
P-450scc (bovine)	$TEMLAGGVNT_{329}TSMTLQW$.

Especially noteworthy is the highly conserved Gly-Gly or Ala-Gly sequence (residues 248 and 249) and the near invariant Thr 252. In P450cam, Thr 252 participates in a local distortion of helix I that forces a local unwinding and kinking of the helix which results in a pocket for an O_2 molecule. Furthermoe, Kunze et al. (1983) used suicide substrate inhibitors to map the topography of microsomal P450s and found that exactly the same region which is masked by helix I in P450cam is blocked in microsomal P450s. Based on these observations, we have suggested that helix I is similarly positioned in all P450s owing to the requirements in forming an O_2 binding site.

Variable Regions

As shown in Fig. 1, the P450cam molecule naturally divides into a helical-rich domain (right side of Fig. 1) and a helical-poor domain (left side of Fig. 1) where most of the beta structure is located. We have already emphasized that helix I very likely is similarly positioned in all P450s owing to the requirements in forming an O_2 binding pocket. This further suggests that the helical packing arrangement surrounding helix I which also forms the heme binding pocket exhibits a similar topography in other P450s. These structural requierments together with the sequence homology between P450cam and microsomal P450s (Black & Coon, 1986) suggests that all P450s exhibit a similar three dimensional structure.

Nevertheless, one of the most interesting aspects of P450 structure is that they *must* differ in the substrate binding region despite these similarities in forming the O_2 pocket and heme ligation environment. We have argued elsewhere that the left domain (Fig. 1) will most likely exhibit the greatest degree of topographical variability (Poulos et al., 1985) owing to the lower content of regular secondary structure and the inherently greater flexibility in beta pairs as opposed to helices. This argument is further underscored by comparing the sequences of P450b and e which differ in sequence by only 14 amino acids (Suwa et al., 1985) yet the relative turnover numbers with various substrates differ significantly (Ryan et al., 1982). Eight of the 14 amino acid differences occur in the left, helical-poor domain with four positions being good candidates for substrate contact points (Poulos et al., 1987). It therefore seems reasonable to suggest that the enzymatic differences between P450s b and e are due primarily to sequence differences in the helical-poor domain.

Finally, potential stereochemical requirements of the actual hydroxylation reaction further suggest that substrate variability resides in the helical-poor domain. Groves and coworkers have provided strong evidence from model heme systems that an aliphatic substrate must approach the iron-linked oxygen atom from the side, parallel to the heme such that the substrate interacts with the oxygen p orbitals (Groves &

Kruper, 1985). As illustrated in Fig. 4, camphor approaches the iron-linked oxygen atom such that the oxygen p orbitals maximally interact with carbon 5 for stereospecific hydroxylation exactly as Groves predicted.

Figure 4. top view showing the p orbitals of a hypothetical ironlinked oxygen atom whould be oriented relative to the camphor molecule. As shown, carbon 5 of the camphor would maximally interact with the oxygen p orbitals.

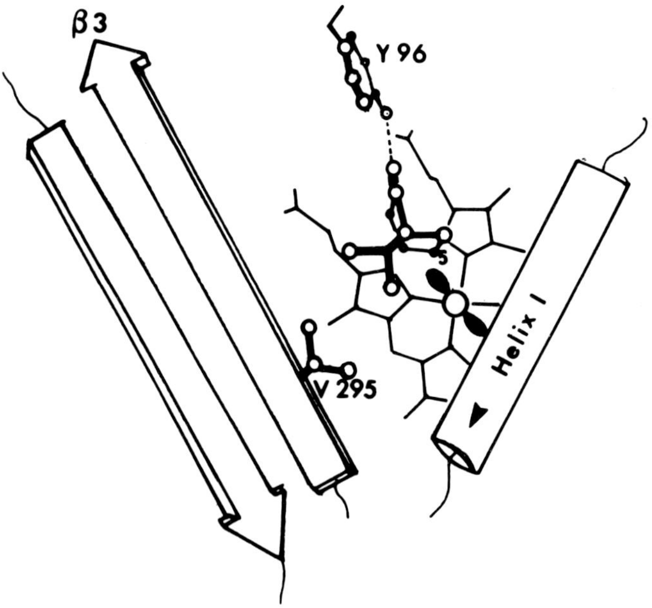

If this reaction trajectory is an essential aspect of P450 chemistry and helix I is similarly positioned in all P450s, substrates that differ in size (espeically larger than camphor) and shape must differ in the beta 3 segment of the helical-poor domain and the region surrounding Tyr 96. An additional complexity arises when we consider how the substrate gains access to the active site since this will require channels of different size, shape, and flexibility involving residues not directly linked to those residues that form specific contacts with the substrate. Nevertheless, the current working hypothesis regarding the potential regions controlling substrate specificity can be used as a guide for further experimental work like site-directed mutagenesis until such time that we obtain at least one additional P450 structure.

References

Black, S.D. and Coon, M.J. (1986). In "Cytochrome P450: Structure, Mechanism, and Biochemistry", (Ortiz de Montellano, P. ed.) pp. 161. Plenum.

Dawson, J.H., Anderson, L.A. and Sono, M. (1982). Journ. Biol. Chem., 257, 3606.

Fisher, M.T. and Sligar, S.G. (1985). Biochemistry, 24, 6896.

Griffin, B.W. and Peterson, J.A. (1975). Journ. Biol. Chem., 250, 6445.

Groves, J.T. and Kruper, W.J. (1985). Journ. Chem. (Israel), 25, 148.

Jung, C., Bendzko, P., Ristau, O. and Gunsalus, I.C. (1985). In "Cytochrome P450, Biochemistry, Biophysics, and Induction (Vereczkey, L. and Magyar, K., eds) pp. 19, Akademiai Kiado, Budapest.

Kunze, K.L., Mangold, B.L.K., Wheeler, C., Beilan, H.S. and Ortiz de Montellano, P.R. (1983). Journ. Biol. Chem., 258 4202.

Philson, S.B., Debrunner, P.G., Schmidt, P.G. and Gunsalus, I.C. (1979). Journ. Biol. Chem., 254, 10173.

Poulos, T.L. and Finzel., B.C. (1984) In "Peptide and Protein Reviews", (Hearn, H.T.W., ed) pp. 115, Marcel Dekker.

Poulos, T.L., Finzel, B.C., & Howard, A.J. 1986, Biochemistry, 25, 5314.

Poulos, T.L., Perez, M., & Wagner, G,C., 1982, Journal of Biol. Chem., 257, 10427.

Poulos, T. L., Finzel, B.C., Gunsalus, I.C., Wagner, G.C. & Kraut, J. 1985, Journal of Biological Chemistry, 260, 16122.

Poulos, T.L., Finzel, B.C., & Howard, A.J. (1987), Journal Molec. Biol, 195, 687.

Ryan, D.E., Thomas, P.E., and Levin, W. (1982). Arch. Biochem. Biophys., 216, 272.

Suwa, Y., Mizukami, Y., Sogawa, K., and Fujii-Kuriyama, Y. (1985). Journ. Biol. Chem., 260, 7890.

Sligar, S.G. and Gunsalus, I.C. (1979). Proc. Natl. Acad. Sci. (USA), 73, 1078.

Sligar, S.G. and Murray, R.I. (1986). In "Cytochrome P450:Structure, Mechanism, and Biochemistry (Ortiz de Montellano, Paul ed.) pp. 429, Plenum.

Probes of P-450cam: Genetic Engineering and Metabolic Chemistry

WILLIAM M. ATKINS and STEPHEN G. SLIGAR

Departments of Biochemistry and Chemistry, University of Illinois, Urbana, IL, USA

The relatively recent evolution of recombinant DNA methods allows the biophysicist and biochemist to contemplate the selective alteration of specific amino acids in a protein structure. To the extent that one can then relate these selective protein modifications to the functioning of the macromolecule, this approach can be utilized to unravel the molecular mechanisms of the cytochrome P-450 monoxygenases. A particularly exciting application of this technique is the molecular engineering of the P-450 active site in order to elucidate the chemical mechanisms of atmospheric dioxygen bond cleavage and substrate functionalization, to understand the timing and linkage of associated electron transfer events, and to rationalize the molecular basis for substrate recognition. With these aims in mind, we have undertaken an intensive effort to alter the enzymatic mechanisms and substrate specificity of the P-450cam active site through modification of substrate and protein. This two pronged attack seeks to correlate the effect of protein modification elicited through site-directed mutagenesis with variations in substrate structure, and thus examines both the "hand" and the "glove" aspects of biological recognition.

Cytochrome P-450cam normally catalyzes the hydroxylation of the monoterpene camphor to form a single regio- and stereo-specific product; 5-exo-hydroxycamphor (Sligar & Murray, 1986). The molecular basis for this selectivity in monoxygenation can be appreciated through the three dimensional structure of the protein (Poulos et. al, this volume) where it appears that specificity in substrate recognition is conferred by specific protein-substrate interactions, including hydrogen bonding between the camphor ketone group and tyrosine-96 in the protein, interdigitation of the substrate's 8,9-*gem* methyl groups with valine-295, and van der Walls contact between the camphor 10-methyl group and the

Leu244-Val247 hydrophobic cleft. The specific role of these various interactions can be elucidated by selective mutagenesis of the appropriate amino acids in the P-450cam gene (Unger *et. al*,1986; Koga *et. al*, 1985). If a particular amino acid side chain is responsible for conveying a high degree of substrate specificity, mutagenesis of this residue could yield a protein with altered stereochemistry or regiospecificty of hydroxylation. Conversely, the corresponding structure of the substrate can be altered to relieve specific contacts, and the resulting effect on metabolic profile assessed. As documented herein, the "goodness of fit" between substrate and protein not only dictates the stereochemistry and regiospecificity of functionalization, but also can profoundly effect the actual chemical reactivity of intermediate iron-oxygen states. We have termed this structure dependent phenomena "metabolic switching", and have utilized the powerful techniques of isotope methods to help define the identity of common intermediates involved. In the tightly coupled production of 5-*exo*-hydroxy camphor, two reducing equivalents originating in NADH are transferred sequentially to the cytochrome via putidaredoxin reductase and putidaredoxin with the subsequent scission of the molecular oxygen bond. Nearly all reducing equivalents are accounted for by substrate hydroxylation and monoxygenase derived water and there is minimal production of hydrogen peroxide. In contrast, the cytochrome P-450 systems isolated from the hepatic endoplasmic reticulum are much less tightly coupled, and as much as half of the pyridine nucleotide supplied reducing equivalents are used to generate hydrogen peroxide or water (Gorsky *et. al*, 1984). Although the existence of a cytochrome oxidase activity has been clearly demonstrated in these P-450 systems, the precise mechanism has remained undetermined. In view of the tight coupling of the P-450cam system we chose this monoxygenase to explore the metabolic fate of dioxygen and substrate under the two pronged attack of substrate and protein modification as discussed above, examining the role of Tyr96 and Val295 contacts through site directed mutagenesis and the effect of substrate structure through the metabolism of norcamphor, a camphor analog which lacks the 8,9, and 10 methyl groups. In the results and discussion sections that follow, we show that P-450cam demonstrates a prolific four electron oxidase activity as well as significant metabolic switching upon deuteration of the norcamphor substrate. This demonstration of an isotopically sensitive branching which affects the monoxygenase/oxidase ratio offers compelling evidence for a common intermediate that is responsible for hydrogen abstraction and which is also susceptible to a two electron reduction to form water in an activity paralleling that of the cytochrome oxidases. In addition, we report the first site-directed mutagenesis of a P-450 active site, with the replacement of the hydrogen bonding tyrosine-96 to a phenylalanine.

NORCAMPHOR HYDROXYLATED PRODUCTS. Cytochrome P-450cam was found to catalyze the hydroxylation of norcamphor to form three distinct products identified by GC-MS: 5-*exo*-hydroxy-norcamphor (45%), 6-*exo*-hydroxy-norcamphor (47%), and 3-*exo*- hydroxy-norcamphor (8%). When deuterium is substituted at the 5- and 6-carbons of norcamphor, there are no new products formed but a significant intramolecular isotope effect which results in a change in the relative and absolute yield of each product:

Regiospecificity Of Norcamphor Hydroxylation

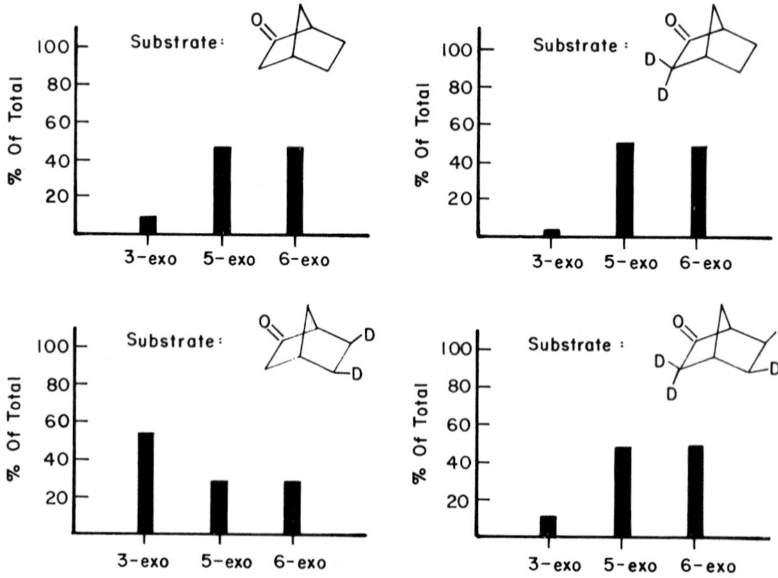

When the sum total of all oxygenated products formed from norcamphor is compared to the number of NADH reducing equivalents consumed in the reaction, a striking unaccountability of electrons is observed. These are shown to reside in excess water produced by the four electron reduction of atmospheric dioxygen by P-450cam. Metabolism of specifically deuterated norcamphor demonstrates a deuterium isotope effect on the branching ratio of substrate hydroxylation to excess water production and suggests that this oxidase activity of P-450cam results from the two electron reduction of a single oxygen-iron intermediate $[FeO]^{3+}$.

STEREOCHEMISTRY OF NORCAMPHOR METABOLISM. The stereochemical course of norcamphor hydroxylation at the 5- and 6- positions was investigated by determining the deuterium content of the corresponding alcohols when 5,6-*exo,exo*-d_2-norcamphor was metabolized. The results are summarized in the Table below.

Solvent	Product	%d_0	%d_1	%d_2
H_2O	5-exo-OH	0.96	94.4	4.6
	6-exo-OH	1.1	95.9	2.8
D_2O	5-exo-OH	98.1	1.9	0.0
	6-exo-OH	98.4	1.6	0.0

Several interesting features of norcamphor hydroxylation contrast the stereochemical course observed with camphor. It is apparent that the 5-alcohol obtained from norcamphor results from nearly complete specificity for *exo*-deuterium abstraction, with only minor *endo* abstraction followed by inversion of configuration. It is likely that due to the primary isotope effect, this stereochemical result significantly underestimates the *exo* specificity that occurs when protium is present at both the *exo* and *endo* face. Thus, with norcamphor as substrate, P-450cam shows a drastic reduction in the fraction of 5-*exo* alcohol produced with inversion of configuration as compared to the substrate camphor (Gelb *et. al*, 1982). The deuterium data for the 6-alcohol reflect a similar preference for *exo* abstraction. The importance of these stereochemical considerations becomes evident when the stoichiometry of norcamphor turnover as a function of substrate deuterium content is determined.

STOCHIOMETRY OF NORCAMPHOR METABOLISM. The complete stoichiometry of the system was determined by precisely quatitating the NADH and O_2 consumption along with formation of H_2O_2 and hydroxy-norcamphor. When the sum totals of oxygenated products are compared to the absolute number of reducing equivalents consumed in the reaction, a striking unaccountability of electrons is observed. These are found to reside in excess water produced by four electron reduction of dioxygen. In the presence of norcamphor, P-450cam displays a remarkable degree of oxidase activity with a significantly less efficient monoxygenase chemistry. A striking trend is observed when increasing amounts of deuterium are present on the substrate as shown below:

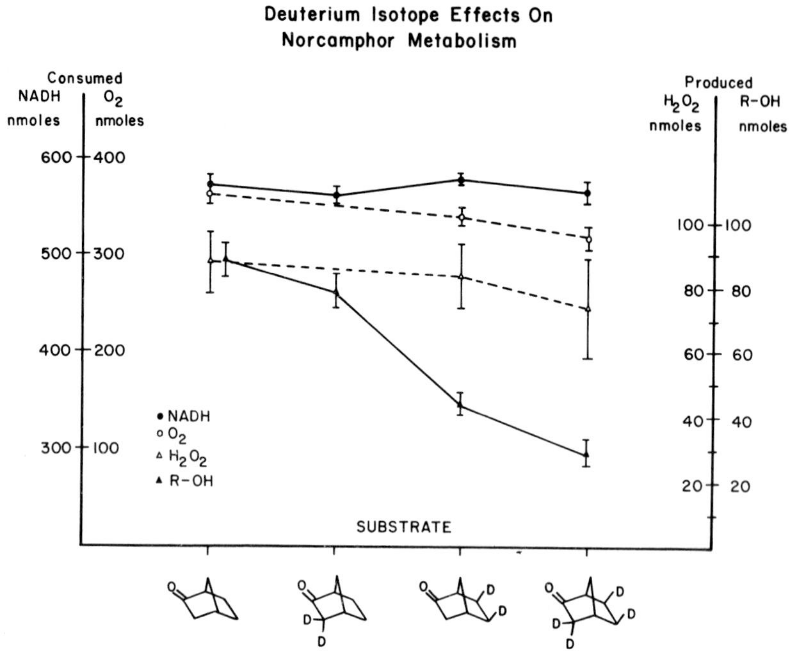

These results demonstrate that, in addition to a deuterium isotope effect on the regioselectivity of hydroxylation, there exists an isotope-dependent switching from monoxygenase to oxygenase activity at the active site of P-450cam. The near complete specificity for the *exo* face abstraction at the 5- and 6-positions should result in a maximal isotope effect on this metabolic switching when the 3,3-(5,6-*exo,exo*)-d_4-norcamphor is processed by the enzyme. Placement of deuterium at the *endo* face would be expected to have a minimal effect, although a secondary isotope effect might be detected.

KINETIC ISOTOPE EFFECTS. The kinetics of hydrogen peroxide formation, as well as NADH and oxygen consumption, have been examined in addition to the stoichiometry results discussed above. The results, which are summarized in the following Table, indicate a significant disparity in the kinetic deuterium isotope effects for these processes when norcamphor is utilized as substrate. The deuterium isotope effects for hydrogen peroxide formation and oxygen consumption are nearly identical and greater than 1.0, whereas the isotope effect on NADH consumption is inverse. This apparent paradox may be understood in terms of a model involving isotope-dependent partitioning of enzyme intermediates as shown in the following Scheme:

Rates and Isotope Effects of Norcamphor Metabolism by P-450cam

	nmoles / min / nmole P-450					
Substrate	$(V_{max})_{O_2}$	$(^HV/^DV)_{O_2}$	$(V_{max})_{H_2O_2}$	$(^HV/^DV)_{H_2O_2}$	$(V_{max})_{NADH}$	$(^HV/^DV)_{NADH}$
Norcamphor	18.6 ± 0.5		5.9 ± 0.7		18.2 ± 0.8	
5, 6-d_2-norcamphor	13.8 ± 0.6	1.34	5.1 ± 0.7	1.16	20.7 ± 0.8	0.87
3, 3, 5, 6-d_4-norcamphor	13.6 ± 1.2	1.36	4.7 ± 0.4	1.25	23.5 ± 0.5	0.77
d-camphor	49.6 ± 1.4				61.4 ± 0.6	

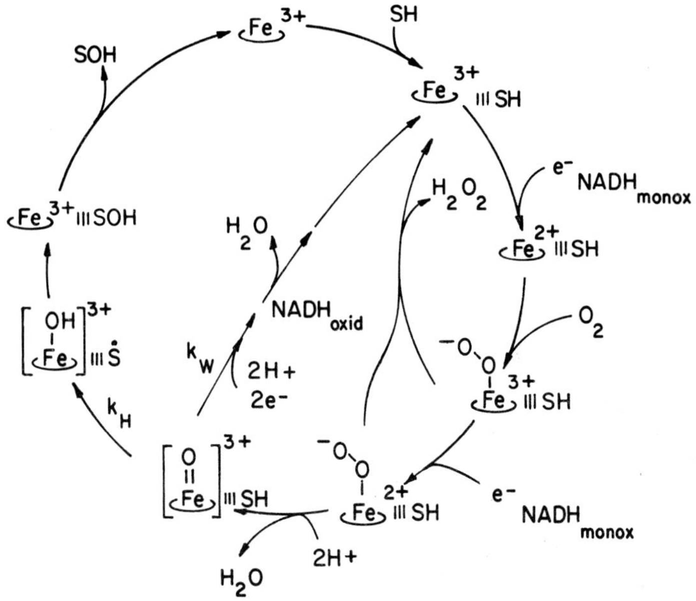

Clearly, there must exist some route for NADH consumption which is kinetically independent of oxygen consumption and hydrogen peroxide production, such that the enzyme intermediates that are substrates for further two electron reduction to form water are present in higher steady state concentrations when the substrate norcamphor is deuterated. Thus there are effectively two competing cycles and an isotope dependent degree of flux through each path. Deuterium substitution results in a slower rate of flux through each cycle, as indicated by the isotope effects on oxygen and hydrogen peroxide metabolism. However, due to the difference in oxidase stoichiometry, NADH is consumed more quickly when the oxidase cycle is more heavily traversed, as would be expected when the substrate is deuterated.

SITE-DIRECTED MUTAGENESIS. In order to examine the structural basis for the substrate specificity of P-450cam and the frequency of oxidase and oxygenase catalytic intermediates, we embarked on a modification of the "glove" via site-directed mutagenesis. As one example of this approach, the Tyr96 residue was changed to Phe96 by single-stranded methods (Taylor *et. al*, 1985) using the oligonucleotide 5'-TGAAGTCGAAGGCTTCGC-3' as the mutagenic primer, and M13mp10 as the single-stranded vector (please refer to Unger *et. al* for the nucleotide sequences). The mutant gene was subcloned into pUC19 and the protein expressed in TB1. From dot blot hybridization, the mutagenic frequency of transformants was greater than 30%. Based on gas chromatographic analysis, the only product obtained with Phe 96 and the substrate camphor is 5-*exo*-hydroxycamphor.

DISCUSSION. Several key points of departure between camphor and norcamphor metabolism have been documented. Based on the deuterium content of the product alcohols obtained with norcamphor, it appears that P-450cam is significantly more selective for hydrogen (deuterium) abstraction from the *exo* face of the norcamphor than it is with camphor. In addition, there is a significant intra-molecular isotope effect on the regioselectivity of hydroxylation when norcamphor is deuterated at various positions. This type of metabolic switching has also been observed when camphor was substituted with fluorine at both the 5-*exo* and *endo* positions (Eble & Dawson, 1984) where the *gem*-diflourosubstrate afforded the 9-hydroxy metabolite. We have also demonstrated that in the presence of norcamphor, P-450cam displays a prolific oxidase activity in which molecular oxygen is reduced by four electrons to afford water, and that there exists a novel type of metabolic switching which results in an isotope-dependent partitioning between monoxygenase and this oxidase activities. This suggests that there is a common intermediate involved in both processes. The kinetic isotope results support this contention that it is the P-450 intermediate which is sensitive to deuterium substitution which undergoes further two electron reduction by NADH. Clearly the multiplicity of metabolic options observed in the presence of norcamphor results from a lack of absolute complimentarity between the enzyme active site and norcamphor. The use of such substrate analogs in conjunction with the high resolution crystal structure allows one to predict which protein/substrate contacts determine the stereo- and regio-specificity of hydroxylation as well as the active site structural features which allow for monoxygenase vs. oxidase activity. These predictions have been tested by site-directed mutagenesis of the protein , or "glove" half of the recognition problem. The results observed with the Tyr96-Phe96 mutation, taken together with the norcamphor metabolic studies, suggest that the hydrophobic and/or steric restrictions associated with the methyl groups of camphor contribute more than the hydrogen bond formed with the substrate carbonyl in determining the optimal substrate complex for efficient monoxygenase reactivity and absolute regiospecificity. As a final comment, it should be noted that the facile reduction of the iron-oxo hydroxylating intermediate to produce a second water molecule suggests that attempts to force stability on this reactive intermediate by providing the enzyme with unmetabolizable substrates is perhaps doomed to failure.

Research supported by NIH grants GM31756 and GM33775.

REFERENCES

Elbe, J. S. & Dawson, J. H.,1984, Journal of Biological Chemistry, 259, 14389-14395.

Gelb, M., Heimbrook, D., Malkonin, P., & Sligar, S.,1982, Biochemistry 21, 370-379.

Gorsky, L. P., Koop, D. R., & Coon, M. J.,1984, Journal of Biological Chemistry, 259, 6812-6817.

Koga, H., Rauchfuss, B., & Gunsalus, I. C.,1985, Biochemical and Biophysical Research Communications, 130, 412-417.

Sligar, S. G., & Murray, R. I.,1986, Cytochrome P-450, Structure, Mechanism, and Biochemistry (P. Ortiz de Montellano, ed.) Plenum Press, New York, pp. 413-479.

Taylor, J. W., Ott J., & Eckstein, F.,1985, Journal of Nucleic Acids Research, 13, 8764-8785.

Unger, B. P., Gunsalus, I. C., & Sligar, S. G.,1985, Journal of Biological Chemistry, 261, 1158-1163.

Rearrangement of Alkyl Hydroperoxides Mediated by Cytochrome P-450: Evidence for the Oxygen Rebound Mechanism

KENNETH M. FISH [1], G.E. AVARIA and JOHN T. GROVES [2]

[1] Department of Biological Chemistry, The University of Michigan, Ann Arbor, MI 48109, USA and [2] Department of Chemistry, Princeton University, Princeton, NJ 08540, USA (address to which correspondence should be sent)

INTRODUCTION

Inorganic ferrous perchlorate catalyzes the oxidation of hydrocarbons by hydroperoxides. This reaction was reported to follow the oxygen rebound mechanism (Groves & Van Der Puy 1974, Groves & McClusky 1976).

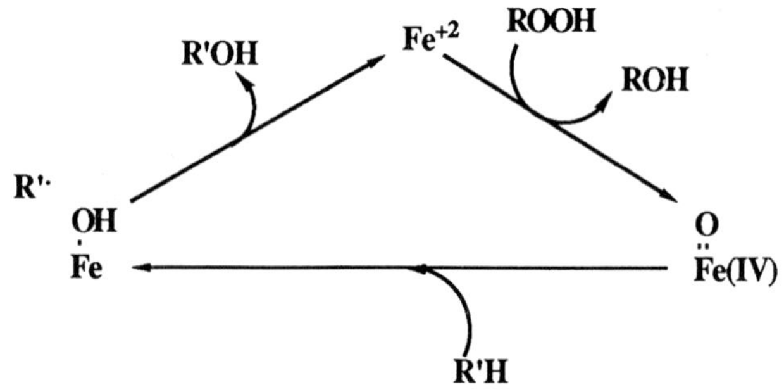

Oxygen is first delivered from hydroperoxide to the ferrous ion forming a ferryl oxene. This reactive intermediate inserts oxygen into C-H bonds by hydrogen atom abstraction and ˙OH transfer to the resulting radical. Therefore oxygen is transferred from a hydroperoxide donor to substrate by rebounding off the iron atom.

Cytochromes P-450 mediate the insertion of oxygen into carbon-hydrogen bonds (for a recent review see K. M. Fish 1986). Saturated hydrocarbons react with oxygen and NADPH to give alcohols, water and NADP+ by the monooxygenase activity of P-450 and P-450 reductase.

Alternatively by the peroxide shunt pathway, alkyl hydroperoxides serve as oxygen donors for cytochromes P-450 producing alcohols and the ferryl oxene porphyrin π cation radical (Nordblom, et al. 1976, Hrycay & O'Brien, 1972, Lichtenberger et. al. 1976, Rahimtula & O'Brien, 1974, Hrycay et. al. 1975). The resulting intermediate inserts oxygen into C-H bonds of unactivated hydrocarbons. Therefore the following overall reaction can be written.

$$ROOH + R'H \xrightarrow{P\text{-}450} ROH + R'OH$$

It follows that the oxygen rebound mechanism might be operating during cytochrome P-450 mediated oxygen transfer from hydroperoxide to saturated hydrocarbon. The reaction of simple alkyl hydroperoxides served as probes to test this hypothesis.

RESULTS AND DISCUSSION

Reaction of simple alkyl hydroperoxides with cytochrome P-450 LM_2 produced vicinal diols as the only products (Table 1). Intramolecular rearrangement via the Barton Reaction would result if homolytic -O-O- cleavage occurred. Thus 1,4-diols would be produced (R. H. Hesse 1969).

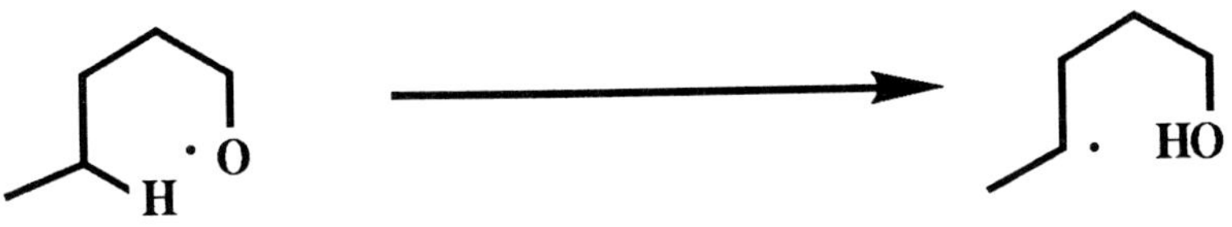

Since only 1,2-diols were produced, it is likely that O-O bond cleavage of the hydroperoxide occured in a heterolytic fasion.

Because rearramgement is regioselective in all cases, it follows that the caged alcohol formed on reaction of hydroperoxide with P-450 must be in a preferred conformation for H$^{\cdot}$ abstraction. Alternatively, the active oxygen species may abstract a nearby hydrogen atom before rearrangement or dissociation can occur.

Reaction of 2-methyl-2-hydroperoxyhexane with cytochrome P-450 in the presence of n-octanol produced rearranged hydroperoxide only. No oxidation of octanol occurred.

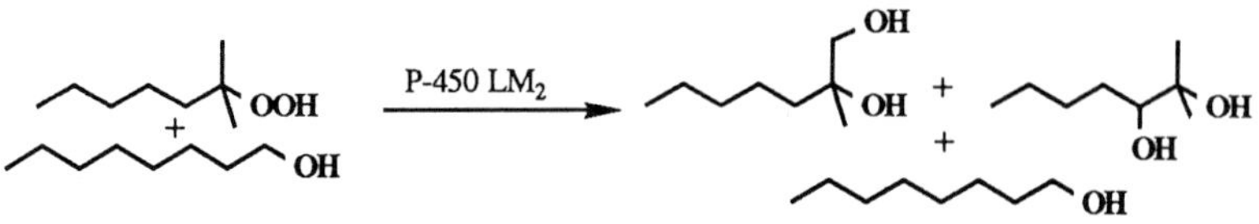

Similarly, a competitive reaction of the above peroxide and cyclohexene produced glycols only.

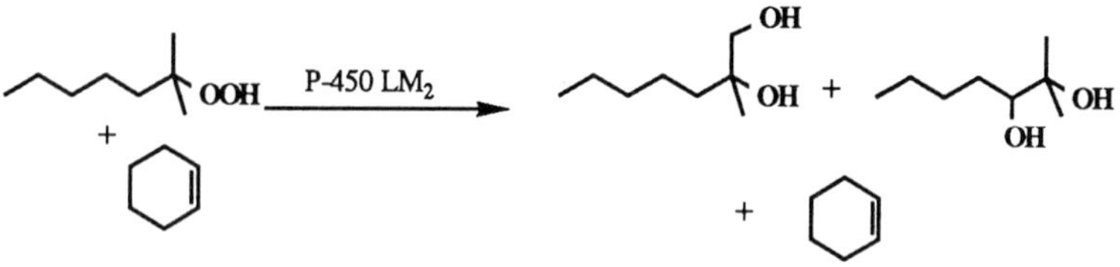

These results indicate that the *caged* alcohol intermediate reacted with the ferryl oxene. Hydrogen atom abstraction was fast with respect to dissociation. If the intermediate alcohols dissociated from the active site, octane diols and oxidized cyclohexenes would have been produced in the respective reactions.

Mixtures of hydroperoxyhexane and doubly labelled ^{18}O-hydroperoxyhexane were rearranged by P-450 to the vicinal diols. Unlabelled 1,2-hexanediol and doubly labelled 1,2-hexanediol were the only products detected.

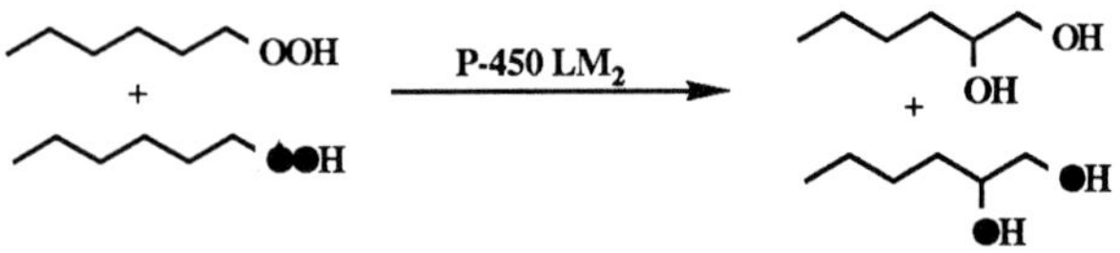

Since no partially labelled diols were produced, intermolecular scrambling can be ruled out. Hydrogen atom atstraction and ˙OH delivery to the radical occured rapidly with respect to dissociation of the caged alcohol supporting the above conclusions.

Hydroxylation of hexanol by the *monooxygenase* activity of P-450 produced 1,3-

hexanediol. The regioselectivity of this reaction was different than the P-450 mediated peroxide rearrangement reaction. Taken together with the doubly labelled ^{18}O peroxide rearrangements, these results require that the caged alcohol conformation in the latter reaction did not change much before H˙ abstraction at the α carbon. Thus hydrogen atom abstractin was fast with respect to rearrangement. The preferred conformation of the caged alcohol was similar to that of the enzyme bound peroxide.

Side chain cleavage of sterols is catalyzed by cytochrome $P\text{-}450_{scc}$. This process occurs via two regioselective and enantioselective hydroxylations prior to C-C bond cleavage (Larroque et. al., 1981).

Cytochrome P-450 LM_2 rearranges simple alkyl hydroproxides regioselectively (vide supra). Hydrogen atom abstraction and/or ˙OH transfer to the caged radical could be enantioselective.

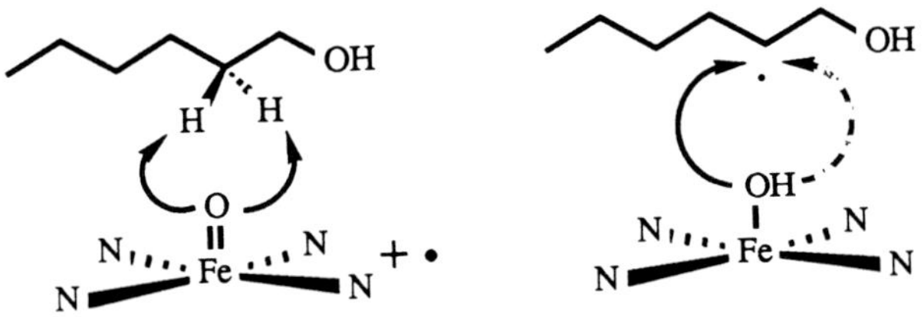

Liver microsomal cytochrome P-450 mediated rearrangement of 2-deuteriohyderperoxyhexanes produced the following diols.

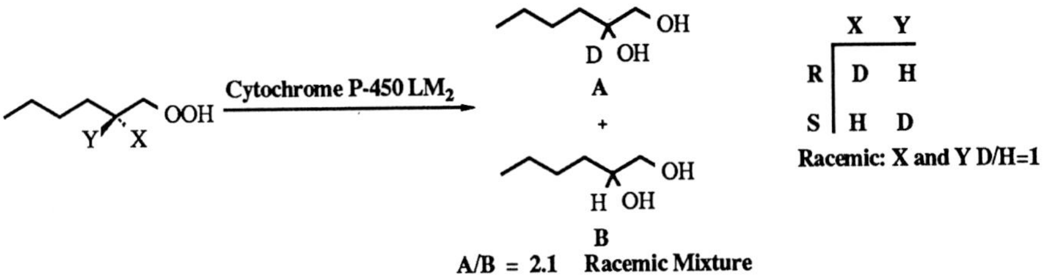

Enantiomerically enriched substrates produced racemic gylcols. Product distributions were identical to that of racemic 2-d-hydroperoxyhexane. Hydrogen atom abstraction and collapse of the caged pair to give products occured without enantioselectivity in either step.

Deuterated and undeuterated products were formed in a 2.1:1 ratio respectively. Transfer of ˙OH must have occurred after racemization of the caged radical. The intramolecular kinetic isotope effect of 2.1 indicated that there was a preference for hydrogen atom vs deuterium atom abstraction. Hydroxylations mediated by P-450 and model systems can be accompanied by k_H/k_D as large as 11.5 (Groves et. al 1978). In the latter case, a linear transition state was inferred.

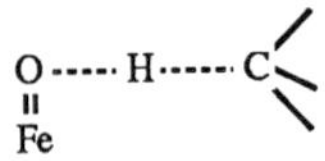

In the former case, the ***caged*** alcohol underwent little rearrangement (vide supra), since hydrogen atom abstraction was rapid. A nonlinear transition state would give rise to less than maximal preference for H˙ vs D˙ abstraction.. Therefore a relatively small kinetic isotope effect is expected. From the magnitude of the observed k_H/k_D, the C-H-oxene angle can be approximated to be 90°-120° (R. A. More O'Ferral 1971).

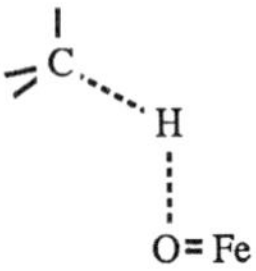

CONCLUSIONS

Cytochrome P-450 LM_2 mediated rearrangement of simple alkyl hydroperoxides **was** consistent with the oxygen rebound mechanism (Scheme 1). The peroxy oxygen was delivered to heme forming the ferryl oxene porphyrin π cation radical. Before rearrangement and/or dissociation of the caged alcohol from the active site, hydrogen atom abstraction occured at the α carbon of the caged radcal. The carbon radical had a long enough lifetime for racemization. Finally, ˙OH was delivered from the iron to radical producing a 1,2-diol. Neither H˙ abstraction nor ˙OH

Scheme 1

Proposed Mechanism For Oxygen Rebound During Cytochrome P-450 Mediated Hydroperoxide Rearrangement

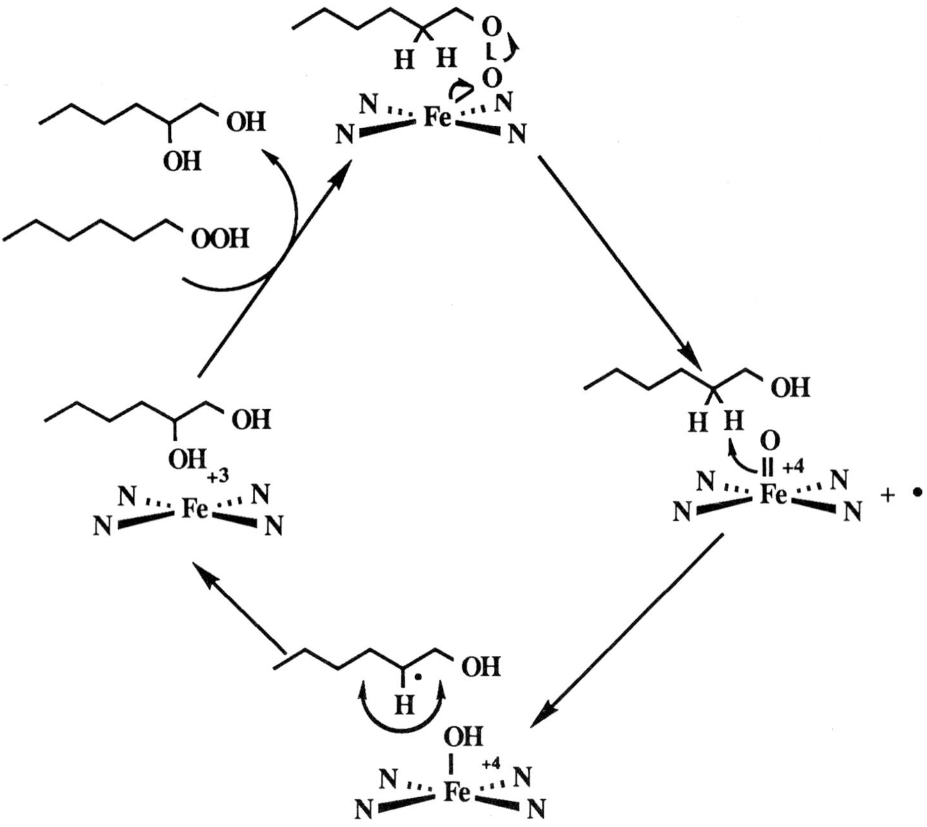

transfer was enantioselective. Oxygen was inserted into the C-H bond to the alpha carbon by rebounding off the iron.

Table 1

Summary of Glycol Formation From Medium Chain Hyderperoxides by Cytochrome P-450 LM_2 -Groves and Avaria-Neisser, unpublished.

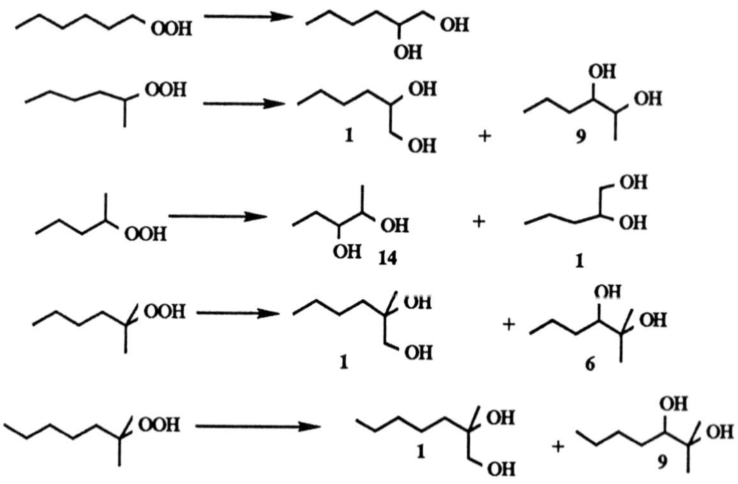

REFERENCES

Fish, K. M., 1986, Thesis: Manganese Horseradish Peroxidase and Cytochrome P-450 Probes into the intermediatesof Oxidized Heme Proteins, The University of Michigan, Ann Arbor Michigan, and references therein.

Groves J. T., & Van Der Puy, M., 1974, Journal of The American Chemical Society, 96, 5274-5275.

Groves, J. T. & McClusky, G. A., 1976 Journal of The American Chemical Society, 98, 859-861.

Groves, J. T,.McClusky, G. A. White R. E., & Coon M. J., 1978 Biochemical and Biophysical Research Communications, 18, 154-160.

Hesse, R. H. 1969, Advances in Free Radical Chemistry, edited by G. H. Williams, (Academic Press), New York, 3, 83-137.

Hrycay, E. G. & O'Brien, J. P., 1972, Archives of Biochemistry and Biophysics, 153, 480-484.

Hrycay, E. G. Gustafsson, J. A., 1975, Ingeleman-Sundberg and L. Ernster, FEBS Letters, 56, 151.

Larroque, C. Rousseau, J., & van Lier, J. E., 1981, Biochemistry, 20, 925-929.

Lichtenberger, F. Nastainczyk, W. & Ullrich, V., 1976, Biochemical and Biophysical Research Communications, 70, 939-946.

Moore O'Ferrall, R. A, 1971, Journal of the Chemical Society, London, Sect. B, 785-790.

Nordblum, G. D., White, R. E., & Coon, M. J., 1976, Archives of Biochemistry and Biophysics, 175, 524-533.

Rahimtula, A. D. & O'Brien, P. J., 1974, Biochemical and Biophysical Research Communications, 60, 440-447.

Reductive Rearrangement of Hydroperoxides: A Role for Cytochrome P-450 in Lipid Peroxidation

A.D.N. VAZ, E.S. ROBERTS and M.J. COON

Department of Biological Chemistry, The University of Michigan Medical School, Ann Arbor, MI 48109-0606, USA

INTRODUCTION

The complex process by which biological membranes react with molecular oxygen to produce peroxidized lipids and their breakdown products has been termed "lipid peroxidation." At least three distinct steps appear to be involved in this process that are analogous to free radical chain reactions, namely, the initiation, propagation, and termination phases. Following the early report by Hochstein & Ernster (1963) that ADP-activated lipid peroxidation, as measured by malonaldehyde formation, is coupled to the NADPH oxidase system in liver microsomes, numerous investigations have dealt with this subject. Attention has been focussed primarily on the mechanism of initiation, and therefore on reactive intermediates produced by the reduction of oxygen, such as superoxide and hydroxyl radicals, and equivalent species derived from lipids, such as peroxy and alkoxy radicals. Although NADPH-cytochrome P-450 reductase may serve as the primary source of superoxide radicals (Aust et al. 1972; Bösterling & Trudell 1981) and hydroxyl radicals (Lai et al. 1979) for initiation of lipid peroxidation, P-450 has been implicated in reports by Hrycay & O'Brien (1971) and Svingen et al. (1979). More recently, Lindstrom & Aust (1984) have reported the reduction of 13-hydroperoxy-9,11-octadecadienoic acid (13-HOO-$C_{18:2}$) to the corresponding 13-hydroxy fatty acid by a reconstituted rat liver microsomal system, and Ekström & Ingelman-Sundberg (1984) have concluded that P-450 may contribute to the initiation of lipid peroxidation in reconstituted membrane vesicles.

While some products obtained during lipid peroxidation such as hydrocarbons, malondialdehyde, and short chain lipid aldehydes have been reported (Poli et al. 1985), and hydrocarbon exhalation is believed to be a measure of this pathophysiological process in vivo (Wendel & Dumelin 1981), the mechanism of formation of such compounds has been largely speculative. In attempts to develop a model for the 4ε-oxidase activity of P-450 (Gorsky et al. 1984) using hydroperoxides as substrates with a reconstituted system, we observed that simple 2ε-reduction of the hydroperoxide did not occur (Vaz & Coon 1985). Instead, such model compounds were found to undergo reductive rearrangement to yield a hydrocarbon and a carbonyl compound (Vaz & Coon 1987a). For example, cumyl hydroperoxide (CHP) yields methane and acetophenone. Since 13-HOO-$C_{18:2}$ yields pentane under our experimental conditions, we have proposed that the known formation of alkanes and aldehydes in membrane lipid peroxidation involves the same P-450-dependent reductive cleavage reaction (Vaz & Coon 1987a).

The present paper is concerned with the role of P-450 in the disposition of both xenobiotic and physiologically occurring organic hydroperoxides. The mechanisms by which P-450 catalyzes other related reactions, including the reduction of O_2 to radical species and the non-reductive cleavage of peroxy compounds, have been reviewed elsewhere (White & Coon 1980; Black & Coon 1987).

RESULTS AND DISCUSSION

In our initial report on the NADPH-dependent reduction of hydroperoxides catalyzed by P-450 form 2, we used an isocratic solvent system consisting of 40% MeOH in water for the separation of CHP, cumyl alcohol, and acetophenone by HPLC (Vaz & Coon 1985). This solvent system, with variations in the MeOH concentration, was inadequate for the other benzylic hydroperoxides and alcohols examined. In subsequent studies (Vaz & Coon 1987a), CH_3CN was found to be the solvent of choice. As shown in Fig. 1, with the use of 30% CH_3CN in water, separation was achieved of all the benzylic hydroperoxides from the corresponding alcohols and carbonyl compounds formed in the reductive reaction. For quantitation, alcohols and hydroperoxides were in the 2-150 nmol range and acetophenone and benzaldehyde in the 5-1,000 pmol range. Also shown in Fig. 1 is the product analysis from the reaction of individual hydroperoxides with P-450 form 2 in the reconstituted system. It is clear from these profiles that the corresponding alcohols are not formed in the reductive reaction, but rather the carbonyl compounds. Quantitation by HPLC indicated that the hydroperoxides are converted in very good yields to the carbonyl

Fig. 1. HPLC analysis of products derived from cumyl, benzyl, and α-methylbenzyl hydroperoxides. In Expt. a the reaction mixture contained CHP, and the standards used were cumyl alcohol (A), acetophenone (B), and CHP (C). In Expt. b the reaction mixture contained benzyl hydroperoxide, and the standards used were benzyl alcohol (A), benzyl hydroperoxide (B), and benzaldehyde (C). In Expt. c the reaction mixture contained α-methylbenzyl hydroperoxide and the standards used were 1-phenylethanol (A), benzaldehyde (B), α-methylbenzyl hydroperoxide (C), and acetophenone (D). An IBM C-18 analytical reversed phase column was used with an isocratic solvent system of acetonitrile/water 30:70. The column was flushed with 90% acetonitrile before each estimation.

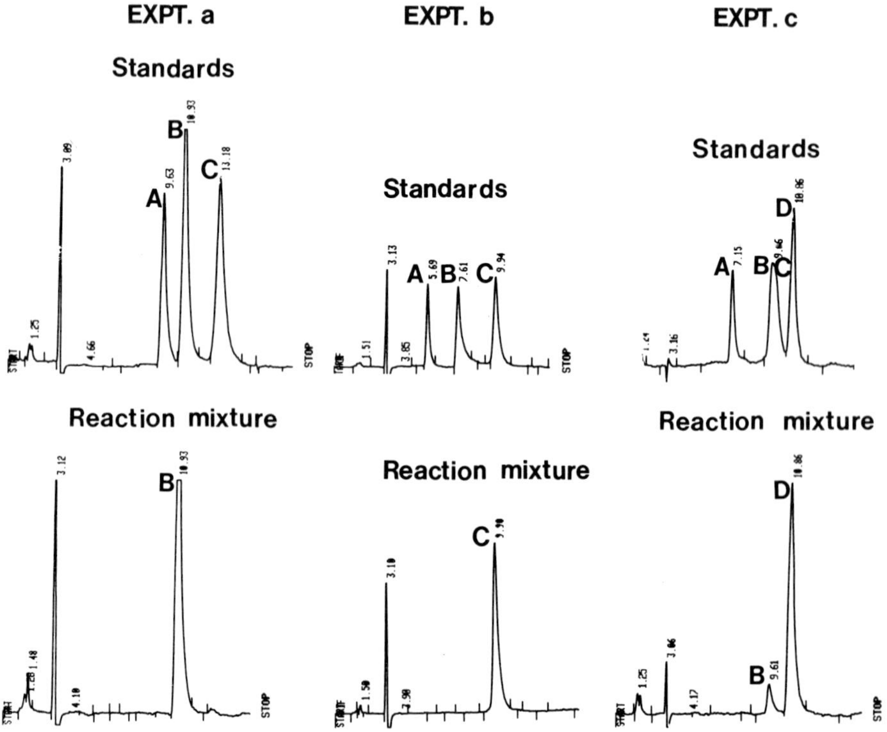

compounds; see Table 1, which is a composite of results reported by Vaz & Coon 1987a. Fragmentation of the hydroperoxide, which is a consequence of reduction of the O-O bond, is diagnostic of homolysis to yield an alkoxy radical

Table 1. Requirements for product formation from hydroperoxides.

System	Hydroperoxide tested	Change (nmol) NADPH	$\phi COCH_3$	ϕCHO	Acetone	CH_4	C_5H_{12}
Complete	Cumyl	-99	91	n.d.	n.d.	80	
	α-Me-benzyl	-157	99	20	n.d.	<25	
	Benzyl	-174		124		n.d.	
	t-Butyl	-211			174	177	
	13-HOO-$C_{18:2}$	-37					25
No NADPH	Cumyl		23	n.d.	n.d.	n.d.	
No P-450	Cumyl		<0.01	n.d.	n.d.	n.d.	
	13-HOO-$C_{18:2}$	-2			n.d.		n.d.
No reductase	Cumyl		40		n.d.	n.d.	
	13-HOO-$C_{18:2}$	-5					n.d.

For product formation, n.d. indicates that none could be detected. The starting amounts of the hydroperoxides, in nmol, were: cumyl (100), α-methylbenzyl (150), benzyl (200), t-butyl (200), and 13-HOO-$C_{18:2}$ (31).

that is inherently unstable and rearranges by β-scission to a carbonyl compound and an alkyl radical (Walling & Padwa 1963; Neta et al. 1984). With CHP as substrate, the β-scission product was identified by GC/MS as methane (Fig. 2, taken in part from Vaz and Coon 1987a). When the reaction was carried out in a D_2O/H_2O mixture, a solvent deuteron was incorporated into methane. Quantitation by GC on a Carbosieve G column indicated a nearly stoichiometric formation of the hydrocarbon. These results are mechanistically diagnostic of a stepwise one-electron reduction, as depicted for CHP:

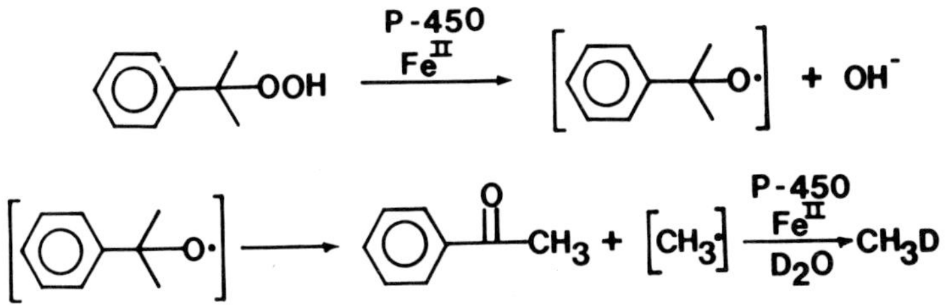

Fig. 2. GC/MS analysis of methane. The electron impact fragmentation pattern is shown for CH_4 enzymatically generated in H_2O (A) and in 70% D_2O (B), with m/e as a function of ion intensity. The intensity at m/e = 17 is a measure of $CH_3{}^2H$. The pattern for a methane standard is also shown (C). In (D) the RIC profile is shown for (a) a CH_4 standard, (b) a reaction mixture with heatdenatured enzymes, (c) a reaction mixture with reductase omitted, and (d) a complete reaction mixture; the vertical arrows indicate where CH_4 was identified.

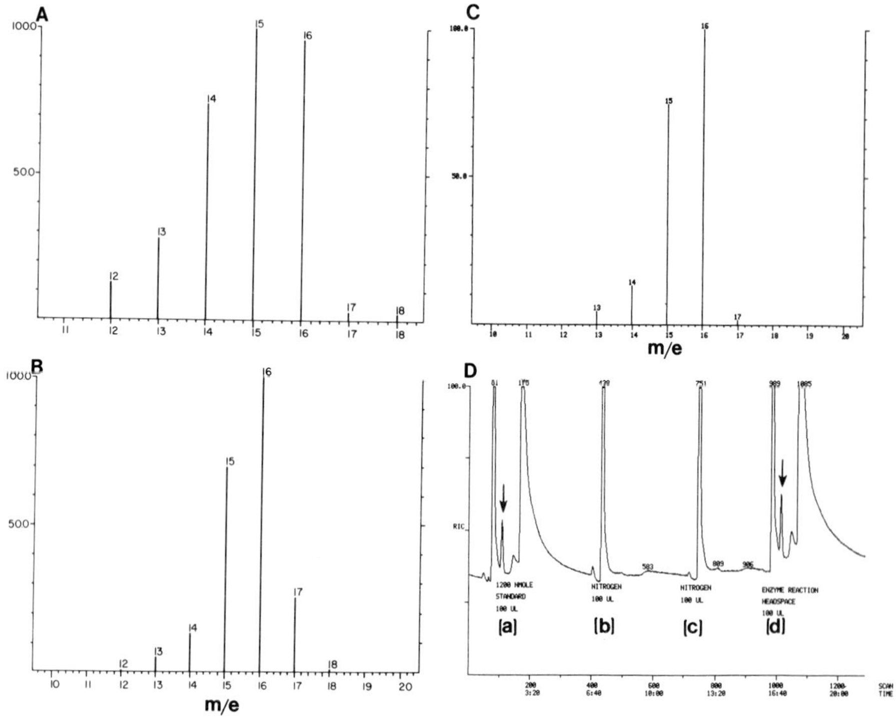

The first electron transfer yields the cumyloxy radica and a hydroxide ion. β-Scission of the former results in a methyl radical which, with uptake of the second electron along with a solvent proton, gives methane. As already seen in Table 1, t-butyl, α-methyl-benzyl, and benzyl hydroperoxides also result in a nearly stoichiometric yield of products that can be accounted for by the proposed homolytic mechanism.

The formation of hydrocarbons from the hydroperoxides requires catalytically competent P-450 and reductase as well as NADPH (Table 1). However, in the absence of NADPH, α-methylbenzyl, benzyl, cumyl, and t-butyl hydroperoxides still form the corresponding carbonyl compounds although in lower yields. This suggests that ferric P-450 is also capable of activating the peroxide bond by a homolytic mechanism, as previously established (Blake & Coon 1981; Coon et al. 1982). The lack of hydrocarbon formation in the absence of NADPH can be accounted for by the failure of the derived radical to be reduced; instead, the radical causes extensive destruction of the P-450 heme.

The products obtained in the P-450-catalyzed reductive cleavage reaction with model hydroperoxides are similar to those obtained during lipid peroxidation in biological membranes (Poli et al. 1985). We examined the possibility, therefore, that the physiological substrates for this reaction might be lipid hydroperoxides. When 13-HOO-$C_{18:2}$ was used as a substrate, pentane was detected (Table 1). Furthermore, we have recently obtained several lines of evidence that the other fragment of the lipid hydroperoxide cleavage is the expected aldehyde acid. The rearrangement and cleavage of hydroperoxides is also effected by other forms of P-450 (Vaz & Coon 1987b), but form 2 is one of the most active. Preliminary studies indicate that the same reactions occur in intact microsomal membranes.

SUMMARY

In the present paper, evidence is presented for the role of P-450 in catalyzing the cleavage of both xenobiotic and biologically occurring hydroperoxides, and the mechanism of the reaction is discussed. The mechanism is believed to involve two one-electron transfers, with homolysis of the peroxide oxygen-oxygen bond and formation of an alkoxyl radical, followed by β-scission of the latter and then reduction of the resulting alkyl radical to the hydrocarbon.

Since 13-HOO-$C_{18:2}$ yields pentane and an aldehyde acid in the reconstituted enzyme system under the same conditions used for model hydroperoxides, we propose that the known formation of alkanes and aldehydes in membrane lipid peroxidation involves reductive cleavage by P-450. Drug induction and other treatments that alter the concentration of the reductase and of specific P-450s in membranes would, therefore, be expected to influence the rate of conversion of unsaturated lipids to hydrocarbons and other products.

Acknowledgment.

This research was supported by Grant DK-10339 from the National Institutes of Health.

REFERENCES

Aust, S.D., Roerig, D.L. & Peterson, T.C., 1972, Biochemical and Biophysical Research Communications, 47, 1133-1137.

Black, S.D. & Coon, M.J., 1987, Advances in Enzymology and Related Areas of Molecular Biology, 60, 35-87.

Blake, R.C., II & Coon, M.J., 1981, Journal of Biological Chemistry, 256, 12127-12133.

Bösterling, B. & Trudell, J.R., 1981, Biochemical and Biophysical Research Communications, 98, 569-575.

Coon, M.J. & Blake, R.C, II, 1982, Evaluation of homolytic and heterolytic mechanisms of peroxide oxygen-oxygen bond cleavage by cytochrome P-450. In Oxygenases and Oxygen Metabolism, edited by M. Nozaki, S. Yamamoto, Y. Ishimura, M.J. Coon, L. Ernster & R.W. Estabrook (Academic Press, New York), pp. 485-495.

Ekström, G. & Ingelman-Sundberg, M., 1984, Biochemical Pharmacology, 33, 2521-2523.

Gorsky, L.D., Koop, D.R. & Coon, M.J., 1984, Journal of Biological Chemistry, 259, 6812-6817.

Hochstein, P. & Ernster, L., 1963, Biochemical and Biophysical Research Communications, 12, 388-394.

Hrycay, E.G. & O'Brien, P.J., 1971, Archives of Biochemistry and Biophysics, 147, 14-27.

Lai, C.-S., Grover, T.A. & Piette, L.H., 1979, Archives of Biochemistry and Biophysics, 193, 373-378.

Lindstrom, T.D. & Aust, S.D., 1984, Archives of Biochemistry and Biophysics, 233, 80-87.

Neta, P., Dizdaroglu, M. & Simic, M.G., 1984, Israel Journal of Chemistry, 24, 25-28.

Poli, G., Dianzani, M.U., Cheeseman, K.H., Slater, T.F., Lang, J. & Esterbauer, H., 1985, Biochemical Journal, 227, 629-638.

Svingen, B.A., Buege, J.A., O'Neal, F.O. & Aust, S.D., 1979, Journal of Biological Chemistry, 254, 5892-5899.

Vaz, A.D.N. & Coon, M.J., 1985, NADPH oxidation by hydroperoxides: catalysis by cytochrome P-450 and evidence for homolytic cleavage of the peroxide bond. In Cytochrome P-450, Biochemistry, Biophysics and Induction, edited by L. Vereczkey & K. Magyar (Elsevier, Amsterdam, and Akademiai Kiado, Budapest), pp. 545-548.

Vaz, A.D.N. & Coon, M.J., 1987a, Proceedings of the National Academy of Sciences USA, 84, 1172-1176.

Vaz, A.D.N. & Coon, M.J., 1987b, Federation Proceedings, 46, 1956.
Walling, C. & Padwa, A., 1963, Journal of the American Chemical Society, 85, 1593-1597.
Wendel, A. & Dumelin, E.E., 1981, Methods in Enzymology, 77, 10-15.
White, R.E. & Coon, M.J., 1980, Annual Review of Biochemistry, 49, 315-356.

HUMAN CYTOCHROME P-450 POLYMORPHISMS AND FUNCTIONAL CORRELATES OF CYTOCHROME P-450 MULTIPLICITY IN MAN

Human Liver Cytochrome P-450 Mephenytoin 4-Hydroxylase and Related Proteins

F. PETER GUENGERICH, DIANE R. UMBENHAUER, CECILE GED, MARTHA V. MARTIN, GRANT R. WILKINSON and R. STEPHEN LLOYD

Departments of Biochemistry and Pharmacology and Center in Molecular Toxicology, Vanderbilt University School of Medicine, Nashville, TN 37232, USA

INTRODUCTION

In the three years since the last Symposium, considerable progress has been made in the characterization of human cytochrome P-450 (P-450) enzymes and their genes. At least seven liver P-450s and one extrahepatic P-450 have been purified from human sources; the genes coding for other human P-450s have also been characterized even when the proteins have not been available (Distlerath & Guengerich 1987; Nebert et al. 1987). Qualitative and quantitative variations in P-450-based activities towards drugs are of considerable interest in clinical settings. In some instances genetic polymorphisms have been described but only very recently have inferences into molecular mechanisms been possible. One of these genetic polymorphisms involves mephenytoin 4-hydroxylation (Wedlund et al. 1984; Küpfer & Preisig 1984) and this chapter will focus specifically on biochemical approaches to understanding the phenomena involved.

METHODS

Details of much of the methodology are contained in the references cited, including enzyme assay (Shimada et al. 1985) and purification (Shimada et al. 1986). cDNA clones were isolated from a bacteriophage λgt11 cDNA library (prepared from a single human liver) and sequenced in M13 mp9 phage using a dideoxy termination method (Umbenhauer et al. 1987). Peptides were prepared by proteolysis with trypsin and separated by high performance liquid chromatography, and amino acid sequences were determined in the Vanderbilt Protein Chemistry Facility using amino acid analysis (Waters Picotag system) and an Applied Biosystems 470A sequenator.

RESULTS AND DISCUSSION

Enzyme purification

The story of purification of these proteins goes back to early efforts in this laboratory to isolate P-450 from human liver autopsy samples (Wang et al. 1980). The preparations obtained in that reference and those subsequently purified from tissues obtained from organ transplant donors and denoted P-450_1 and P-450_8 (Wang et al. 1983) proved to be related to the P-450 mephenytoin hydroxylase. In subsequent work we monitored S-mephenytoin hydrolase activity during purification and isolated a protein termed P-450_{MP} (Shimada et al. 1986). Some individuals contained either a low M_r (P-450_{MP-1}) or high M_r (P-450_{MP-2}) variant of the protein and many individuals had both proteins. Reconstitution and immunoinhibition experiments implicated a role for cytochrome $\underline{b}_5$ in catalysis. P-450_{MP-1} and P-450_{MP-2} have similar catalytic activities (V_{max} and K_m) and the existence of the two proteins is not the basis of the polymorphism. To date we have been unable to find a clear relationship of either V_{max} or V_{max}/K_m with K_m in liver microsomal samples prepared from different individuals.

cDNA clone mp-8

Polyclonal and monoclonal antibodies raised to P-450_{MP} were used to isolate clone mp-8 from a human liver λgt11 cDNA library (Umbenhauer et al. 1987). The cDNA clone, only fourteen 5' nucleotides short of containing the entire coding sequence, yielded a predicted N-terminal amino acid sequence identical with those experimentally determined for both P-450_{MP-1} and P-450_{MP-2} proteins (Shimada et al. 1986). *In vitro* translation experiments indicate that distinct mRNAs code for P-450_{MP-1} and P-450_{MP-2}. Blotting of genomic DNA endonuclease fragments with non-overlapping 5' and 3' mp-8 cDNA probes suggest that at least 7 gene-like sequences are in the family.

Regulation

Blotting experiments (done in collaboration with J. Hamerton and D.C. Riddell, Winnipeg) utilizing human-hamster cell hybrids indicate localization of all P-450_{MP} genes (P-450 IIC9 family, Nebert et al. 1987) on chromosome 10. No correlation between catalytic activity and levels of either mRNA (probed with mp-8 cDNA probes) or immunochemically detectable P-450_{MP} has been found to date. Neither detectable P-450_{MP} protein, catalytic activity, nor mRNA is present in human fetal liver (although other P-450s are) (Cresteil et al. 1985; Shimada et al. 1986; Knodell et al. 1987). The enzyme(s) has a rather diffuse localization pattern in liver, in contrast to the centrilobular pattern of P-450_{NF} (Ratanasavanh et al. 1986). Levels of P-450_{MP} are decreased in primary liver tumors (El Mouelhi et al. 1987). While the effects of disease and drug treatment on P-450_{MP} levels have not been addressed in vivo, alterations in hexobarbital metabolism (an apparent marker) suggest that disease (cirrhosis) may decrease P-450_{MP} function and that rifampicin treatment may elevate activity (Breimer et al. 1977; Zilly et al. 1977).

Catalytic specificity

Only the S-enantiomer of mephenytoin is hydroxylated (at the 4-position) by P-450_{MP}. S-Nirvanol (desmethyl mephenytoin, phenylethylhydantoin) is also 4-hydroxylated but the K_m appears to be higher (Shimada et al. 1986). Some in vitro experiments provide evidence that ethotoin, mephobarbital, methsuximide, and phensuximide are also substrates (Hall et al. 1987). Both in vitro and in vivo studies suggest that hexobarbital 3'-hydroxylation is also catalyzed by P-450_{MP}. In vitro experiments indicate that tolbutamide methyl hydroxylation is catalyzed by P-450_{MP} but phenotypic poor metabolizers of mephenytoin show normal clearance of tolbutamide (Knodell et al. 1987). Although a rabbit P-450 1 cDNA clone shows about 80% sequence identity to the mp-8 clone, the characteristic estradiol 2-hydroxylase and progesterone 21-hydroxylase activities of this protein are not seen with human liver P-450_{MP} (Umbenhauer et al. 1987). A rat P-450 cDNA sequence is also about 80% identical (Yoshioka et al. 1987) but the function and regulation of the related protein (P-450_{UT-A}) are quite different from human P-450_{MP}, particularly in regard to hormonal regulation, and another rat P-450 (P-450_{PCN-E}) is the major one responsible for mephenytoin 4-hydroxylation (Shimada & Guengerich 1985). In summary, some but not all hydantoins and barbiturates appear to be substrates for P-450_{MP} hydroxylation. The literature on hexobarbital metabolism may be applied to the mephenytoin polymorphism. Caution is needed in the interpretation of catalytic activity assignments in this multi-gene family.

Further sequence studies

Screening of the (single liver) cDNA library has been extended to isolate four clones which all differ in their restriction maps. They appear to be of two types, and some clones in each type may be fidelity artifacts of cloning. However, some clones closely related in their coding regions

Figure 1. Amino acid sequences derived from proteins ($P\text{-}450_{MP\text{-}1}$, $P\text{-}450_{MP\text{-}2}$, $P\text{-}450_{MP\text{-}3}$) and predicted from cDNA clones (mp-8, mp-20).

```
MP-1      M D S L V V L V L   L S   L L L L S L W R Q S S G
MP-2      M D S L V V L V L   L S   L L L L S L W R Q S S G   G   L
cDNA 8            V L V L C L S C L L L L S L W R Q S S G R G K L P P G P T P L P V I G N I L Q I G I K D I S K S L T
cDNA 20                       S F M L L F S L W R Q S C R R R K L P P G P T P L P I I G N M L Q I D V K D I C K S F T
MP-3

cDNA 8    N L S K V Y G P V F T L Y F G L K P I V V L H G Y E A V K E A L I D L G E E F S G R G I F P L A E R A N R G F
cDNA 20   N F S K V Y G P V F T V Y F G M N P I V V F H G Y E A V K E A L I D N G E E F S G R G N S P I S Q R I T K G L
MP-3              V Y G P V F T V Y F G M N P I V V F H G Y E A V K

cDNA 8    G I V F S N G K K W K E I R R F S L M T L R N F G M G K R S I E D R V Q E E A R C L V E E L R K T K A S P C D
cDNA 20   G I I S S N G K R W K E I R R F S L T T L R N F G M G K R S I E D R V Q E E A H C L V D E L R K T K A S P C D
MP-3                                                                                                      A S P C D

cDNA 8    P T F I L G C A P C N V I C S I I F H K R F D Y K D Q Q F L N L M E K L N E N I K I L S S P W I Q I C N N F S
cDNA 20   P T F I L G C A P C N V I C S V V F Q K R F D Y K D Q N F L T L M K R F N E N F R I L N S P W I Q V C N N F P
MP-3      P T F I L G C A P C N V I C               F D Y K D Q N F               F N E N F R I L N S P W I Q V C N N F P

cDNA 8    P I I D Y F P G T H N K L L K N V A F M K S Y I L E K V K E H Q E S M D M N N P Q D F I D C F L M K M E K E K
cDNA 20   L L I D C F P G T H N K V L K N V A L T R S Y I R E K V R E H Q A S L D V N N P R D F I D C F L I K M E Q E K
MP-3      L L I D C F

cDNA 8    H N Q P S E F T I E S L E N T A V D L F G A G T E T T S T T L R Y A L L L L L K H P E V T A K V Q E E I E R V
cDNA 20   D N Q K S E F N I E N L V G T V A D L F V A G T E T T S T T L R Y G L L L L L K H P E V T A K V Q E E I D H V
MP-3              S E F N I E N L V G T V A D L F V A

cDNA 8    I G R N R S P C M Q D R S H M P Y T D A V V H E V Q R C I D L L P T S L P H A V T C D I K F R N Y L I P K G T
cDNA 20   I G R H R S P C M Q D R S H M P Y T D A V V H E I Q R Y S D L V P T G V P H A V T T D T K F R N Y L I P K G T
MP-3                                                                                                  N Y L I P K

cDNA 8    T I L I S L T S V L H D N K E F P N P E M F D P H H F L D E G D N F K K S K Y F M P F S A G K R I C V G E A L
cDNA 20   T I M A L L T S V L H D D K E F P N P N I F D P G H F L D K N G N F K K S D Y F M P F S A G K R I C A G E G L
MP-3

cDNA 8    A G M E L F L F L T S I L Q N F N L K S L V I P K N L D T T P V V N G F A S V P P F Y Q L C F I P V
cDNA 20   A R M E L F L F L T T I L Q N F N L K S V D L L K N L N T T A V T K G I V S L P P S Y Q I C F I P V
MP-3
```

differ markedly in their 3' non-coding regions. Clone mp-8 and its close relative mp-4 are about 80% identical to clone mp-20 (and its close relative mp-12) (Fig. 1). The N-terminal sequences of $P\text{-}450_{MP-1}$ and $P\text{-}450_{MP-2}$ both correspond to mp-8. An immunochemically-related protein (designated $P\text{-}450_{MP-3}$) was isolated and partially sequenced--it appears to correspond to mp-20 and mp-12 (and also to a protein (P-450HLg) recently isolated by Wrighton et al. (1987)) but not to mp-8 or mp-4, as shown by the sequences of the isolated peptides (Fig. 1). A monoclonal antibody has been isolated that recognizes $P\text{-}450_{MP-1}$ (after electrophoresis and immunoblotting) but not $P\text{-}450_{MP-2}$ or $P\text{-}450_{MP-3}$. Purified $P\text{-}450_{MP-3}$ did not have mephenytoin 4-hydroxylase, hexobarbital 3'-hydroxylase, or tolbutamide methyl hydroxylase activity; catalytic activities remain unknown for this closely-related protein. In order to further elucidate the basis of catalytic specificity in this multi-gene family, clone mp-8 has been extended to fill the 5' region and inserted into shuttle vector pJDB207, containing a phosphoglycerate kinase promoter and ampicillin and leucine selection sites, for expression in yeast.

More than 30 restriction endonucleases were used to screen for restriction fragment length polymorphisms (RFLPs). These are found frequently in human populations but most are unrelated to differences in catalytic activity. A difference in SacI patterns was detected in a panel of 9 blood cell DNA preparations obtained from individuals who had been phenotyped *in vivo*. When the code in the blind study was broken, all of the phenotypic differences could be seen in the altered banding patterns. The genetic basis of this RFLP and the genomic location are under study.

At this time it appears that the S-mephenytoin 4-hydroxylase is a P-450 which is part of a complex multi-gene family--related proteins with other functions are also expressed. Some influences on this P-450, $P\text{-}450_{MP}$, may be exerted by drugs and disease; preliminary evidence suggests that an RFLP may be the basis of the mephenytoin polymorphism but more extensive studies are required.

REFERENCES

Breimer, D.D., Zilly, W. & Richter, E., 1977, *Clinical Pharmacology and Therapeutics*, 21, 470.

Cresteil, T., Beaune, P., Kremers, P., Celier, C., Guengerich, F.P. & Leroux, J.P., 1985, *European Journal of Biochemistry*, 151, 345.

Distlerath, L.M. & Guengerich, F.P., 1987, Enzymology of human liver cytochromes P-450. In *Mammalian Cytochromes*

P-450, Vol. 1, edited by F.P. Guengerich (CRC Press), p. 133.

El Mouelhi, M., Didolkar, M.S., Elias, E.G., Guengerich, F.P. & Kauffman, F.C., 1987, Cancer Research, 47, 460.

Hall, S.D., Guengerich, F.P., Branch, R.A. & Wilkinson, G.R., 1987, Journal of Pharmacology and Experimental Therapeutics, 240, 216.

Knodell, R.G., Hall, S.D., Wilkinson, G.R. & Guengerich, F.P., 1987, Journal of Pharmacology and Experimental Therapeutics, 241, 1112.

Küpfer, A. & Preisig, R., 1984, European Journal of Clinical Pharmacology, 26, 753.

Nebert, D.W., Adesnik, M., Coon, M.J., Estabrook, R.W., Gonzalez, F.J., Guengerich, F.P., Gunsalus, I.C., Johnson, E.F., Kemper, B., Levin, W., Phillips, I.R., Sato, R. & Waterman, M.R., 1987, DNA, 6, 1.

Ratanasavanh, D., Beaune, P., Baffet, G., Rissel, M., Kremers, P., Guengerich, F.P. & Guillouzo, A., 1986, Journal of Histochemistry and Cytochemistry, 34, 527.

Shimada, T. & Guengerich, F.P., 1985, Molecular Pharmacology, 28, 215.

Shimada, T., Misono, K.S. & Guengerich, F.P., 1986, Journal of Biological Chemistry, 261, 909.

Shimada, T., Shea, J.P. & Guengerich, F.P., 1985, Analytical Biochemistry, 147, 174.

Umbenhauer, D.R., Martin, M.V., Lloyd, R.S. & Guengerich, F.P., 1987, Biochemistry, 26, 1094.

Wang, P.P., Beaune, P., Kaminsky, L.S., Dannan, G.A., Kadlubar, F.F., Larrey, D. & Guengerich, F.P., 1983, Biochemistry, 22, 5375.

Wang, P., Mason, P.S. & Guengerich, F.P., 1980, Archives of Biochemistry and Biophysics, 199, 206.

Wedlund, P.J., Aslanian, W.S., McAllister, C.B., Wilkinson, G.R., and Branch, R.A., 1984, Clinical Pharmacology and Therapeutics, 36, 773.

Wrighton, S.A., Thomas, P.E., Willis, P., Maines, S.L., Watkins, P.B., Levin, W. and Guzelian, P.S., 1987, Federation Proceedings, 46, 2143.

Yoshioka, H., Morohashi, K., Sogawa, K., Miyata, T., Kawajiri, K., Hirose, T., Inayama, S., Fujii-Kuriyama, Y. & Omura, T., 1987, Journal of Biological Chemistry, 262, 1706.

Zilly, W., Breimer, D.D. & Richter, E., 1977, European Journal of Clinical Pharmacology, 11, 287.

Molecular Mechanisms of the Debrisoquine- and Mephenytoin-Polymorphisms: Studies at the Protein and DNA Level and with Circulating Autoantibodies against P-450 bufl (db1) and P-450 meph

U.A. MEYER, T. CATIN, A. DEMIERRE, R. FONNE, T. KRONBACH, U.T. MEIER, R. SKODA, F. VILBOIS, U. ZANGER and F.J. GONZALEZ[1]

Department of Pharmacology, Biocenter of the University, 4056 Basel, Switzerland and [1] Laboratory of Molecular Carcinogenesis, NCI, NIH, Bethesda, MD 20892, USA

INTRODUCTION

In the last few years several genetically determined polymorphisms of cytochrome P450 (P450)-mediated drug oxidation have been discovered. These polymorphisms cause impaired biotransformation of certain drugs in socalled "poor metabolizers" (PM) as compared to "extensive metabolizers" (EM, Franklin & Parke,1986).

The debrisoquine/sparteine-type polymorphism occurs in 5-10 % of individuals in Caucasian populations and affects the metabolism of over 20 drugs including bufuralol and dextrometorphan. PM-subjects are homozygous for a recessive gene (for review, see Price-Evans, 1986).

Deficient metabolism of the anticonvulsant mephenytoin (mephenytoin polymorphism) is observed in 2-8 % of Caucasian and over 20 % of Japanese subjects (Küpfer and Preisig, 1984). The deficiency is inherited as an autosomal recessive trait and is distinct from the debrisoquine polymorphism.

The present paper summarizes our recent studies designed to elucidate the molecular mechanisms of these two polymorphisms at the microsomal, purified protein and RNA/DNA level.

RESULTS AND DISCUSSION

Debrisoquine Polymorphism

Studies at the microsomal level: Two functionally different cytochrome P450 isozymes both able to catalyze bufuralol 1'-hydroxylation have been purified from human livers (Gut et al, 1984; 1986a). Indeed, with (+)- and (-)-bufuralol as substrate a high affinity/high stereoselectivity isozyme (P450bufI) and a low affinity/low stereoselectivity enzyme (P450bufII) appear to be responsible for bufuralol metabolism in human liver microsomes. It has been difficult to evaluate the relative contribution of these two isozymes in EM- and PM-microsomes. We have now observed that purified P450bufI and the corresponding microsomal activity can be uniquely assessed by using the peroxygenase function of this P450 in a cumene hydroperoxide (CuOOH)-supported reaction. The microsomal CuOOH mediated activity of EM-microsomes and of purified P450bufI had almost identical kinetic properties and stereospecificity. Thus by kinetic analysis of microsomes from liver biopsies EM- and PM-subjects, a striking selectivity of the CuOOH-mediated reaction in identifying the PM condition was achieved, namely a drastically reduced Vmax paralleld by an increased Km and a loss of stereoselectivity (Zanger et al, submitted).

Further purification of P450 buf I: Using the CuOOH-mediated reaction specific for the P450bufI enzyme to screen column eluates and by a number of other modifications we have markedly improved the purification of human liver P450-bufI. The sA of the final preparation was 29.5 nmol 1'-OH-bufuralol/nmol P450/min, by far the highest so far reported. The N-terminal sequence of the isolated protein was determined to be GLEALVPLAVIVAIFLLLVDLM (Vilbois et al, unpublished). This N-terminal sequence lacked the first methionine (M) residue, but it was in complete agreement with the nucleotide sequence of the recently cloned human P450db1 cDNA (Gonzalez et al, see below).

Development and discovery of antibodies specifically recognizing P450bufI (P450db1): Three different types of antibodies all specifically recognizing P450bufI were developed or discovered. First, the purified P450bufI described above was used for immunizations of mice for the development of <u>monoclonal antibodies</u> (Fig. 1). Secondly, in collaboration with L. Kiffel and J.C. Homberg (Hôpital Necker and Hôpital St. Antoine, Paris), we have discovered that circulating antibodies which occur in some children with chronic active hepatitis (<u>anti-LKM1-autoantibodies</u>, Homberg et al, 1985) uniquely recognize P450bufI-type activity, as demonstrated by inhibition and immunoprecipitation. We have used anti-LKM1 IgG covalently linked to Protein A-Sepharose to isolate the microsomal anti-LKM1 antigen and it was identical with P450bufI in its Mr and its N-terminal aminoacid sequence. The anti-LKM1 antigen was isolated from solubilized microsomes of 17 human livers including liver biopsies from EM- and PM-subjects. The amount of immunoisolated P450bufI closely corelated with the Vmax of CuOOH-mediated bufuralol-1'-hydroxylation (rs = 0.96, $p < 0.001$). No protein could be isolated from PM-livers (Zanger et al., submitted). Thirdly, rabbit antibodies raised against purified rat P450db1, an isozyme with high debrisoquine 4-hydroxylase activity (Gonzalez et al, 1987) also specifically recognized P450bufI activity. Western blots with <u>anti-rat P450db1 antibodies</u> of liver microsomes from 14 different subjects revealed again a single protein and the relative density of this band correlated with the Vmax-values for CuOOH-mediated bufuralol 1'hydroxylation (Fig. 1).

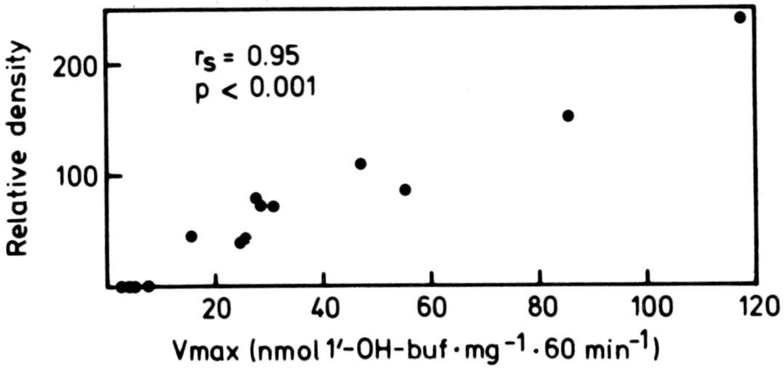

<u>Fig. 1</u>. Relationship of density of immunoblots with anti-rat P450db1-IgG and CuOOH-mediated bufuralol 1'-hydroxylation in microsomes of 14 human livers.

No protein was detectable in two PM samples and in organ transplant donor livers with equally low bufuralol-1'-hydroxylase activity. Fig.2 shows the recognition in Western blots of purified P450bufI and the anti-LKM1 antigen by the newly developed anti-P450bufI monoclonal antibody BI-114/2 as well as decreased (P72) or undetectable (P76) protein in livers with markedly decreased activity.

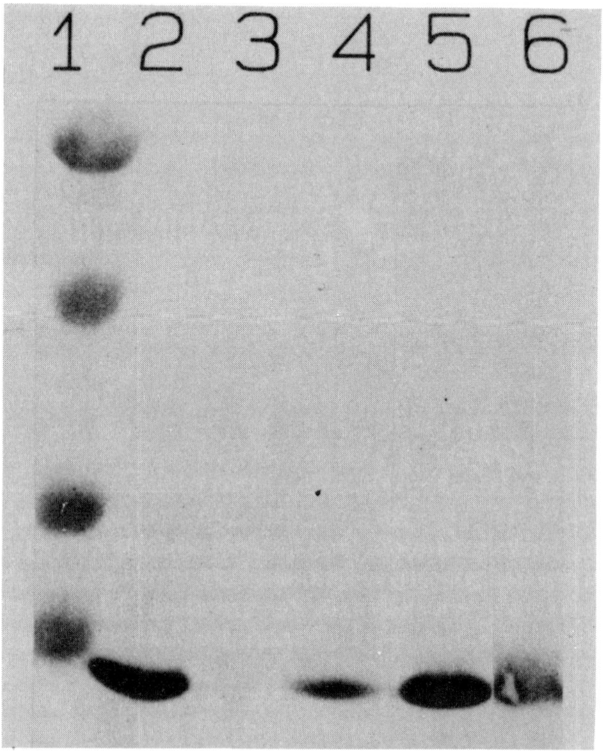

Fig. 2. Immunoblots with monoclonal antibody BI-114/2 of human liver microsomes K16 (lane 2, Vmax 118 nmol/mg/h), P76 (lane 3, Vmax 12 nmol/mg/h), P72 (lane 4, Vmax 15 nmol/mg/h), purified P450bufI (lane 5, 2 µg), purified LKM1-antigen (lane 6). Lane 1: Marker proteins

Collectively, the immunological data provide convincing evidence for the specific absence rather than a functional alteration of P450bufI (P450db1) in the PM-subjects examined as cause of the debrisoquine polymorphism. They are in contrast to previous studies with polyclonal antibodies raised against the purified human enzyme (Gut et al, 1986a; Distlerath et al, 1985) where no correlation between immunoreaction and activity was detected. These antisera possibly recognized additional proteins of identical molecular weight.

Studies at the DNA level: A full length human cDNA (db1-cDNA) was cloned using the rabbit anti-rat P450db1 antibody. The relationship of the db1-cDNA to P450bufI was convincingly demonstrated by identical N-terminal sequence and functional expression in COS-1 cells. Thus P450bufI corresponds to P450db1 (or P450IID1). In livers with markedly decreased bufuralol-1'-hydroxylase activity, different aberrant pre-mRNA splicing defects were discovered which would explain the absence of the P450bufI protein in PM-livers (Gonzalez et al, this Symposium).

We have used the db1-cDNA to analyze leucocyte DNA from in vivo phenotyped EM and PM subjects in order to identify mutant alleles of the db1 gene locus (Skoda et al, submitted). Restriction fragment length polymorphisms (RFLP's) generated by the endonuclease *Xba*I were found to be associated with the PM-phenotype when DNA from 24 unrelated PMs and 29 unrelated EMs was compared. A polymorphic 44 kb band was present in 58 % of PMs, but only in 3.4 % of EMs and a polymorphic 11.5 kb fragment was present in 33 % of PMs but in none of the EMs examined. The segregation of these RFLPs was studied in 6 families of PM propositi in which obligate heterozygote carriers of the recessive gene can be identified. These studies revealed that each of the two fragments is allelic with a 29 kb fragment present in all EMs and suggests that the 44 kb and 11.5 kb RFLPs identify two independent mutated alleles of the dbl gene. Moreover, at least a third (and probably more) mutated alleles not detected by these RFLPs must be present in the population. The mutations leading to these RFLPs and their relationship to the aberrant splicing defects are presently being investigated.

Mephenytoin Polymorphism

Studies at the microsomal level. The inherited deficiency of mephenytoin metabolism affects the stereoselective 4-hydroxylation of S-mephenytoin. Initial studies revealed that PM-microsomes could be distinguished from those of EMs by a decreased Vmax and increased Km for S-mephenytoin hydroxylation and a loss in stereoselectivity for the S-enantiomer (Meier et al, 1985a, 1985b). P450 isozymes which catalyze mephenytoin hydroxylation have been purified from human livers by this (P450meph, Gut et al., 1986b) and another laboratory (P450 MP-1 and MP-2; Shimada et al., 1986).

Discovery of human autoantibodies against P450meph. In sera of patients suffering from tienilic acid-induced hepatitis, high titres of "anti-liver kidney microsome (anti-LKM2)" antibodies have been detected (Homberg et al, 1985). These autoimmune antibodies strongly inhibit the microsomal oxidation of tienilic acid and recognize specifically $P450_8$ on Western blots of human liver microsomes (Beaune et al, 1987). $P450_8$ is immunologically related to P450 meph and P450 MP (Gut et al, 1986b; Shimada et al, 1986). We have accumulated compelling evidence that the anti-LKM2 antibodies specifically recognize the P450 responsible for microsomal S-mephenytoin hydroxylation (Meier & Meyer, 1987). They potently inhibited S-mephenytoin 4-hydroxylation and immunoprecipitated the same single protein as rabbit anti-P450meph. The immunoprecipitated protein had the same N-terminal sequence as P450 MP.

Immunoaffinity purification of P450meph from EM and PM liver microsomes. Immunopurification of P450meph was performed on Protein A-Sepharose columns to which anti-LKM2 IgG was covalently coupled. The same amount of protein was isolated from solubilized liver microsomes of EMs and one in vivo phenotyped PM-subject. Comparison of the EM-type P450 to that isolated from PM-liver revealed no difference in regard to Mr, isoelectric point, relative content in microsomes, peptide maps, amino acid composition, and N-terminal protein sequence (Meier & Meyer, 1987).

CONCLUSIONS

Debrisoquine-Polymorphism: We have improved the specificity of the assay for P450bufI (= P450db1) and have characterized immunological probes which uniquely recognize P450bufI. Our data provide strong evidence for the specific absence of P450bufI (rather than a functional alteration of this isozyme) in PM subjects as the cause of the debrisoquine polymorphism. The development of a db1-cDNA has led to the discovery of pre-mRNA splicing defects in PM-livers and of RFLP's defining independent mutant alleles of the db1 gene. Further studies are now aimed at the mechanism of these mutations and the development of DNA-probes for their easy identification.

Mephenytoin-Polymorphism: Our data with anti-LKM2 autoantibodies suggest that the mephenytoin 4-hydroxylase deficiency is caused by a minor structural change leading to a functionally altered P450meph isozyme.

Acknowledgment: Supported by the Swiss National Science Foundation.

REFERENCES

Beaune, P., Dansette, P.M., Mansuy, D., Kiffel, L., Finck, M., Amar, C., Leroux, J.P. & Homberg, J.C., 1987. Proceedings of the National Academy of Sciences, 84, 551.

Distlerath, L.M., Reilly, P.E.B., Martin, M.V., Davis, G.C., Wilkinson, G.R., and Guengerich, F.P., 1985. Journal of Biological Chemistry, 260, 9057.

Franklin, R.A. & Parke, D.V., 1986 Eds. Human Genetic Variations in Oxidative Drug Metabolism. Special Issue, Xenobiotica, 16, 359-509.

Gonzalez, F.J., Matsunaga, T., Nagata, K., Meyer, U.A., Nebert, D.W., Pastewka, J., Kozak, C., Gillette, J., Gelboin, H.V., Hardwick, J.P., 1987, DNA, 6, 149.

Gut, J., Gasser, R., Dayer, P., Kronbach, T., Catin, T., and Meyer, U.A., FEBS Letters, 173, 387.

Gut, J. Catin, T., Dayer, P., Kronbach, T., Zanger, U., and Meyer, U.A. 1986a, Journal of Biological Chemistry, 261, 11734.

Gut, J., Meier, U.T., Catin, T. and Meyer, U.A., 1986b, Biochimica Biophysica Acta, 884, 435.

Homberg, J.-C., Abuaf, N., Helmy-Khalil, S., Biour, M., Poupon, R., Islam, S., Darnis, F., Levy, V.G., Opolon, P., Beaugrand, M., Toulet, J., Danan, G. and Benhamous, J.-P., 1985, Hepatology, 5, 722.

Kupfer, A. and Preisig, R., 1984, European Journal of Clinical Pharmacology, 26, 753.

Meier, U.T., Dayer, P., Male, P.-J., Kronbach, T. and Meyer, U.A., 1985a, Clinical Pharmacology and Therapeutics, 38, 488.

Meier, U.T., Kronbach, T. and Meyer, U.A., 1985b, Analytical Biochemistry, 151, 286.

Meier, U.T. and Meyer, U.A., 1987, Biochemistry, (in press).

Price Evans, D.A., 1986, In Ethnic Differences in Reactions to Drugs and Xenobiotics. Progress in Clinical and Biological Research, 214, edited by W. Kalow, H.W. Goedde and D.P. Agarwal (Alan R. Liss Inc., N.Y.) 491.

Shimada, T., Misono, K.S. and Guengerich, F.P., 1986, Journal of Biological Chemistry, 261, 909.

Human and Rat Debrisoquine 4-Hydroxylase and Ethanol-Inducible P-450 Gene Families: Structure, Regulation and Polymorphisms

F.J. GONZALEZ [1], R. SKODA [2], J.P. HARDWICK [3], B.J. SONG [1], M. UMENO [1], O.W. McBRIDE [1], C. KOZAK [1], E. MATSUNAGA [1], T. MATSUNAGA [1], S. KIMURA [1], S.S. PARK [1], C.S. YANG [4], D.W. NEBERT [1], H.V. GELBOIN [1] and U.A. MEYER [2]

[1] National Institutes of Health, Bethesda, MD, USA, [2] Biocenter, Basel, Switzerland, [3] Argonne National Laboratory, Argonne, IL, USA and [4] New Jersey Medical School, Newark, NJ, USA

Introduction and Strategy for Isolating Human cDNAs

The molecular biology of human P450 genes is a new frontier in pharmacology research. Analysis of human P450s by direct cloning, sequencing, and expression of their cDNAs can circumvent problems associated with enzyme purification from human tissue samples. These problems include 1) the difficulty in procurement of valuable and rare sources of tissues, 2) genetic and pathological variability in the tissue resulting in lack of reproducibility of the purification schemes that are carried out on a single tissue sample, and 3) the recent widespread prevalence of HTLVIII contaminated human tissues. Purified human P450 proteins, however, have been obtained and antibodies developed (Gut et al., 1986; Shimada et al., 1986).

Our approach to the study of human P450s (Fig. 1) is to use antibodies generated against purified rat P450 proteins to screen gt11 cDNA expression libraries prepared from human liver mRNA. Full-length cDNAs are isolated, completely sequenced and expressed into active proteins by use of cDNA expression systems. In this way the rat and orthologous human P-450s can be compared both structurally and catalytically. The rat and human cDNAs can then be used to study gene regulation in rodents, as well as human genetic variability in expression of P450s. In the remaining part of this chapter we discuss the structure, regulation and polymorphism of rat and human debrisoquine 4-hydroxylase (P450dbl) and ethanol-inducible P450 (P450j). P450dbl is a known drug metabolizing enzyme of clinical significance; a polymorphism of "extensive" and "poor" metabolizers (EM and PM phenotypes) associated with this enzyme has been described (Idle and Smith, 1979). P450j, on the other hand, is highly active in the metabolism of certain nitrosamines, such as N-nitrosodimethylamine (Patten et al., 1986; Thomas et al., 1987).

Figure 1. Strategy for studying human P450s.

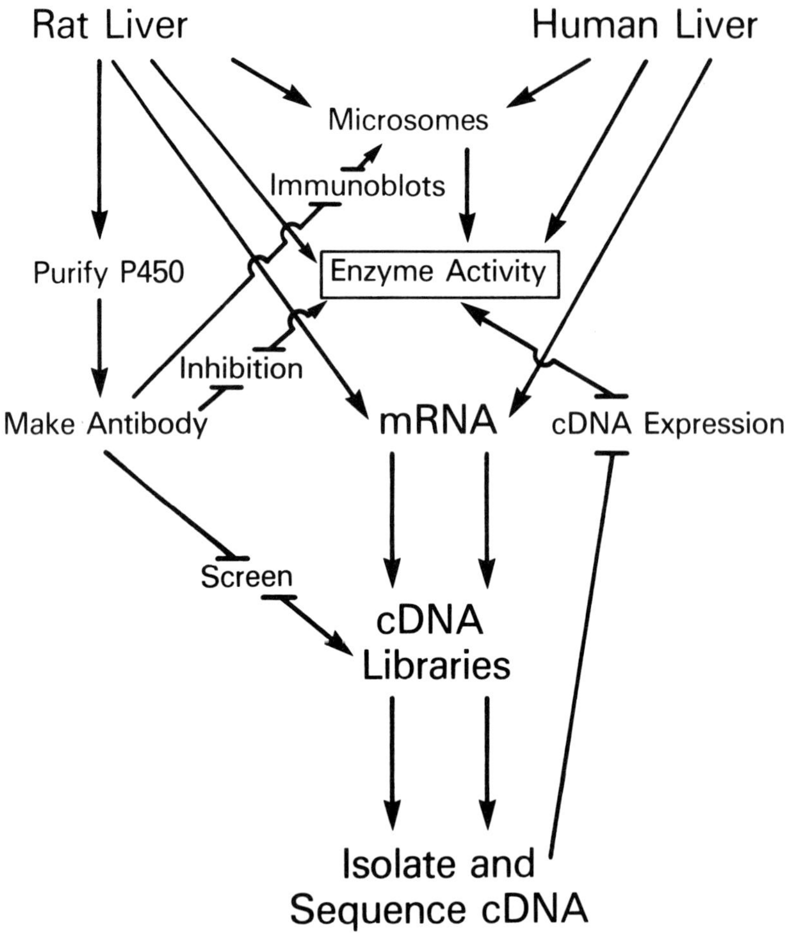

Structure and Evolution of P450IID1 (db1)

Two P450s that are immunologically similar were purified from rat liver. One of them (db1) has a high catalytic activity toward debrisoquine 4-hydroxylation and bufuralol 1'-hydroxylation. Using antibody against db1, we isolated and sequenced rat db1 and db2 cDNAs (Gonzalez et al.,1987). Rat db1 and db2 share 78% nucleotide and 73% deduced amino acid similarities. Comparison of the db1 amino acid sequence with other P450 sequences established that it belongs in a separate subfamily in the P450II gene family (P450IID, Nebert et al., 1987). At least four rat genes were isolated from a genomic library and sequenced. These genes all share about 70% to 90% nucleotide similarities across their exonic regions, and evidence of recent gene conversion events was uncovered. The dbl cluster is located on mouse chromosome 15 (Gonzalez et al., 1987).

By screening human liver cDNA libraries with anti-(rat) dbl we isolated and sequenced a human cDNA (hP450dbl). The hP450dbl shares 71% and 73% respective nucleotide and amino acid similarity with rat dbl. Two tandom, highly similar genes that are located on the long arm of human chromosome 22, were isolated; one of which corresponds to the P450dbl gene. We do not have evidence that the second gene is expressed. All the rat and human genes contain nine exons.

Structure and Evolution of P450IIE (P450j)

By use of polyclonal antibody against a rat acetone-induced P450 (Patton et al., 1986) we isolated both rat and human cDNAs against P450j (Song et al., 1986). Only a single gene product could be found in rat and human liver by use of antibody and cDNA probes. Human P450j cDNA shares 75% nucleotide and 78% deduced amino acid similarities to the respective rat cDNA. The P450j gene is located on mouse chromosome 7 and human chromosome 10, and contains nine exons in both rat and human. Comparison of the rat and human P450j protein sequences with other known P450 sequences established that P450j is a separate subfamily in the P450II gene family (P450IIE subfamily).

Regulation of P450dbl and P450j

P450dbl and P450j proteins are increased soon after birth and in the case of the latter enzyme, this increase is due to transcriptional activation. Both enzymes are present in adult male and female animals and their levels are not modulated by the common P450 inducers, 3-methylcholanthrene, phenobarbital or dexamethasone. In fact, we have not found a compound that will induce P450dbl. P450j, on the other hand, is induced by a variety of treatments and the mechanism by which this enzyme is increased is different from that described for the 3-methylcholanthrene-, phenobarbital- and dexamethasone-inducible P450 proteins.

P450j is induced by a novel post-transcriptional mechanism. Administration of acetone, ethanol or pyrazole to rats results in a 3- to 6-fold increase in cellular levels of P450j protein. P450j mRNA, however, is not increased;

Figure 2. Regulation of P450j expression.

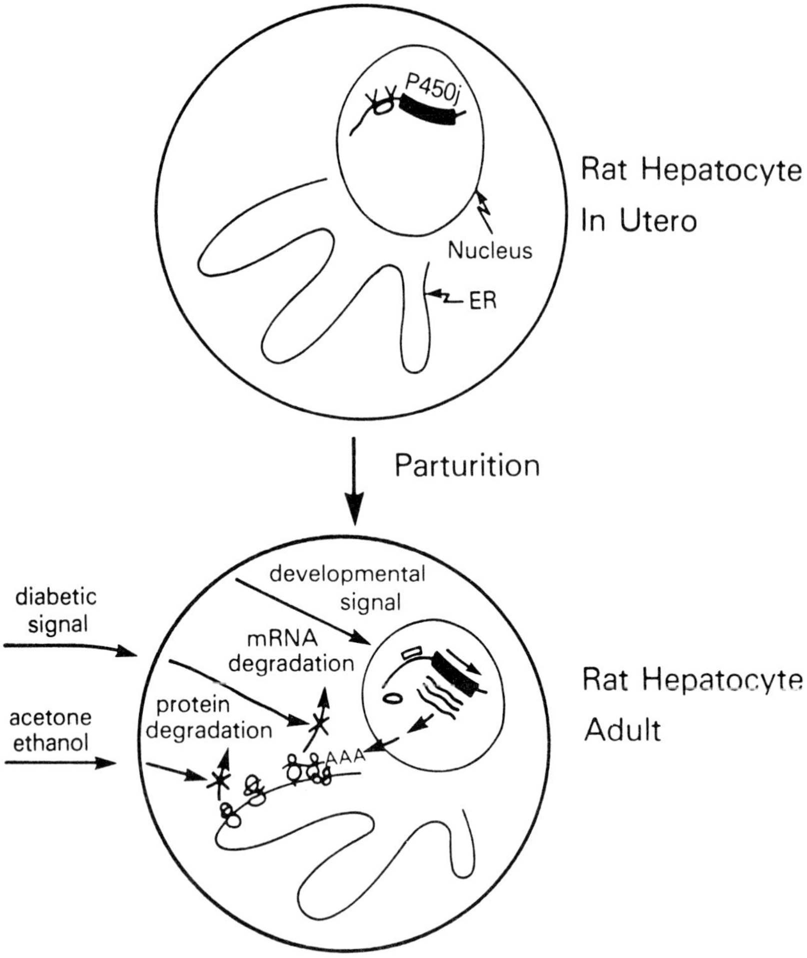

suggesting a post-transcriptional mechanism of induction (Son et al., 1986, 1987). In vivo labeling studies using [^{14}C]-sodium bicarbonate, revealed that the rate of synthesis of P450j and total cellular protein is suppressed by treatment with acetone and the turnover rate of P450j protein is markedly decreased. These data suggest that P450j protein is specifically stabilized against degradation by treatment of the rat with acetone. It has been known for several years that P450j activity is elevated in rats that had been made diabetic by administration of alloxan or streptozotocin

(Past and Cook, 1982; Peng et al., 1983). Surprisingly, we found that P450j mRNA levels are elevated about 10-fold in chemically-induced diabetic rats (song et al., 1987). This increase was not due to transcriptional activation of the P450j gene suggesting that P450j mRNA is preferentially stabilized (Song et al., 1987). Thus, P450j induction under abnormal nutritional or pathophysiologic states is the result of protein and/or mRNA stabilization. The mechanisms of specificity of these stabilization phenomena are currently not understood. Regulation of P450j is summarized in Figure 2

Alterations in levels of human P450 proteins may be due to defects in P450 regulatory molecules (e.g. receptors), defects in P450 genes, or dietary or therapeutic imbalances. For instance, in the latter case, P450p protein levels in human was shown to be elevated by treatment with dexamethasone (Molowa et al., 1986)

Expression and Restriction Fragment Length Polymorphisms Associated with the P450j Gene

Analysis of P450j levels in several human liver microsome samples revealed that all livers contained the protein with a maximum 2- to 4-fold difference in level between samples. Fetal liver microsomes, however, do not contain detectable P450j protein, as was expected from developmental studies in rat liver. The marginal differences among liver specimens and the fact that all adult liver samples so far examined contain the enzyme would suggest that P450j is not polymorphic in humans. Further analysis of more human livers and extrahepatic samples, at the protein and enzymatic activity level, should establish if concentrations of this enzyme show marked variation between individuals.

A restriction fragment length polymorphism (RFLP) was found linked to the P450j gene. Digestion of human lymphocyte DNA with Taq I revealed an RFLP with a relatively high allelic frequency. Twenty other restriction enzymes were tested that did not show any RFLPs. Although this Taq I RFLP may be useful for linkage studies on chromosome 10, it is probably not associated with an altered P450j gene.

Expression and Polymorphisms in the P450db1 Gene

Production of the metabolite 4-hydroxydebrisoquine from the parent drug is impaired in 5% to 10% of the white population of Europe and North America (Idle and Smith, 1979; Eichelbaum, 1984). Livers from kidney transplant donors and surgical resections have been collected and

Figure 3. Strategy for studying P450db1 gene mutations.

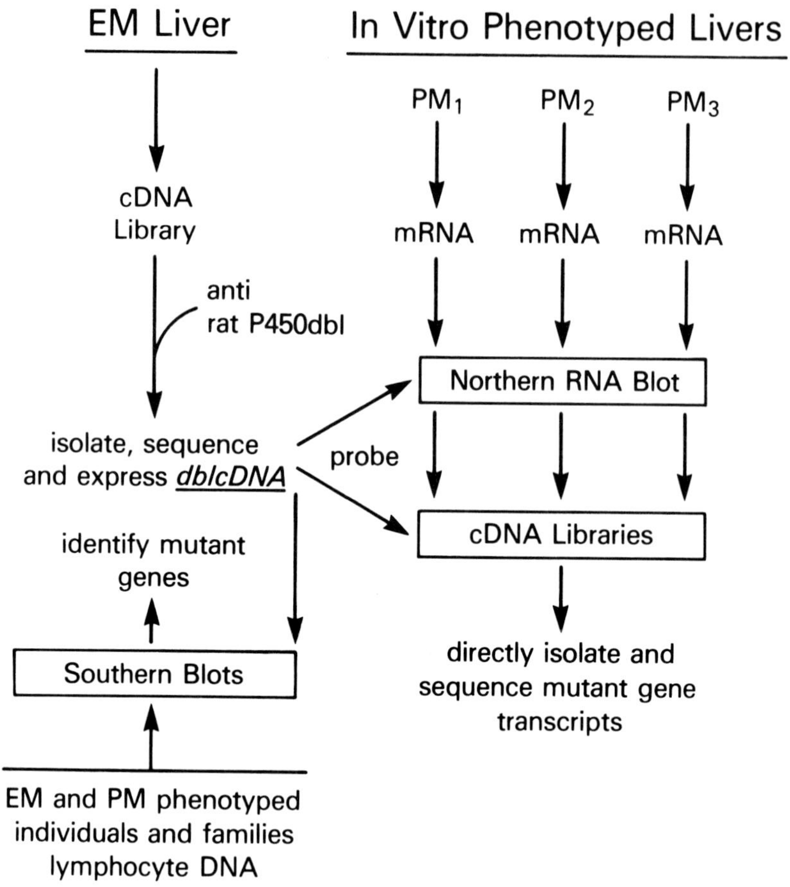

analyzed as being either PM or EM. Western blot analysis of several of these "in vitro phenotyped" livers, with antibody to rat P450db1, revealed the absence of the enzyme in several livers phenotyped as PM while all EM livers contained the protein (see Meyer et al., this Symposium). Examination of P450db1 mRNAs by Northern blots with the db1 cDNA showed an 1800 nucleotide message in EM livers that was absent in several PM livers. In addition, low levels of abnormal sized mRNAs were also identified in PM livers. Direct analysis of these "varient" mRNAs by construction of cDNA libraries from PM livers and isolating and sequencing cDNAs that hybridized with the db1 probe (Fig. 3) revealed the presence of incorrectly spliced db1 mRNAs. One mRNA contained the db1 gene's 5th intron and mRNA from another PM liver contained the 6th intron.

These studies have established that the molecular basis of the debrisoquine polymorphism in several samples we have studied is due to low levels or absence of the P450db1 protein. Therefore, the widespread human defect in debisoquine metabolism is the result of mutant P450db1 alleles. Further haplotype analysis and family studies should reveal the total number of different mutant P450db1 genes.

REFERENCES

Gonzalez, F. J., Matsunaga, T., Nagata, K., Meyer, V. A., Nebert, D. W., Pastewka, J., Kozak, C. A., Gillette, J., Gelboin, H. V. & Hardwick, J. P., 1987, DNA, 6, 149.

Gut, J., Catin, T., Dayer, P., Kronback, T., Zanger, V. & Meyer, U. A., 1986, Journal of Biological Chemistry, 261, 11734.

Idle, J. R. & Smith, R. L., 1979, Drug Metabolism Reviews, 9, 301.

Molowa, D. T., Schuetz, E. G., Wrighton, S. A., Watkins, P. B., Kremers, P., Mendez-Picon, G., Parker, G. A. & Guzelian, P. S., 1986, Proceedings of the National Academy of Sciences U.S.A., 83, 5311.

Nebert, D. W., Adesnik, M., Coon, M. J., Estabrook, R. W., Gonzalez, F. J., Guengerich, F. P., Gunsalus, I. C., Johnson, E. F., Kemper, B., Levin, W., Phillips, I. R., & Waterman, M. R., 1987, DNA, 6, 1.

Past, M. R. & Cook, D. E., 1982, Biochemical Pharmacology, 31, 3329.

Patten, C. J., Ning, S. M., Lu, A. Y. H., & Yang, C. S., 1986, Archives of Biochemistry and Biophysics, 251, 629.

Peng, R. P., Tennant, P., Lorr, N. A. & Yang, C. S., 1983, Carcinogenesis, 4, 703.

Shimada, T., Misono, K. S., & Guengerich, F. P., 1986, Journal of Biological Chemistry, 261, 909.

Song, B. J., Gelboin, H. V., Park, S. S., Yang, C. S. & Gonzalez, F. J., 1986, Journal of Biological Chemistry, 261, 16689.

Song, B. J., Matsunaga, T., Hardwick, J. P., Park, S. S., Veech, R.L., Yang, C. S., Gelboin, H. V. & Gonzalez, F. J., 1987, Molecular Endocrinology, in press.

Thomas, P. E., Bandiera, S., Maines, S. L., Ryan, D. E. & Levin, W., 1987, Biochemistry, 26, 2280.

The Contribution of Polymorphic Isozymes of Cytochrome P-450 to the Pharmacokinetics and Toxicity of Foreign Compounds in Man

ALAN R. BOOBIS, ROB J. EDWARDS, ALISON SINGLETON, DOROTHEA SESARDIC, BERNARD MURRAY, CHRISTOPHER J. SPEIRS, STEPHEN MURRAY, SHINICHI KOBAYASHI and DONALD S. DAVIES

Department of Clinical Pharmacology, Royal Postgraduate Medical School, DuCane Road, London W12 0HS, UK

INTRODUCTION

Since the first report that the 4-hydroxylation of debrisoquine is polymorphic in man (Mahgoub et al. 1977) numerous other drugs have been shown to exhibit impaired metabolism in the poor metaboliser (PM) phenotype (see Idle & Smith 1984). This was shown to be due to an absence or deficiency of a specific isozyme of cytochrome P-450 catalysing these reactions (Davies et al. 1981) (cytochrome P-450dbl). Recently Gonzalez et al. (in press) have found that in the PM subjects studied to date there is an absence of cytochrome P-450dbl, due to instability of the mRNA.

Impaired metabolism in the PM phenotype might give rise to altered susceptibility to the biological effects of foreign compounds in such individuals. As a consequence, much effort has been expended on devising methods for identifying those drugs and other xenobiotics that are subject to polymorphic oxidation in man. In particular, methods for assessing the impact that polymorphic oxidation has on the pharmacokinetics of such compounds are considered here, followed by some approaches to studying any role that polymorphic oxidation might have on their toxicity.

METHODS

Studies in vitro

Human liver samples were from the Human Tissue Bank at the Royal Postgraduate Medical School (Boobis et al. 1983). Debrisoquine 4-hydroxylase, bufuralol 1'-hydroxylase and phenacetin O-deethylase activities were determined as previously described (Speirs et al. 1986). Mutagenicity assays were performed using modifications of the Ames/Salmonella test, with strain TA98, as previously reported (Plummer et al. 1986). Methods for the synthesis of peptides and production of polyclonal antibodies against them will be presented elsewhere. Other immunological and chromatographic techniques were as previously described (Sesardic et al. 1986). The methods used for epitope identification have recently been summarised (Edwards et al. 1987) and will be published in full elsewhere.

Studies in vivo

The effects of quinidine on the metabolic ratio (MR) for debrisoquine were determined in healthy volunteers, using modifications of our previous protocol (Speirs et al. 1986). All subjects gave their written informed consent to participate in the studies, which had local Research Ethical Committee permission. Debrisoquine and 4-hydroxy-debrisoquine were assayed by capillary column gas chromatography, negative ion chemical ionisation mass spectrometry (Boobis et al. 1983).

RESULTS

Determination of the influence of polymorphic oxidation of a drug on its pharmacokinetics

Given that there is evidence from studies in vitro that cytochrome P-450db1 contributes to the oxidation of a drug, how can that contribution to the overall pharmacokinetics of the compound be determined in vivo. Amongst several possibilities, we have concentrated on evaluating the "inhibitor approach" (Speirs et al. 1986), in which subjects act as their own controls.

Table 1. Effect of quinidine on the 4-hydroxylation of debrisoquine in vivo.

Dose of quinidine		Mean	Median	Std dev
5 mg	Pre-dose MR	1.8	0.6	3.1
(n=6)	Post-dose MR	4.8	2.9	6.1
	Post/pre	4.7	3.8	3.0
50 mg	Pre-dose MR	1.6	0.4	2.2
(n=8)	Post-dose MR	17.7	15.5	12.0
	Post/pre	25.8	12.7	24.7

Studies in vitro have shown that several compounds inhibit debrisoquine 4-hydroxylase activity (Boobis et al. 1983). Many of these, not surprisingly, are alternative substrates for cytochrome P-450db1. The most potent inhibitor, with a K_i of less than 0.1 μM, is quinidine (Speirs et al. 1986), although this is apparently not a substrate for the isozyme that it inhibits so effectively (Guengerich et al. 1986). Nevertheless, inhibition at low concentrations is highly specific for cytochrome P-450db1 (Speirs et al. 1986). Based on these observations, the effects of quinidine on the oxidation of debrisoquine in vivo were determined. Low doses of quinidine (50 mg) caused a profound elevation in the metabolic ratio for debrisoquine (Speirs et al. 1986).

Fig. 1. Time course of inhibition of debrisoquine 4-hydroxylase by quinidine in vivo.

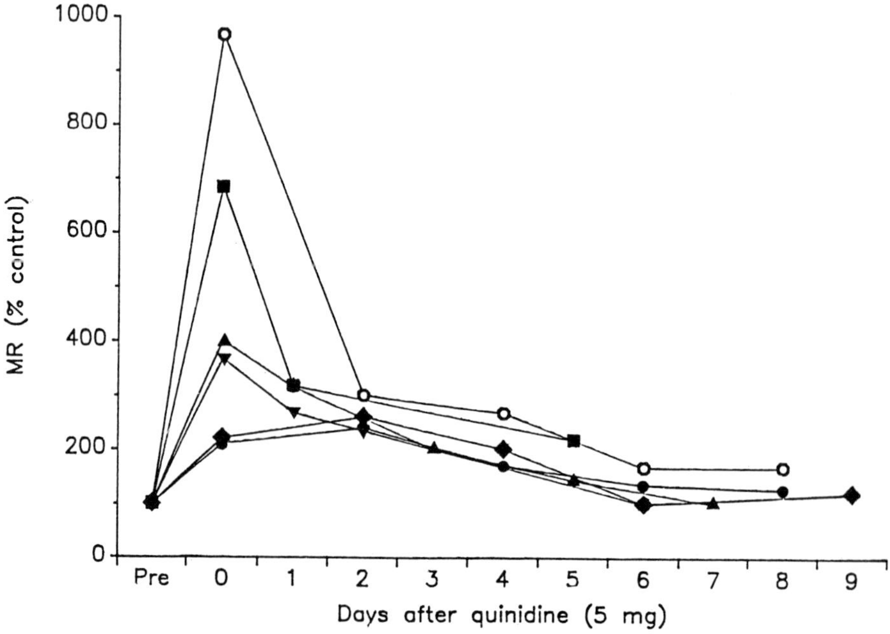

The metabolic ratio (MR) was determined from a 6-hour urine sample collected following 1 mg debrisoquine every 48 h.

The effects of 50 mg and 5 mg of quinidine on the debrisoquine MR have now been compared (Table 1). The MR was increased by 26-fold following 50 mg and by 5-fold following the 5 mg dose. All subjects showed an increase in their MR following quinidine, even at the lower dose.

The duration of the inhibitory effect of quinidine in vivo was determined following a dose of 5 mg (Fig. 1). It took at least 6 days before the MR returned to pre-treatment values, and in some subjects the MR was still elevated after 8 days.

The nature of this inhibition was investigated in vitro. Both re-isolating the microsomal fraction by ultracentrifugation and diluting by 10-fold the microsomal suspension containing quinidine reversed the inhibition. In both instances, only that quinidine not irreversibly bound would have been removed.

Determination of the involvement of polymorphic oxidation in the toxicity of a compound

When investigating a toxic reaction, it is obviously not possible to perform some of the studies that might be carried out on the metabolism of a drug in normal volunteers. Amongst the alternatives possible, both correlation (McManus et al. 1984) and inhibition (Plummer et al. 1986) studies in vitro have proved helpful. For example, there is no correlation between debrisoquine 4-hydroxylase activity and the mutagenic activation of Trp-P-2, IQ, MeIQ, 2-aminofluorene and 2-acetylaminofluorene (Table 2). Activities catalysed by the same isozyme do correlate, such as debrisoquine 4-hydroxylase and bufuralol 1'-hydroxylase, whereas activities known to be catalysed by different isozymes, such as debrisoquine 4-hydroxylase and phenacetin O-deethylase, do not correlate. Inhibition studies, either with alternate substrates such as

Table 2. Spearman rank correlation matrix for debrisoquine 4-hydroxylase activity of human liver with some oxidation reactions of toxicological importance.

	POD	Trp-P-2	IQ	2-AAF	2-AF	B1'OH
D4H	0.246	0.298	0.112	-0.043	-0.107	0.794
POD		0.888	0.606	0.625	0.471	0.443
Trp-P-2			0.567	0.637	0.421	0.422
IQ				0.740	0.506	0.297
2-AAF					0.608	0.241
2-AF						0.256

Abbreviations used are: POD, phenacetin O-deethylase; Trp-P-2, IQ, 2-AAF and 2-AF, the mutagenicity of Trp-P-2, IQ, 2-acetylaminofluorene and 2-aminofluorene in the Ames test; B1'OH, bufuralol 1'-hydroxylase; D4H, debrisoquine 4-hydroxylase. For values of $R_s >= 0.384$, $p <= 0.05$ (n=27).

debrisoquine, or with quinidine, have shown that cytochrome P-450dbl does not contribute significantly to the mutagenic activation of aflatoxin B_1, IQ, MeIQ, MeIQx or DiMeIQx.

Use of specific antibodies

Although inhibition studies, such as those described above, can provide much useful information on the contribution of cytochrome P-450dbl to a particular reaction, this depends upon the availability of specific inhibitors. For many isozymes, no such inhibitors are known. Alternatively, antibodies could provide the required specificity. However, due to the degree of homology between members of the same sub-family of cytochrome P-450, it is often difficult to obtain antibodies mono-specific for one isozyme (Thomas et al. 1984, 1985). One solution is to produce antibodies directed against a unique site on the isozyme of interest. This could be accomplished by identifying the epitope with which an inhibitory (monoclonal) antibody to an isozyme in the rat, say, reacts and synthesising the peptide corresponding to this epitope for the orthologous isozyme in man. Alternatively, unique peptide regions could be identified from published nucleotide sequences for the human isozymes. Such a site would obviously have to be part of a functional domain of the protein for such an antibody to be inhibitory. Identification of such domains of cytochrome P-450 is not trivial, but several approaches currently underway in different laboratories are likely to provide useful information on this in the near future.

The epitope with which a specific monoclonal antibody to rat cytochrome P-450c reacts was identified (Edwards et al. 1987). This epitope is also present on the human orthologues of both forms c and d. The epitope for this antibody was synthesised, and its identity confirmed by the reaction of the antibody with the peptide.

A 9 amino acid sequence common to rat cytochromes P-450c and d (residues 174-182 of form c), but different in the orthologues of these isozymes in other species, was identified. A polyclonal antibody was raised to this region, in the form of a synthetic peptide conjugated to lysozyme, in a rabbit. In solid phase enzyme linked immunosorbent assay (ELISA) this antibody reacted not only with the immunogenic peptide conjugated to haemocyanin (KLH), but also with the KLH-conjugated peptides from the human orthologues of c and d, and with KLH-cysteine, but not with KLH alone (Table 3).

ELISA assay with the pure peptides revealed that the antibody reacted with only the immunogenic peptide, and that the cross-reactivity with the conjugated peptides was due entirely to the presence of antibody species against the coupling group. The antibody also recognised both purified isozymes containing the peptide, cytochromes P-450c and d, in ELISA assay. Western blotting confirmed that the antibody recognised both forms c and d within microsomes, and that there was no apparent reactivity with forms b, e, j or any major constitutive form. There was also no reaction with mouse forms P1 or P3, rabbit forms 4 or 6, or with the human orthologue of form d.

Table 3. EC_{50} values for the binding of anti-peptide $c_{174-182}$ to different ligands determined by ELISA

Ligand	EC_{50} (g/l)
Keyhole limpet haemocyanin (KLH) -rat $c_{174-182}$	0.025
KLH-human $d_{172-180}$	0.059
KLH-human $c_{170-178}$	0.066
KLH-cysteine	0.100
KLH	>1
Rat $c_{174-182}$	0.027
Human $d_{172-180}$	>1
Human $c_{170-178}$	>1
Hepatic microsomes from	
Control rat	>1
3-Methylcholanthrene-induced rat	~0.32

The antibody was raised against a synthetic peptide corresponding to residues 174-182 of rat cytochrome P-450 form c, conjugated to lysozyme. Peptides human $d_{172-180}$ and $c_{170-178}$ represent the orthologous sequences for the human isozymes, d and c, respectively.

DISCUSSION

The "inhibitor approach" to determining the contribution of the polymorphic isozyme of cytochrome P-450, cytochrome P-450dbl, to the overall pharmacokinetics of a new drug in vivo has been further characterised, using quinidine. Doses of quinidine as low as 5 mg are sufficient to inhibit debrisoquine 4-hydroxylase activity in vivo. Thus, it should be possible to convert the phenocopy of any subject from EM to PM with between 5 and 100 mg of quinidine. Although quinidine is a substrate for cytochrome $P\text{-}450_{NF}$ (Guengerich et al. 1986), its K_m is such that no inhibition of this isozyme would be expected at the doses of quinidine used here. Recent studies (DD Breimer, personal communication) would suggest that this is indeed so.

The effects of quinidine in vivo are relatively long lasting. However, this is not due to irreversible inhibition, as shown by studies in vitro. Thus, the long lasting inhibition observed in vivo is due probably to the persistence of inhibitory concentrations of quinidine in the liver for longer than the published data on the pharmacokinetics of this compound would suggest (Moffat et al. 1986). Brosen et al (personal communication) have shown that the half-life of quinidine is far longer than previously supposed.

Although the kinetics of drugs can be studied in vivo using the "inhibitor approach", the majority of toxic pathways are obviously not suited to this method. Alternative approaches are therefore required, such as correlation and inhibition studies, both in vitro. Thus, it has been shown that cytochrome P-450db1 does not contribute significantly to the metabolic activation of IQ, MeIQ, DiMeIQ, DiMeIQx, Trp-P-2, 2-aminofluorene, 2-acetylaminofluorene or aflatoxin B_1.

Although studies with quinidine and alternate substrates can provide much useful information on the specificity of cytochrome P-450db1, alternative methods are required for other polymorphic isozymes, where selective chemical inhibitors are not available. Specific inhibitory antibodies would provide an ideal alternative. Whereas polyclonal antibodies are usually inhibitory, they are often not specific (Thomas et al. 1985), whereas monoclonal antibodies, although often specific, are frequently not inhibitory (Thomas et al. 1984). The use of site-specific antibodies would overcome these difficulties, provided a suitable site could be identified.

Methods have been developed that have enabled the identification of the specific epitope with which a monoclonal antibody reacts. This approach could be used to identify the epitope with which an inhibitory antibody interacts, if this were linear. The corresponding epitope for the human orthologue of the isozyme in question could then be synthesised and used to raise antibodies that should be inhibitory.

The possibility of obtaining antibodies against linear regions of an isozyme of cytochrome P-450 predicted from its nucleotide sequence has been demonstrated. A polyclonal antibody raised against such a peptide showed no cross-reactivity with the corresponding regions of orthologous isozymes, as predicted from their dissimilar nucleotide sequences at this site. However, the amount of immunoglobulin in such antibody preparations directed against redundant epitopes makes the production of monoclonal antibodies essential in future studies of this kind.

ACKNOWLEDGEMENTS
This work was supported by grants from the Cancer Research Campaign, Medical Research Council and Health & Safety Executive

REFERENCES

Boobis, A.R., Murray, S., Kahn, G.C., Robertz, G.-M. & Davies, D.S., 1983, Molecular Pharmacology, 23, 474.

Davies, D.S., Kahn, G.C., Murray, S., Brodie, M.J. & Boobis, A.R., 1981, British Journal of Clinical Pharmacology, 11, 89.

Edwards, R.J., Sesardic, D., Boobis, A.R. & Davies, D.S., 1987, Biochemical Society Transactions, 15, 455.

Guengerich, F.P., Müller-Enoch, D. & Blair, I.A., 1986, Molecular Pharmacology, 30, 287.

Idle, J.R. & Smith, R.L., 1984, In Proceedings of the Second World Conference on Clinical Pharmacology and Therapeutics, edited by L. Lemberger & M.M. Reidenberg (American Society for Experimental Pharmacology and Therapeutics), p. 148.

Mahgoub, A., Idle. J.R., Dring, L.G., Lancaster, R. & Smith, R.L., 1977, Lancet, ii, 584.

McManus, M.E., Boobis, A.R., Minchin, R.F., Schwartz, D.M., Murray, S., Davies, D.S. & Thorgeirsson, S.S., 1984, Cancer Research, 44, 5692.

Moffatt, A.C., Jackson, J.V., Moss, M.S. & Widdop, B., (editors), 1986, In Clarke's Isolation and Identification of Drugs, 2nd edn (The Pharmaceutical Press), p. 952.

Plummer, S., Boobis, A.R. & Davies, D.S., 1986, Archives of Toxicology, 58, 165.

Sesardic, D., Boobis, A.R., McQuade, J., Baker, S., Lock, E.A., Elcombe, C.R., Robson, R.T., Hayward, C. & Davies, D.S., 1986, Biochemical Journal, 236, 569.

Speirs, C.J., Murray, S., Boobis, A.R., Seddon, C.E. & Davies, D.S., 1986, British Journal of Clinical Pharmacology, 22, 739.

Thomas, P.E., Reik, L.M., Ryan, D.E., Bandiera, S. & Levin, W., 1985, In Drug Metabolism; Molecular Approaches and Pharmacological Implications, edited by G. Siest (Pergamon Press), p. 131.

Thomas, P.E., Reik, L.M., Ryan, D.E. & Levin, W., 1984, Journal of Biological Chemistry, 259, 3890.

Pharmacokinetic and Pharmacodynamic Consequences of Beta-Adrenoceptor Blocker Oxidation Polymorphism in Man

G.T. TUCKER, M.S. LENNARD, V.S. OTTON, P.R. JACKSON, L.E. RAMSAY and H.F. WOODS

University of Sheffield Department of Therapeutics, Royal Hallamshire Hospital, Sheffield S10 2JF, UK

INTRODUCTION

Following anecdotal reports claiming that the oxidative metabolism of some beta-blockers is under polymorphic control of the debrisoquine-type, controlled studies have established the pharmacokinetic and pharmacodynamic consequences of this phenomenon (Lennard et al 1986a). In this review the findings of the initial investigations with metoprolol, timolol, propranolol and atenolol are summarised, followed by a brief mention of further studies of biochemical, stereochemical, theoretical, ethnic and clinical aspects.

PATHWAYS OF METABOLISM OF BETA-BLOCKERS

It is well-established that the extent of beta-blocker metabolism is related to their lipophilicity. Hydrophilic compounds, such as atenolol, undergo minimal metabolism and are recovered largely unchanged in the urine and faeces whereas the more lipid-soluble ones, such as metoprolol, timolol and propranolol, are almost completely metabolised. The main routes are: metoprolol - side-chain O-demethylation (followed by conversion to a carboxylic acid)(65% of dose), side-chain alpha-hydroxylation (10%) and deamination (10%); timolol - cleavage of the morpholino ring (40%), via hydroxylation and O-dealkylation; propranolol - aromatic 4'-hydroxylation (40%), side-chain oxidation (40%) and glucuronic acid conjugation (15%). The potential for co-segregation of the oxidative pathways with the 4-hydroxylation of debrisoquine is, therefore, apparent.

PANEL AND POPULATION STUDIES

Comparisons of the pharmacokinetics of metoprolol, timolol, propranolol and atenolol in panels of extensive (EM)

and poor (PM) metabolisers of debrisoquine have shown significant interphenotypic differences for the first two compounds. No differences in overall kinetics were seen with propranolol, although plasma concentrations of its 4'-hydroxy metabolite were significantly lower in PM's. As expected, atenolol kinetics were similar in the two phenotypes (Table 1).

Table 1. Interphenotypic comparisons of the pharmacokinetic properties of 4 beta-blockers (Lennard et al 1986a)

Drug	Oral AUC ratio PM/EM	Peak plasma concentration ratio PM/EM	Elimination $t_{\frac{1}{2}}$ (h) PM	EM
Metoprolol	6	3	2.8	7.6
Timolol	2-4	2	3.2	5.5
Propranolol	1	1	4.0	4.4
Atenolol	1	1	6.7	6.4

The pharmacokinetic findings for the 4 beta-blockers were reflected in the time-courses of the mean percentage reduction in post-exercise tachycardia. Thus, significant interphenotypic differences in the intensity and duration of beta-blockade were demonstrated for metoprolol and timolol, but not for propranolol and atenolol (Lennard et al 1986a). The results with propranolol suggested that, contrary to common belief, the 4-hydroxy metabolite may not contribute significantly to its pharmacological effect.

Following these panel studies a population study has confirmed the polymorphic nature of metoprolol metabolism by showing significant correlations between urinary debrisoquine/4-hydroxy debrisoquine ratios and those of metoprolol/alpha-hydroxy metoprolol and of metoprolol/the end-product of metoprolol O-demethylation (McGourty et al 1985). Furthermore, data from the families of probands were shown to be compatible with poor oxidation of metoprolol being inherited as an autosomal recessive trait. More recently, a population study of timolol has revealed a strong correlation between the urinary debrisoquine/4-hydroxy debrisoquine ratio and that of timolol/an end-product of its morpholino ring cleavage (Lennard et al 1986b).

BIOCHEMICAL STUDIES

In vitro support for similar genetic control of the metabolism of metoprolol and debrisoquine comes from the demonstration of mutual competitive inhibition in human and rat liver microsomes (Otton et al 1984; Crewe et al 1984).

We have also used quinidine and quinine as 'specific' inhibitors to probe the possibility of the involvement of more than one isozyme in the O-demethylation of metoprolol. These studies were prompted by the observation that, whereas the frequency distribution of the urinary metoprolol/alpha-hydroxy metoprolol ratio is clearly bimodal, that for metoprolol/the O-demethylation product is not, showing some overlap between EM's and PM's (McGourty et al 1985). This could arise if O-demethylation was effected by two enzymes, only one of which is linked to debrisoquine oxidation. Accordingly, when quinidine and quinine were added to human liver microsomes only part of the O-demethylase activity towards metoprolol could be inhibited, whereas all of the alpha-hydroxylation was blocked (Fig.1). The existence of an O-demethylating enzyme unrelated to debrisoquine hydroxylase was also suggested by results with microsomes from a human liver with negligible alpha-hydroxylase activity and little capacity for debrisoquine and sparteine metabolism (a PM liver?). These microsomes showed only residual O-demethylation of metoprolol, which could not be inhibited by quinidine/quinine (Fig.2).

STEREOCHEMICAL ASPECTS

The in vitro studies with human microsomes (Figs.1&2) indicated that R(+)-metoprolol is more rapidly O-demethylated by the high affinity 'isozyme' than S(-)-metoprolol, whereas the low affinity 'isozyme' appeared to be selective for the S(-)-isomer. Alpha-hydroxylation was also greater for the S(-)-form. These findings are compatible with in vivo observations of higher plasma concentrations of S(-)- than R(+)-metoprolol in EM's and the reverse in PM's (Lennard et al 1986a).

Most of the beta-blocking activity of metoprolol is thought to reside in the S(-)-enantiomer. Therefore, the kinetic data predict that for a given total ((S) + (R)) plasma concentration of metoprolol, the drug should produce more effect in EM's than in PM's. Accordingly, it has been shown that the log plasma drug concentration-response curve for EM's is shifted to the left of that for PM's (Lennard et al 1986a). Therefore, differential stereoselective

Fig.1. The effect of quinidine and quinine on the alpha-hydroxylation (upper panel) and O-demethylation (lower panel) of the isomers of metoprolol by human liver microsomes (EM liver).

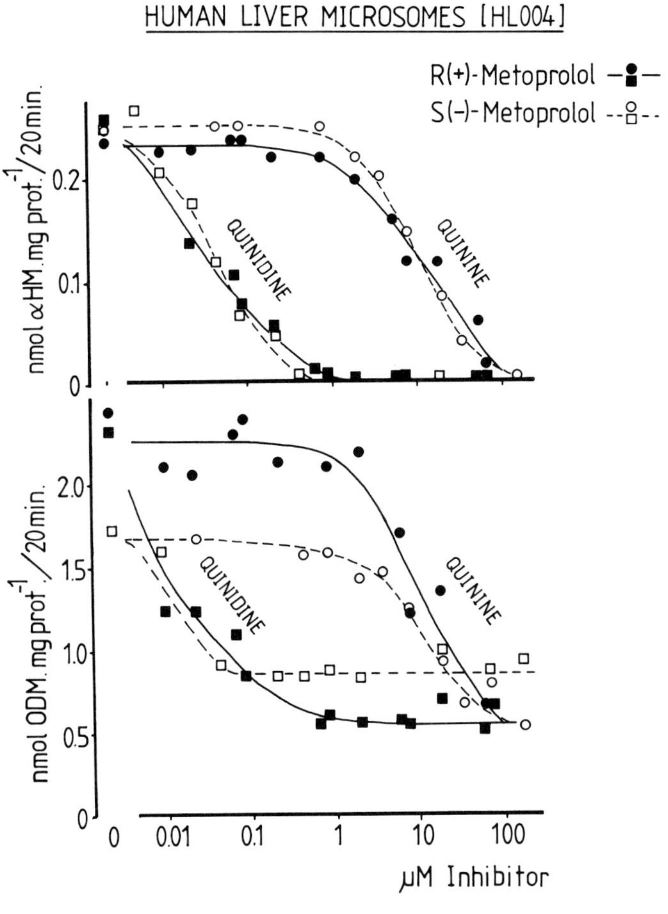

elimination between phenotypes contributes to interindividual variation in plasma concentration-response relationships.

THEORETICAL ASPECTS

Metoprolol has been used as a model drug to investigate the pharmacokinetic basis and discriminatory power of var-

Fig.2. The effect of quinidine and quinine on the O-demethylation of the isomers of metoprolol by human liver microsomes (PM liver).

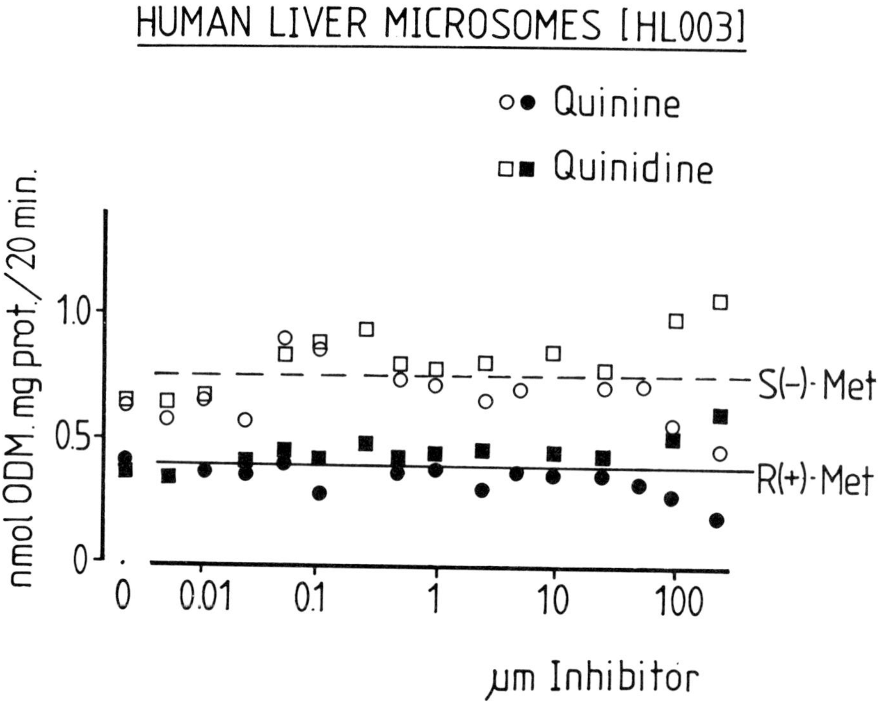

ious indirect experimental indices (urinary drug/metabolite ratio, AUC, urinary metabolite recovery etc) for detecting polymorphic drug oxidation (Jackson et al 1986). Computer simulations of frequency distributions based on a first-pass model of metoprolol kinetics, including the description of clearance via the polymorphic pathway according to the Hardy-Weinberg Law, were similar to those observed experimentally. They illustrated, for example, why the AUC value is a relatively poor index of polymorphism when the frequency of the gene defect is low. Further simulations have examined the effects of changing the number of subjects in a population study, the drug absorption rate or renal clearance, and the influence of urine collection period on the ability to discriminate bimodality.

ETHNIC ASPECTS

Having established that the oxidation of metoprolol co-segregates with that of debrisoquine in white British subjects, these studies were extended to investigate the metabolism of both drugs in a Nigerian population (Iyun et al 1986).

Although significant correlations between urinary drug/metabolite ratios of the two compounds were found in Nigerians, in contrast to the British data, there was no evidence of bimodality in the frequency distributions of the ratios for either drug and the medians were shifted significantly to the right. It was concluded that the poor hydroxylation trait may be present at a much lower frequency than in whites, absent altogether, or obscured by environmental factors. A suggestion that liver residues of anti-malarial aminoquinolines might inhibit metoprolol and debrisoquine metabolism, thereby perturbing the frequency distributions of their metabolic ratios, is supported by preliminary in vitro data (Lancaster et al 1987).

The findings of Iyun et al (1986) were in marked contrast to those of an earlier report of a 9% incidence of the PM phenotype for debrisoquine in a Nigerian population (Mbanefo et al 1980). However, inspection of these data also showed no clear evidence of bimodality and the assignment to the PM phenotype was based upon the antimodes defined in Caucasians. Clearly, different ethnic groups should be examined separately for evidence of polymorphisms and antimodes should not be extrapolated from one to another. This practice calls in to question the interpretation of much of the published data for the debrisoquine/sparteine oxidation polymorphism across ethnic groups.

CLINICAL IMPLICATIONS?

The clinical consequences of the polymorphic oxidation of beta-blockers are unresolved but focus on the choice of drug, the choice of dosage form (especially in relation to slow-release preparations) and the need to titrate dosage in the individual patient.

Angina Pectoris

Control of angina is thought tc correlate with beta-blockade and, therefore, maintenance of effect over 24h is desirable. The large interphenotypic differences in beta-blocking activity at 24h after dosing with metoprolol suggest that PM's may require lower doses than EM's and also that the drug may need to be given only once daily to PM's. Slow-release preparations of metoprolol may not be required in

PM's, whereas the currently available ones do not sustain beta-blockade over 24h in most EM's (Lennard et al 1986a).

Hypertension

The antihypertensive effect of beta-blockers correlates poorly with their plasma concentrations. However, in view of the lack of a significant hypotensive effect of metoprolol in EM's at 24h after administration, once-daily dosing with metoprolol and its available slow-release preparations may be unsatisfactory in this phenotypic group.

Adverse Reactions

There is no doubt that beta-blockers have a wide therapeutic margin and that potentially fatal adverse reactions are rare. On the other hand, more than 20% of patients discontinue these drugs and up to 30% suffer from fatigue. Nevertheless, most reports (Clark et al 1984) have not shown an association between the PM phenotype and susceptibility to adverse effects of beta-blockers. However, these are retrospective studies and controlled data are needed to evaluate clinical significance.

A small controlled study in PM's and EM's has assessed the effects of phenotype on fatigue (Lewis et al - in prep) Subjects were given placebo, metoprolol (100mg) and atenolol (100mg) in random order and the time to complete an exercise step test was determined. The degree of exhaustion was also measured by a visual analogue scale. Although differences between either drug and control were not significant, highly significant correlations were found between the differences in both exercise times and VAS scores for the drugs and the debrisoquine metabolic ratio.

On the occasions when beta-blockers have to be prescribed to patients with asthma, metoprolol is one of the drugs recommended because of its cardioselectivity. However, there is evidence that $beta_1$-selectivity is lowered at high plasma concentrations of metoprolol, suggesting that obstructive airways disease could be exacerbated in PM's. This receives some support from a study of 12 asthmatic patients (6EM,6PM) comparing the effects of metoprolol (100mg) and atenolol (50mg) on FEV_1,FVC and response to inhaled salbutamol (McGourty et al - in preparation).

CONCLUSIONS

For certain beta-blockers some of the interindividual variability in pharmacokinetics and pharmacodynamics can now be accounted for by monogenic control of oxidative metabolism. Whether or not this is clinically significant, the differences produced are generally much larger than those imposed by environmental and physiological factors. Routine phenotyping prior to commencing therapy with metoprolol or timolol is not justified. However, careful titration of the dose by assessment of exercise heart rate and side-effects is prudent. Unfortunately, it is naive to believe that this is or will be universal clinical practice. Alternatively, it may be easier to use a drug whose elimination is independent of metabolism.

REFERENCES

Clark, D.W.G., Morgan, A. & Waal-Manning, H., 1984, British Journal of Clinical Pharmacology, 18, 965.

Crewe, H.K., Lennard, M.S., Tucker, G.T. & Woods, H.F., 1984, British Journal of Pharmacology, 83, 965.

Iyun, A.O., Lennard, M.S., Tucker, G.T. & Woods, H.F., 1986 Clinical Pharmacology and Therapeutics, 40, 387.

Jackson, P.R., Tucker, G.T., Lennard, M.S. & Woods, H.F., 1986, British Journal of Clinical Pharmacology, 22, 541.

Lancaster, D.L., Lennard, M.S., Tucker, G.T. & Woods, H.F., 1987, Abstracts 7th International Symposium on Microsomes and Drug Oxidation, Adelaide.

Lennard, M.S., Tucker, G.T. & Woods, H.F., 1986a, Clinical Pharmacokinetics, 11, 1.

Lennard, M.S., Lewis, R.V., Braun, L.A., Ramsay, L.E., Jackson, P.R., Tucker, G.T. & Woods, H.F., 1986b, British Journal of Clinical Pharmacology, 23, 644P.

Mbanefo, C., Bababunmi, R.A., Mahgoub, A., Sloan, T.P., Idle, J.R. & Smith, R.L., 1980, Xenobiotica, 10, 811.

McGourty, J.C., Silas, J.H., Lennard, M.S., Tucker, G.T. & Woods, H.F., 1985, British Journal of Clinical Pharmacology, 20, 555.

Otton, S.V., Inaba, T. & Kalow, W., 1984, Life Sciences, 34, 73.

ACKNOWLEDGEMENTS

M.S.L. is a Wellcome Trust Lecturer and V.S.O. is a M.R.C. of Canada Fellow.

Assessment of In Vivo Oxidative Drug Metabolizing Enzyme Acxtivity in Man by Applying a Cocktail Approach

D.A. BREIMER, J.H.M. SCHELLENS and P.A. SOONS

Center for Bio-Pharmaceutical Sciences, Division of Pharmacology, University of Leiden, PO Box 9503, 2300 Leiden, The Netherlands

INTRODUCTION

Several different approaches have been undertaken to assess drug metabolizing enzyme activity in humans and to predict the rate of metabolism of a drug in vivo in individual subjects. The use of marker drugs (substrate probes) has found relatively wide application, in particular in determining changes in enzyme activity as caused by induction or inhibition and assessed in longitudinal studies in the same subjects (Vesell 1979). One of the major shortcomings however of any marker drug currently used to measure oxidative enzyme activity in vivo, is that the results obtained with one drug do only rarely possess predictive value towards other drugs. For example several attempts have been made to correlate plasma elimination rates (half-lives) or plasma clearance values of antipyrine with those of another oxidized drug in the same panel of subjects. In most studies the correlations were low or non-existent (Breimer et al. 1984). A number of possible causes may explain these findings, of which the multiplicity of the cytochrome P-450 system is the most important one. Multiplicity implies that different compounds may be oxidized by different isozymes (inter-substrate selectivity) and/or that from a single compound different metabolites may be produced by different isozymes (intra-substrate selectivity) (Breimer 1983). The latter is for instance likely to be the case for antipyrine. The formation of major primary metabolites, 3-hydroxymethylantipyrine (HMA), norantipyrine (NORA), 4-hydroxyantipyrine (OHA), is selectively inducible in rats and to a lesser extent also in man (Teunissen 1984). In pigs OHA is the only metabolite, that is found in appreciable quantities, whereas HMA and NORA are almost totally absent (Van den Broek et al. 1981). It has also been suggested that the formation of

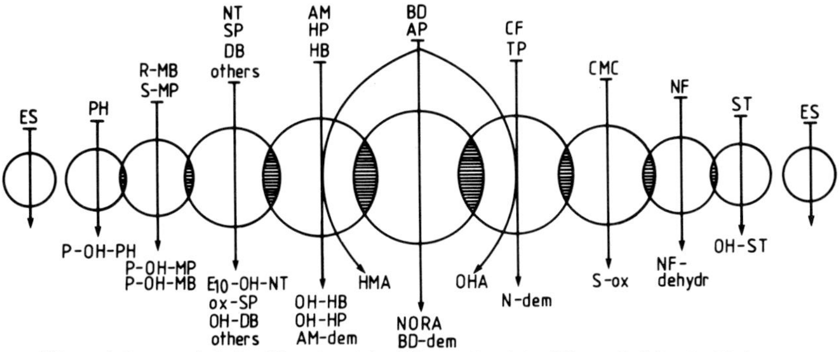

ES = endogenous substrate; PH = phenytoin; MP = mephenytoin; MB = methylphenobarbital DB = debrisoquine; SP = sparteine; NT = nortriptyline; HB = hexobarbital; HP = heptabarbital; AM = aminopyrine; AP = antipyrine; BD = benzodiazepine; TP = theophylline; CF = caffeine; CMC = carboxymethylcysteine; NF = nifedipine; ST = steroid.

Figure 1. Multiplicity of the cytochrome P-450 system with substrate and product selectivity (reproduced from Breimer (1983) and Breimer & Schellens 1986).

these 3 metabolites is under a different genetic regulatory control in humans (Penno et al. 1981). In any case has the occurrence of genetic deficiences in oxidative drug metabolizing enzyme activity provided substantial evidence for the fact that certain isozymes may exhibit considerable selectivity in substrate binding and metabolite formation (Guengerich et al. 1986).

In Figure 1 a schematic representation is given of some isozymes of the cytochrome P-450 system (circles) with different, but in many cases also overlapping, substrate and product selectivities. The arrows represent single primary metabolic pathways with metabolite formation from some frequently used substrates. There is clear evidence that some substrates are indeed predominantly metabolized by the same isozyme(s), for example debrisoquine and sparteine and also that one metabolic pathway of nortriptyline (hydroxylation at the E-10-position) co-segregates with debrisoquine and sparteine oxidation (Eichelbaum 1982). On the other hand antipyrine oxidation occurs independently of debrisoquine or sparteine metabolism (Danhof et al. 1981). The multiplicity of the cytochrome P-450 system as depicted in Figure 1 implies that several substrate probes (or a limited number of polyfunctional substrates like antipyrine), of which the rate of metabolite formation is assessed, are required to obtain a total picture of oxidative drug metabolizing enzyme activity. We have developed an experimental strategy along these lines: the cocktail approach.

Cocktail approach

This approach is characterized by giving combinations of substrate probes simultaneously (in relatively low doses) and measuring either plasma kinetics of unchanged drugs or urinary excretion of metabolites or both. The kinetic parameters that most closely reflect enzyme activity in vivo are the intrinsic clearance of unchanged drug or the clearance for production of separate metabolites (Breimer et al. 1984). Intrinsic clearance can be determined by measuring the area under the plasma concentration versus time curve (AUC) after oral administration of the drug: CL(int) = Dose/AUC and clearance of formation of metabolite is excreted into urine as that metabolite: CL(→ met) = f(m)· CL(int). So-called metabolic ratios are often used in phenotyping panels of subjects for the existence of genetically determined deficiences in drug oxidizing enzyme activity in order to differentiate between poor and rapid metabolizers. They are however not parameters which reflect enzyme activity quantitatively and are therefore less suitable in metabolic correlation studies, which are meant to elucidate common regulation (co-segregation) of metabolic pathways of different substrates (Breimer 1983).

There are in principle at least two major advantages associated with the cocktail approach, a practical one and a conceptual one. It is obviously of great practical advantage when for example phenotyping for different types of polymorphism can be achieved in one session. Conceptually this approach is of advantage in metabolic correlation studies, because it does exclude the influence of intraindividual variability in metabolizing activity with time. In the past in such studies two or more drugs were given to the same panel of subjects, but on different occasions. Although oxidative drug metabolizing enzyme activity seems to be primarily controlled by genetic factors, there may still be a considerable influence of environmental factors which potentially may cause intraindividual differences in activity from one point in time to the other (Rowland 1980). A complication of the cocktail approach is that several drugs and metabolites have to be analysed in the same biological samples, which requires a large degree of selectivity and sensitivity of the analytical methodology required for their determination.

We have in recent years performed several studies along the lines of the cocktail approach and shown that it is feasible both in rats and in humans. The first experiments were done in rats, where for example the correlation was studied between the oxidation of antipyrine and hexobarbital (Van der Graaff et al. 1983). Four groups of rats were pretreated with either phenobarbital or 3-methylcholanthrene to create broadly ranging enzyme activities. Intrinsic clearance values of antipyrine and hexobarbital did not correlate significantly, however, when individual metabolites were taken into consideration much better results were obtained. Correlation coefficient between intrinsic clearance of hexobarbital and 3-hydroxymethylantipyrine was 0.89, hence it is likely that the isozyme(s) responsible for HMA formation is the same or closely related to that or those oxidizing hexobarbital. No significant correlation were found between the rate of formation of the other metabolites of antipyrine and hexobarbital oxidation.

In a similar study design the correlation between antipyrine oxidation and theophylline metabolism was investigated in rats (Teunissen et al. 1986). The highest correlation coefficient (0.99) was found between intrinsic clearance of theophylline and the rate of formation of OHA. The correlation coefficient between intrinsic clearances of unchanged drugs was also high (0.92), which suggests a considerable overlap in substrate selectivity for the isozyme(s) involved. Correlations between theophylline clearance and HMA and NORA formation were considerably lower (0.54 and 0.67 respectively), suggesting that these metabolites may at least partly be formed independently of theophylline.

In a first study in man antipyrine and theophylline were given simultaneously to nine healthy subjects orally and on another occasion by rectal infusion to steady-state (Teunissen et al. 1985). Correlation coefficient between clearance values of theophylline and antipyrine was reasonably high (0.72), but as in rats the best correlation was found between the rate of formation of OHA and theophylline clearance (0.91). The correlation results did not improve when theophylline metabolite formation was taken into consideration. This study clearly showed that the cocktail approach is feasible in humans as well and that it represents a powerful tool in elucidating co-segregation of metabolic pathways of different drugs.

In this chapter the results of two recent studies in man will briefly be reviewed. The first is an extension of the correlation study with antipyrine and theophylline in a greater number of subjects and hexobarbital as a third probe drug. The objective of the second study was to assess in one session the incidence of the poor metabolizing phenotype of sparteine, mephenytoin and nifedipine.

Metabolic correlations between antipyrine, hexobarbital and theophylline

These probe drugs were administered to 26 healthy subjects on two different occasions: in the first experiment a combination of antipyrine (250 mg) and hexobarbital (250 mg) was given and in the second experiment theophylline (150 mg) was added to this combination. The reason for this dual experiment was to make sure that the addition of theophylline would not interfere with the metabolism of antipyrine and hexobarbital. Plasma concentrations of the parent drugs and urine concentrations of unchanged theophylline and the 3 major metabolites of antipyrine were determined. No differences in these parameters were found for antipyrine and hexobarbital between experiment 1 and 2.

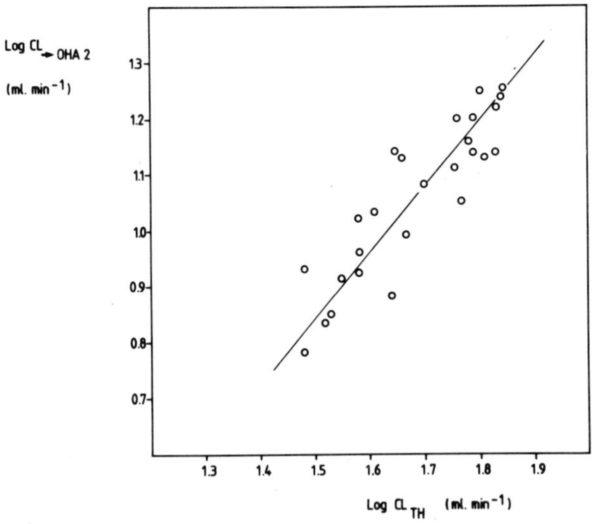

Figure 2. Orthogonal regression analysis of theophylline clearance and clearance of formation of OHA in 26 healthy subjects.

Least squares correlation coefficients were calculated and orthogonal regression analysis was performed for intrinsic clearance values and clearance of formation of metabolites. Interindividual variability in clearance was about 3-fold for antipyrine and theophylline and 5-fold for hexobarbital. High correlation coefficients were found between clearance values of the 3 drugs (0.8 - 0.9), indicating that similar isozymes are involved in their oxidation. With regard to antipyrine metabolite formation the highest correlations existed between HMA, NORA and hexobarbital on the one hand and between OHA and theophylline on the other. Regression analysis gave similar results and in Figure 2 this is shown for OHA and theophylline. The isozymes of the cytochrome P-450 system responsible for the oxidation of HB and formation of HMA and NORA are probably very closely related and the same is the case for theophylline oxidation and OHA formation. This type of experimental approach and the results obtained represent the basis for the information that is given in Figure 1. As an alternative strategy the cocktail approach offers the possibility to pursue enzyme inhibition in vivo by giving combinations of different drugs in order to identify co-segregation of metabolic pathways, which is an experimental procedure widely applied in vitro for the same purpose.

Poor metabolizer incidence of sparteine, mephenytoin and nifedipine in a Dutch population

A cocktail of sparteine sulfate (25 mg), mephenytoin (100 mg) and nifedipine (20 mg) was administered to 172 healthy subjects. In 0-8 and 8-24 hours' urine samples excreation of sparteine, dehydrosparteine, para-hydroxy-mephenytoin and the "M-I"-metabolite of nifedipine (oxidized to the pyridine derivative and ester hydrolyzed) were measured. The frequency histogram of the log metabolic ratios of sparteine is shown in Figure 3, of mephenytoin metabolite excretion in 0-8 hours' urine in Figure 4 and of M-I-excretion in 0-8 hours' urine in Figure 5. For sparteine 1: poor metabolizers were detected (log metabolic ratio > 2) and for mephenytoin 4 (less than 0.5% of doseexcreted into 0-8 hours' urine). These data are well in agreement with the frequencies detected in other Caucasian populations (Kalow 1982). However the frequency distribution of the M-I-metabolite of nifedipine did not show bimodality, whic is in contrast to the results of a previous investigation (Kleinbloesem et al. 1984). In that study the area under the plasma concentration time curve of unchanged drug (AUC was bimodally distributed among a population of 53 subjects

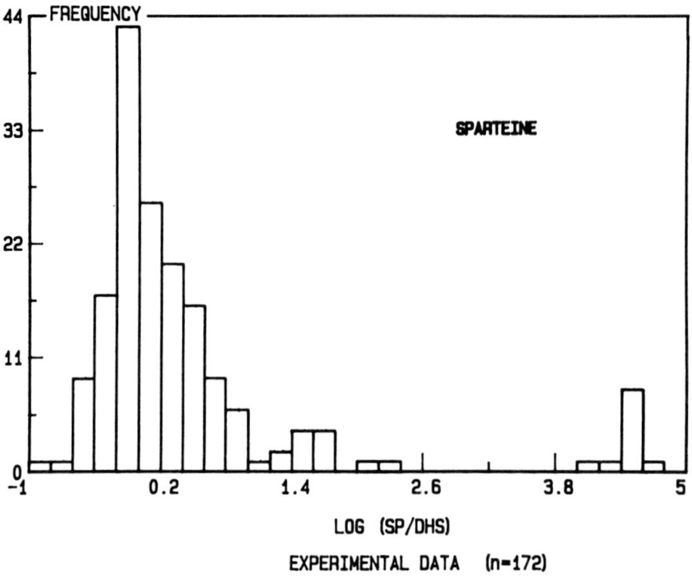

Figure 3. Frequency distribution of log sparteine metabolic ratio in 172 healthy subjects.

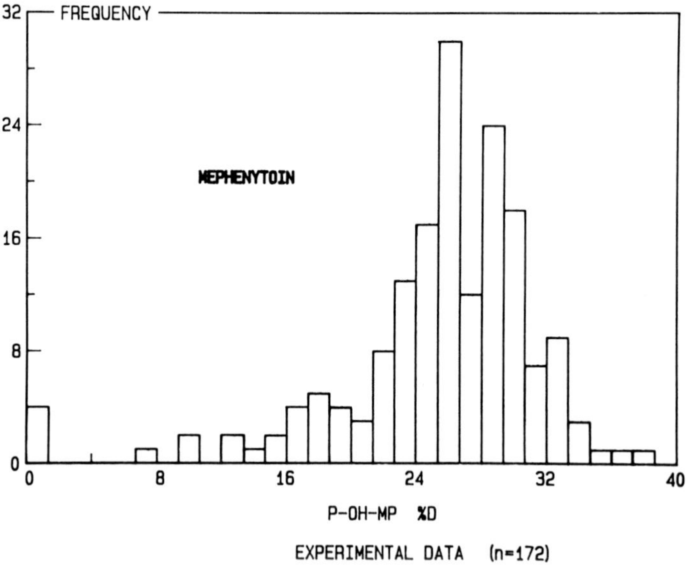

Figure 4. Frequency distribution of p-hydroxy-mephenytoin excretion in 0-8 hours' urine (% dose) in 172 healthy subjects.

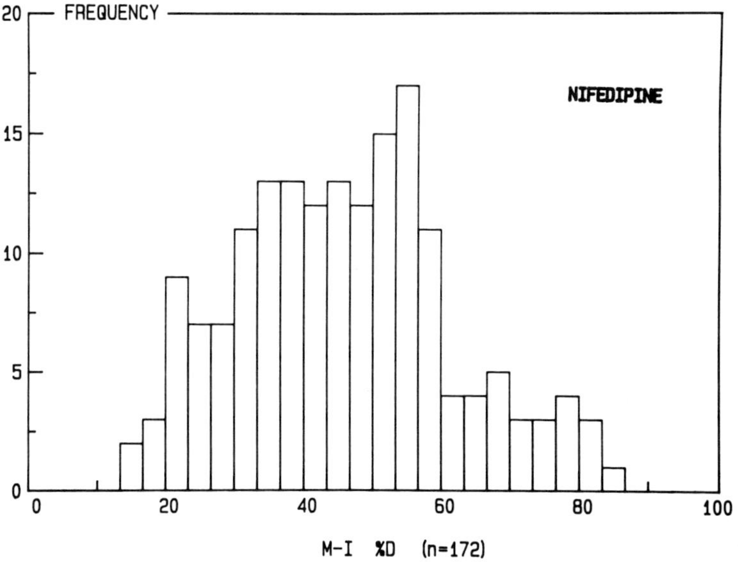

Figure 5. Frequency distribution of the excretion of M-I-metabolite of nifedipine in 0-8 hours' urine (% dose) in 172 healthy subjects.

with 9 "poor metabolizers". The subjects with the highest AUC-values showed the lowest 0-8 hours' urinary excretion of M-I. In a currently ongoing new population study with nifedipine in 130 subjects, in which AUC of unchanged drug and of the pyridine metabolite are measured as well as urinary excretion of M-I, preliminary results indicate that the frequency distribution of the measured parameters is not clearly bimodal. Several other studies with nifedipine have in the mean time shown that the rate of nifedipine metabolism is quite sensitive to enzyme inhibition and induction and also that non-linearity (dose dependence) exists in its kinetics (Schellens & Breimer, unpublished investigations). Furthermore there are indications that parallel or alternative metabolic pathways exist in addition to aromatic oxidation to the pyridine derivative of nifedipine, whereas this metabolite is not only further metabolized to M-I, but also to a side chain hydroxylated product (M-II). Taking all these factors together it becomes clear that the metabolic profile of nifedipine is quite complex and that its rate of metabolism is influenced by several environmental factors. With such a drug it is very difficult to unambiguously prove the existence of polymorphism, if not all metabolic pathways can be assessed quantitatively and simultaneously.

Coming back to the objective and conclusions of the present study with sparteine, mephenytoin and nifedipine: the cocktail approach seems to be quite feasible to assess the poor metabolizer incidence of the independent oxidation polymorphisms of sparteine and mephenytoin.

REFERENCES

Breimer, D.D., 1983, Clinical Pharmacokinetics, 8, 371.

Breimer, D.D. & Schellens, J.H.M., 1986, Acta Pharmacologica Toxicologica, 59, Supplement V, 252.

Breimer, D.D., Vermeulen, N.P.E., Danhof, M., Teunissen, M.W.E., Joeres, R.P. & Van der Graaff, M., 1984, Assessment and prediction *in vivo* oxidative drug metabolizing enzyme activity. In Pharmacokinetics: A Modern View, edited by L.Z. Benet, G. Levy & B.L. Ferraiolo (Plenum Press, New York & London), p. 191.

Danhof, M., Idle, J.R., Teunissen, M.W.E., Sloan, T.P., Breimer, D.D. & Smith, R.L., 1981, Pharmacology, 22, 349.

Eichelbaum, M., 1982, Clinical Pharmacokinetics, 7, 1.

Guengerich, F.P., Distlerath, L.M., Reilly, P.E.B., Wolff, T., Shimada, T., Umbenhauer, D.R. & Martin, M.V., 1986, Xenobiotica, 16, 367.

Kalow, W., 1982, Clinical Pharmacokinetics, 7, 373.

Kleinbloesem, C.H., Van Brummelen, P., Faber, H., Danhof, M., Vermeulen, N.P.E. & Breimer, D.D., 1984, Biochemical Pharmacology, 33, 3721.

Penno, M.B., Dvorchik, B.H. & Vesell, E.S., 1981, Proceedings of the National Academy of Sciences (U.S.A.), 78, 5193.

Rowland, M., 1980, Intra-individual variability in pharmacokinetics. In Towards Better Safety of Drugs and Pharmaceutical Products, edited by D.D. Breimer (Elsevier/North Holland Biomedical Press, Amsterdam), p. 143.

Teunissen, M.W.E., 1984, Factors and Conditions Affecting Antipyrine Clearance and Metabolite Formation. Ph.D.-thesis, University of Leiden, The Netherlands.

Teunissen, M.W.E., De Leede, L.G.J., Boeijinga, J.K. & Breimer, D.D., 1985, Journal of Pharmacology and Experimental Therapeutics, 233, 770.

Teunissen, M.W.E., Brorens, I.O.N., De Langen, H.J., Geerlings, A.M. & Breimer, D.D., 1986, Pharmaceutical Research, 3, 156.

Van den Broek, J.M., Teunissen, M.W.E. & Breimer, D.D., 1981, Drug Metabolism and Disposition, 9, 541.

Van der Graaff, M., Vermeulen, N.P.E., Joeres, R.P., Vlietstra, J. & Breimer, D.D., 1983, Journal of Pharmacology & Exp. Therap., 227, 459.

Vesell, E.S., 1979, Clinical Pharmacology & Therapeutics, 26, 275.

Theophylline and Tolbutamide as Model Substrates for Cytochrome P-450 Isozymes in Animals and Man

D.J. BIRKETT, J.O. MINERS, M.E. McMANUS, I. STUPANS, M. VERONESE and R.A. ROBSON

Department of Clinical Pharmacology, Flinders University of South Australia, Flinders Medical Centre, Adelaide, SA 5042, Australia

INTRODUCTION

The Cytochrome P-450 system is the major enzyme system involved in the metabolic oxidation of drugs and other xenobiotics. Cytochrome P-450 exists as a gene superfamily of enzymes which can be divided into a series of gene families and subfamilies on the basis of structural similarities (Nebert et al. 1987). Individual isozymes show distinct substrate and inhibitor selectivity and can be subject to differential induction and/or repression. The intrinsic metabolic clearance of a drug along a particular oxidative pathway is the sum of the activities (Vmax/Km) of the isozymes that mediate that pathway. In monitoring the effects of genetic and environmental influences on the activity of particular isozymes, it is desirable to have substrates which are metabolised mainly or completely by a single isozyme and which can be used both in vitro and in vivo. This has been achieved for two human isozymes which are subject to polymorphic regulation (human cytochrome P450dbl, P450IID1; and human cytochrome P450mp, P450IIC9). We have been investigating the suitability of the methylxanthines particularly theophylline, and tolbutamide as selective model substrates both in vivo and in vitro for inducible forms of human cytochrome P450.

In Vivo Differentiation of Theophylline and Tolbutamide Metabolism

The structures of theophylline and tolbutamide are shown in Figure 1. The major metabolic pathways for theophylline are 1- and 3-demethylations to 3-methylxanthine (3MX) and 1-methylxanthine (1MX) respectively and 8-oxidation to 1,3-dimethyluric acid (13DMU) (Cornish & Christman 1957; Grygiel et al. 1979). These are cytochrome P450 mediated (Robson et al. 1987). 1MX, but not 3MX, is a good substrate in vitro for xanthine oxidase (Bergmann & Dikstein 1956; Krenitsky et

Figure 1. Structure of theophylline and tolbutamide.

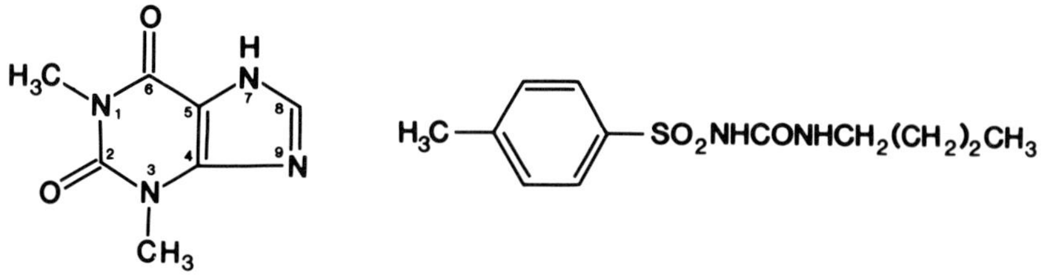

al. 1972) and is also further converted in vivo by this enzyme to 1-methyluric acid (1MU) (Birkett et al. 1983a). Tolbutamide (TB) is metabolised along a single pathway to hydroxymethyl tolbutamide (OHTB) which is further converted by alcohol and aldehyde dehydrogenases to carboxymethyl tolbutamide (Nelson & O'Reilly 1961, Thomas & Ikeda 1966). As this pathway accounts for more than 90% of tolbutamide elimination, plasma unbound clearance of parent drug approximates the partial metabolic clearance.

The results of a number of in vivo drug interaction studies with theophylline and tolbutamide are summarised in Table 1. The effects on theophylline and tolbutamide metabolism are strikingly different, the differences being most marked with sulphinpyrazone which enhanced all three pathways of theophylline metabolism but inhibited tolbutamide methylhydroxylation. None of the pathways was linked in vivo with the debrisoquine phenotype. TP 8-oxidation was less induced by cigarette smoking and less inhibited by cimetidine and propranolol than the 1- and 3-demethylations suggesting that at least two isozymes are involved in theophylline metabolism. As the human in vivo data suggested there were at least three P450 forms involved respectively in tolbutamide methylhydroxylation, the theophylline demethylations and theophylline 8-oxidation, animal models were investigated to determine whether particular P450 forms carrying out these pathways could be defined.

Animal Models for Theophylline and Tolbutamide Metabolism

The metabolic profiles and effects of inducers on theophylline and tolbutamide metabolism by rat and rabbit liver microsomes are shown in Table 2. Tolbutamide hydroxylase was induced by phenobarbitone in the rat suggesting metabolism by forms b and/or e (gene subfamily P450IIB), but was induced by rifampicin in the rabbit suggesting the involvement of Form 3c (gene subfamily P450IIICA). Thus different isozymes appear to be responsible for tolbutamide metabolism in rat and rabbit.

Table 1. Differentiation of theophylline and tolbutamide metabolism in vivo in man.

	Effect on TP1DM	TP3DM	TP8OX	TBOH	References
Cigarette smoking	↑↑	↑↑	↑	-	Grygiel & Birkett 1981 Scott & Poffenbarger 1979
Cimetidine	↓↓	↓↓	↓	-	Grygiel et al. 1984 Dey et al. 1983
Propranolol	↓↓	↓↓	↓	-	Miners et al. 1985b Miners et al. 1984
Sulphaphenazole	-	-	-	↓↓↓	Grygiel 1981 Pond et al. 1977
Sulphinpyrazone	↑	↑	↑	↓↓	Birkett et al.1983b Miners et al. 1982
Debrisoquine	NL	NL	NL	NL	Dahlqvist et al. 1984 Idle et al. 1979 Miners et al. 1985a

TP1DM, theophylline 1-demethylation; TP3DM, theophylline 3-demethylation; TP8OX, theophylline 8-oxidation; TBOH, tolbutamide methylhydroxylation.
↑ induction; ↓ inhibition; - no change; NL - not linked.

Theophylline 8-hydroxylation by rabbit liver microsomes was increased by pretreatment with acetone (an inducer mainly of Form 3a) and phenobarbitone (Forms 2 and 5), and rifampicin also had a small but statistically signficant effect. TCDD increased the 1-demethylation. In the rat, all three pathways of theophylline metabolism were increased by Aroclor 1254, dexamethasone, phenobarbitone and 3-methylcholanthrene. Thus, the effects of inducers in rats and rabbits suggest that theophylline is a "promiscuous" substrate capable of being metabolised by a range of cytochrome P450 isozymes. In agreement with

Table 2. Effects of inducers on theophylline and tolbutamide metabolism by rat and rabbit liver microsomes.

Inducer	Activity pmol/mg microsomal protein/min			
	TP1DM	TP3DM	TP8OX	TBOH
Rat				
Control	0.34±0.09	0.27±0.22	11.11±0.77	1180±330
Aroclor 1254	2.32±0.68*	15.41±6.23*	94.46±24.41*	1160±440
3-Methyl-cholanthrene	1.76±0.19*	1.36±0.74*	70.98±4.5*	620±100*
Phenobar-bitone	1.90±0.23*	1.41±0.37*	42.54±0.74*	3190±680*
Dexameth-asone	2.63±0.51*	1.60±0.82*	46.93±6.65*	960±190
Rabbit				
Control	0.44±0.15	2.72±1.63	10.3±0.6	1730±460
Acetone	0.86±0.32	1.85±0.63	37.8±14.7*	1140±350
Phenobar-bitone	0.74±0.23	1.57±0.58	28.9±3.2*	1880±240
Rifampicin	0.20±0.10*	0.98±0.22	12.1±1.1*	2650±200*
TCDD	0.90±0.25*	5.11±1.59	13.3±4.1	1020±490

*$p < 0.05$ compared to control; n = 3 animals in each group. Substrate concentrations were theophylline 200 uM and tolbutamide 1.5 mM.

this, rabbit Forms 4, 6 and 3b each catalysed all three theophylline pathways in reconstituted systems. None of these cytochromes P450 catalysed tolbutamide methylhydroxylation. Thus, the animal induction studies did not consistently indicate the involvement of a particular isozyme or group of isozymes in the metabolism of either theophylline or tolbutamide.

In Vitro Studies with Human Liver Microsomes

Table 3 shows the kinetic constants for theophylline and tolbutamide metabolism by human liver microsomes. The theophylline metabolic profile in vitro was similar to that observed in vivo. In terms of intrinsic metabolic clearance (Vmax/Km), 3MX formation accounted for 20.0% of the total metabolism in vitro compared to about 20-25% in vivo, 1MX formation accounted for 17.5% compared to 15-20% in vivo, and 8-oxidation accounted for 62.4% compared to 50-60% in vivo. The Vmax/Km value for tolbutamide methylhydroxylation was about 10-fold greater than the sum of those for the three theophylline pathways. There was a highly significant correlation between the Vmax values for the 1- and 3-demethylations of theophylline ($r = 0.98$, $p<0.05$), but not between the Vmax values for 8-oxidation and those for the demethylations. Vmax values for tolbutamide hydroxylation did not significantly correlate with those for any of the theophylline metabolic pathways.

Table 3. Kinetics of theophylline and tolbutamide metabolism by human liver microsomes (n = 5).

	Apparent Km µM	Vmax $pmol.min^{-1}.mg^{-1}$
Theophylline (n = 5)		
1-demethylation	573.0±185.5	2.4±1.5
3-demethylation	722.4±295.3	2.7±1.3
8-oxidation	912.6±360.2	12.0±4.8
Tolbutamide (n = 5)		
methylhydroxylation	120.0±41.0	273.0±66.0

The effects of various compounds on the metabolism of theophylline and tolbutamide are shown in Table 5.

There was no cross inhibition between theophylline and tolbutamide confirming the involvement of different isozymes of P450 in the metabolism of these substrates.

Table 5. Effects of inhibitors on theophylline and tolbutamide metabolism by human liver microsomes.

	% Control Activity			
Inhibitor	TP1DM	TP3DM	TP8OX	TBOH
Caffeine	60±2	61±7	78±15	98±4
Debrisoquine	88±6	82±1	90±5	99±5
Mephenytoin	87±10	88±5	89±10	76±3
Phenylbutazone	84±3	84±6	84±6	9±3
Sulphinpyrazone	89±4	87±3	82±10	32±4
Sulphaphenazole	97±4	96±7	97±3	7±3
Cimetidine	55±8	58±9	61±7	32±3
Verapamil	64±20	67±21	65±16	32±2
Nifedipine	22±13	35±20	188±157	10±4
Propranolol	8±2	9±4	28±5	86±09
Theophylline	-	-	-	101±4
Tolbutamide	96±2	94±1	97±1	-

Substrate concentrations were theophylline 200 μM and tolbutamide 100 μM n = 3 in each case. Inhibitor concentration was 500 μM except for sulphaphenazole where it was 50 μM.

This was further supported by the differential inhibitory effects observed with other compounds. Phenylbutazone, sulphinpyrazone and sulphaphenazole inhibited the metabolism of tolbutamide but not that of theophylline, and the reverse was the case with propranolol. The differential effect was most marked with sulphaphenazole where at least three orders of magnitude separated the concentrations required for inhibition of theophylline and tolbutamide metabolism (Figure 2).

Figure 2. Effect of sulphaphenazole on theophylline and tolbutamide metabolism by human liver microsomes. ○ TP1DM; △ TP3DM; □ TP8OX; ● TBOH.

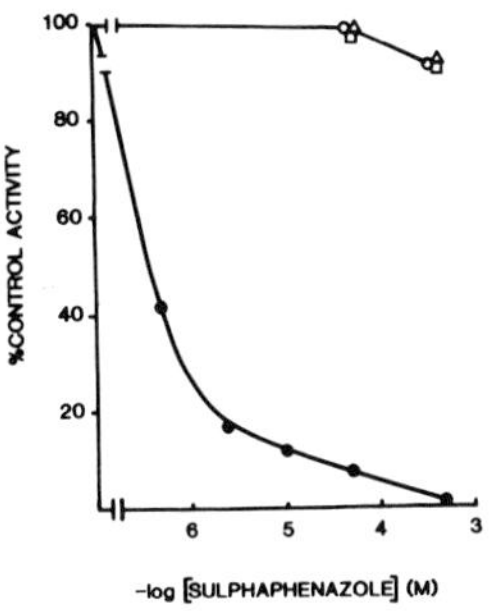

The inhibition studies also confirmed that at least two P450 isozymes are involved in theophylline metabolism. Inhibition of 1-demethylation was always similar to that of 3-demethylation which, together with the high correlation between Vmax values for these pathways, suggests that both demethylations are carried out by the same isozyme or group of isozymes. By contrast, the effects of caffeine, cimetidine, nifedipine and propranolol on the 8-oxidation were clearly different from those on the demethylations. For example, the Ki values for caffeine inhibition were 0.69 ± 0.19mM (1-demethylation), 0.73 ± 0.48mM (3-demethylation) and 2.04 mM (8-oxidation) (ANOVA; $p < 0.05$). The Ki for cimetidine was also higher for the 8-oxidation than for the demethylations which agrees with a similar effect in vivo (Grygiel et al. 1984). Propranolol even at high concentrations did not inhibit the 8-oxidation more than 60-70% whereas the demethylations were essentially totally inhibited.

Nifedipine almost totally inhibited the demethylations in microsomes from two livers, but had relatively little effect on the 8-oxidation (20-30% inhibition). With microsomes from two other livers, nifedipine caused a 2-fold to 4-fold activation of 8-oxidation whereas the demethylations were still inhibited (Figure 3).

Figure 3. Effect of nifedipine on theophylline metabolism by human liver microsomes from subject 5.

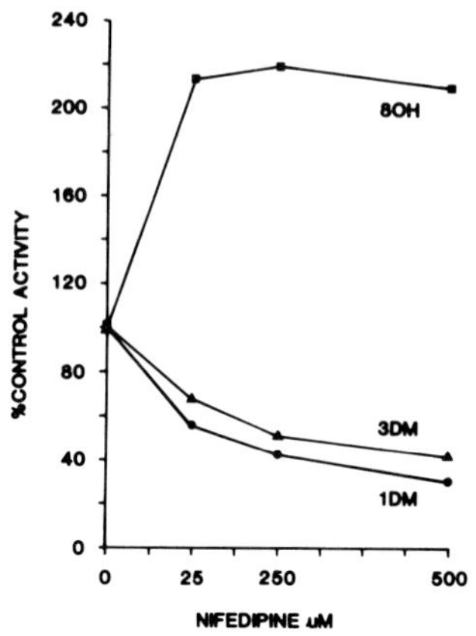

In man, theophylline metabolism is extensively induced by cigarette smoking (Grygiel & Birkett 1981) and it has been suggested that polycyclic hydrocarbon inducible forms of cytochrome P450 are involved to a major extent in methylxanthine metabolism (Campbell et al. 1987). To examine this further, a polyclonal antibody was raised in the goat against the polycyclic aromatic hydrocarbon inducible rabbit cytochrome P450 Form 4. These antibodies recognised orthologous forms in human liver microsomes which could be quantitated by Western blot techniques. The human liver microsomal content of the Form 4 orthologue correlated with theophylline 1-demethylation (r = 0.82) and 3-demethylation (r = 0.81) but not with theophylline 8-oxidation (r = 0.32). The Form 4 antibody, which was inhibitory against the rabbit enzyme, inhibited all pathways of theophylline metabolism by 30-40% with human liver microsomes, but had no effect on tolbutamide methylhydroxylation.

CONCLUSIONS

Predictive capacity in the area of human oxidative drug metabolism will depend on the ability to determine which cytochrome P450 isozymes are involved in the metabolism of a particular drug (its "isozyme profile"), and how the activities of these isozymes are regulated by genetic and environmental factors. This will require identification of selective substrates and inhibitors which can preferably be used both in vivo and in vitro, and the availability of banks of purified isozymes and selective inhibitory antibodies.

Thus far, predictive capacity has been achieved only for drugs metabolised mainly by isozymes subject to polymorphic regulation (human cytochromes P450db and P450mp). In the case of the debrisoquine isozyme the use of an animal model contributed substantially to identification and purification of the human enzyme but no satisfactory animal model exists for the human mephenytoin isozyme.

The results presented above clearly establish that at least three isozymes, or groups of isozymes, are involved in tolbutamide and theophylline metabolism in man and some progress has been made in the identification of these isozymes. Animal models (rat and rabbit) were investigated but found not to be useful in either case.

In vitro studies in man suggested that the mephenytoin isozyme was responsible for tolbutamide methylhydroxylation but this was not borne out by human in vivo studies which showed no linkage between the mephenytoin hydroxylation polymorphism and tolbutamide metabolism (Knodell et al. 1987), illustrating the importance of having model substrates which can be used both in vivo and in vitro. Our results suggest that sulphaphenazole may be a high affinity, selective inhibitor for the human tolbutamide isozyme, as quinidine is for the debrisoquine isozyme, and may, therefore, be useful for identifying other substrates metabolised by this form of cytochrome P450.

The in vitro studies with theophylline have confirmed our previous suggestion based on in vivo results (Grygiel & Birkett 1981, Grygiel et al. 1984, Miners et al. 1985b) that different isozymes are involved in the 8-oxidation and demethylations of theophylline. To date, no distinction has been found between the 1- and 3-demethylations, and it is likely that these pathways are mediated by the same isozyme or groups of isozymes. Preliminary results suggest that this may be the human orthologue to the polycyclic hydrocarbon inducible rabbit cytochrome P450 Form 4. The 8-oxidation of theophylline is at least partially carried out by other isozymes, the identities of which are uncertain. Indeed, the differing effects of nifedipine on theophylline 8-oxidation with microsomes from different individuals suggests that several forms can be involved, and that in an individual the predominant form may depend on environmental factors such as exposure to inducing agents.

ACKNOWLEDGEMENTS

This work was supported by grants from the National Health and Medical Research Council of Australia and the Anti-Cancer Foundation of the Universities of South Australia.

REFERENCES

Bergmann, F. & Dikstein, S., 1956, Journal of Biological Chemistry, 223, 765.

Birkett, D.J., Miners, J.O. & Attwood, J., 1983a, British Journal of Clinical Pharmacology, 15, 117.

Birkett, D.J., Miners, J.O. & Attwood, J., 1983b, British Journal of Clinical Pharmacology, 15, 567.

Campbell, M.E., Grant, D.M., Inaka, T. & Kalow, W., 1987, Drug Metabolism and Disposition, 15, 237.

Cornish, H.H. & Christman, A.H., 1957, Journal of Biological Chemistry, 228, 315.

Dahlqvist, R., Bertilsson, L., Birkett, D.J., Eichelbaum, M., Sawe, J. & Sjoqvist, F., 1984, Clinical Pharmacology and Therapeutics, 35, 815.

Dey, N.G., Castleden, C.M., Ward, J., Cornhill, J. & McBurney, A., 1983, British Journal of Clinical Pharmacology, 16, 438.

Grygiel, J.J., 1981, PhD Thesis, Flinders University of South Australia.

Grygiel, J.J. & Birkett, D.J., 1981, Clinical Pharmacology and Therapeutics, 30, 491.

Grygiel, J.J., Wing, L.M.H., Farkas, J. & Birkett, D.J., 1979, Clinical Pharmacology and Therapeutics, 26, 660.

Grygiel, J.J., Miners, J.O., Drew, R. & Birkett, D.J., 1984, European Journal of Clinical Pharmacology, 26, 335.

Idle, J.R., Sloan, T.P., Smith, R.L. & Wakile, L.A., 1979, British Journal of Pharmacology, 66, 430P.

Knodell, R.G., Hall, S.D., Wilkinson, G.R. & Guengerich, F.P., 1987, Journal of Pharmacology & Experimental Therapeutics, 241, 1112.

Krenitsky, T.A., Neil, S.M., Elion, G.B. & Hitchings, G.H., 1972, Archives of Biochemistry and Biophysics, 150, 585.

Miners, J.O., Foenander, T., Wanwimolruk, S., Gallus, A.S. & Birkett, D.J., 1982, European Journal of Clinical Pharmacology, 22, 321.
Miners, J.O., Wing, L.M.H., Lillywhite, K.J. & Smith, K., 1984, British Journal of Clinical Pharmacology, 18, 853.
Miners, J.O., Wing, L.M.H. & Birkett, D.J., 1985a, Australian and New Zealand Journal of Medicine, 15, 348.
Miners, J.O., Wing, L.M.H., Lillywhite, K.J. & Robson, R.A., 1985b, British Journal of Clinical Pharmacology, 20, 219.
Nebert, D.W., Adesnick, M., Coon, M.J., Estabrook, R.W., Gonzalez, F.J., Guengerich, F.P., Gunsalus, I.C., Johnson, E.F., Kemper, B., Levin, W., Phillips, I.R., Sato, R. & Waterman, M.R., 1987, DNA, 6, 1.
Nelson, E. & O'Reilly, I., 1961, Journal of Pharmacology and Experimental Therapeutics, 132, 103.
Pond, S.M., Birkett, D.J. & Wade, D.N., 1977, Clinical Pharmacology and Therapeutics, 22, 573.
Robson, R.A., Miners, J.O., Matthews, A.P., Stupans, I., Meller, D., McManus, M.E. & Birkett, D.J., 1987, Biochemical Pharmacology, (in press).
Scott, J. & Poffenbarger, P.L., 1979, Diabetes, 28, 41.
Thomas, L.C. & Ikeda, G.J., 1966, Journal of Medical Chemistry, 9, 507.

Isozyme Specificity of Drug Oxidation by Animal and Human Tissues

DONALD S. DAVIES, DOROTHEA SESARDIC, ROBERT J. EDWARDS and ALAN R. BOOBIS

Department of Clinical Pharmacology, Royal Postgraduate Medical School, DuCane Road, London W12 0HS, UK

INTRODUCTION

The discovery of a genetic polymorphism in drug oxidation in man indicated that the cytochrome P450 system might exhibit a high degree of substrate specificity, where a single isozyme would be exclusively responsible for the oxidation of certain compounds. We now know that debrisoquine is almost exclusively oxidised by a single isozyme of cytochrome P450, P450dbl. This degree of specificity has obvious implications for susceptibility to the toxic or carcinogenic effect of drugs and chemicals.

Although there is interest in determining the substrate specificity of the different isozymes of cytochrome P450, would it not be better to determine the isozyme specificity of key oxidation reactions? In order to illustrate this point, we have the used model substrate, phenacetin, to evaluate isozyme specificity towards its O-deethylation in animal and human tissues.

METHODS

Human and animal tissues

Human liver samples were from the Human Tissue Bank at the Royal Postgraduate Medical School. Thirteen samples were biopsies from patients undergoing abdominal surgery, and 28 samples were from organ transplant donors (Boobis et al. 1983, 1985). The use of human tissue for this purpose had local Research Ethical Committee approval. Male Wistar rats (Olac), untreated and treated with 3-methylcholanthrene (MC), 80 mg/kg in corn oil by i.p. injection 48h before sacrifice, were used. Microsomal fractions were prepared from liver, kidney, lung and proximal small intestine from rats and from human liver samples by differential centrifugation as previously described (Boobis et al. 1983).

Isozymes and antibodies

Cytochromes P450c and P450d, and NADPH-cytochrome c reductase were purified as previously described, from livers of rats pretreated with MC (Sesardic et al. 1986, Edwards et al. in preparation). Monoclonal antibodies (MAb's) to rat cytochrome P450c, 3/4/2 (Boobis et al. 1985b), C8 and C9 (Thomas et al. 1984), and a one specific to both P450c and d, 12/2 (Sesardic et al. 1986), were prepared and purified as previously described. A specific polyclonal antibody, αP450d(-c), was as previously described (Reik et al. 1982).

Immunological and metabolic assays

Western blotting was performed as previously reported (Sesardic et al. 1986) and immunoquantification of human cytochrome P-450d was achieved by computerised integration of the peaks obtained by densitometric scanning of Western blots probed with MAb 3/4/2. Phenacetin O-deethylase (POD) activity was determined using the method of Murray and Boobis (1986). In reconstitution experiments, metabolic activity was determined using saturating amounts of reductase and phosphatidylcholine,which were preincubated with cytochrome P450 (20 pmol), prior to addition of NADPH. In immunoinhibition studies, purified Ig fraction of an antibody was incubated with either purified cytochrome P450 or microsomal fraction before addition of cofactor. The reaction was started by the addition of substrate.

RESULTS AND DISCUSSION

The O-deethylation of phenacetin has previously been shown to exhibit biphasic kinetics with both rat and human liver microsomes (Boobis et al. 1981). Enzyme velocity is best described as the sum of two components, one with high affinity, (K_m = 4-6 μM) and one with low affinity (K_m = approximately 400 μM). At substrate concentrations likely to be encountered in vivo, only the high affinity component of activity would be expected to make the significant contribution to POD activity. Thus, only high affinity POD was studied, by using a low substrate concentration.

<u>Phenacetin O-deethylase activity of rat tissues</u>

Pretreatment of rats with polycyclic aromatic hydrocarbons (PAH), such as (MC), resulted in induction of POD activity in both hepatic and extrahepatic tissues. Full kinetics were determined for the activity in the liver. This confirmed (Kahn et al. 1987) that only the high affinity component of POD activity is induced in rat liver by MC, by 25-fold, demonstrating the involvement of PAH-inducible isozymes (P450c and/or P450d). Kinetic studies with P450c and P450d showed that the two isozymes catalyzed high affinity POD activity, with P450d slightly more active (Vmax = 1293 ± 156 pmol/nmol/min and K_m = 2.49 ± 0.72 μM) than P450c (V_{max} = 859 ± 144 pmol/nmol/min and K_m = 5.00 ± 1.62 μM). Given these kinetic constants, POD activity of microsomes could be accounted for by either of isozyme, particularly after MC treatment, where the levels of P450c exceed the levels of P450d by at least 3-fold.

A panel of antibodies was analysed for their ability to inhibit POD activity catalyzed by reconstituted cytochromes P450c and P450d (Table 1.).

Table 1. Effect of antibodies on phenacetin O-deethylase activity of reconstituted rat cytochromes P450c and P450d.

	Antibody			
Isozyme	12/2	C8	αP450d(-c)	C9
P450c	12.9	4.2	125.0	99.7
P450d	17.1	100.0	7.0	105.0

Results are expressed as percent of control activity, with 10 μM phenacetin, which was 568 pmol/nmol/min for P450c and 1290 pmol/nmol/min for P450d.

MAb 12/2 was equally effective at inhibiting POD activity of reconstituted forms c and d, MAb C8 inhibited only P450c, and a monospecific polyclonal antibody αP450d(-c), inhibited only P450d. MAb C9 was non-inhibitory.

The specific inhibitory antibodies to cytochromes P450c and P450d were than used to determine the contribution of these isozymes to the POD activity of liver microsomes from untreated rats (Fig. 1A) and from MC-treated rats (Fig. 1B).

Contrary to the expectation, cytochrome P450c appear to make no contribution to hepatic POD activity. MAb C8, although capable of inhibiting reconstituted P450c activity (Table 1) and microsomal aryl hydrocarbon hydroxylase activity (Thomas et al. 1984) was completely without effect

Figure 1. Effect of inhibitory antibodies on high affinity POD activity of rat liver.

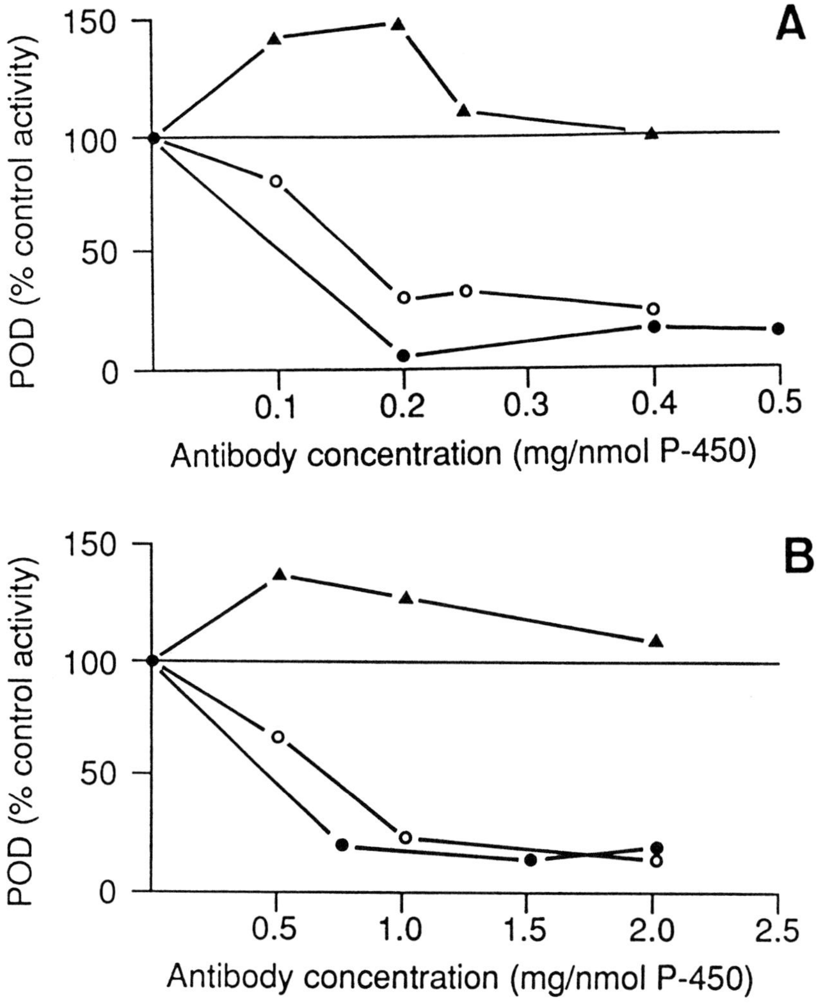

Immunoinhibition was performed with antibodies specific to P450c, C8, (▲), P450d, αP450d(-c), (●) or both isozymes, 12/2, (○). The concentrations of αP450d(-c) were ten times higher than those shown for the MAb's. Results are expressed as percent of control activity, with 4 μM phenacetin, which was 20.3 pmol/mg/min for microsomes from untreated rats and 1250 pmol/mg/min for MC-treated rats.

against microsomal POD activity. An antibody specific to P450d and MAb 12/2, both inhibited completely, and to the same extent, all of POD activity attributable to the high affinity component, in liver microsomes from both untreated and MC-treated rats. Thus, only cytochrome P450d is involved in the O-deethylation of phenacetin in the livers of both untreated and MC-treated rats.

Can we extrapolate these data to other tissues? Although cytochrome P450d is not inducible by PAH in extrahepatic tissues of rats (Boobis et al.1987) there is clearly induction of POD activity. The inhibitory antibodies were used to determine the contribution of PAH-inducible forms of cytochrome P450 to extrahepatic POD activity (Table. 2).

Table 2. Effect of inhibitory antibodies on microsomal POD activity of extrahepatic tissues in the rat.

Tissue	Treatment	POD activity (pmol/mg/min)	Antibody 12/2	C8	αd(-c)
Lung	None	4.1±0.8	118±19	99±10	102±7
Lung	MC	9.3±1.2	49±27	54±5.0	100±10
Kidney	None	<0.2	-	-	-
Kidney	MC	10.9±2.5	28.2±0.3	16.6±2.3	100±10
Gut	None	<0.2	-	-	-
Gut	MC	15.4±4.5	31.5±2.9	25.6±6.4	102±7

The results have been expressed as percent of control activity, with 20 μM phenacetin (data also shown) and are means ± SEM from three different pools, each of tissues from three animals.

Thus, it appears that isozyme specificity in one tissue cannot be extrapolated to that in another even in the same species. From the studies with inhibitory antibodies, it is apparent that, following treatment with MC, which increases the P450c content of all of the extrahepatic tissues studied (data not shown), microsomal POD activity is predominantly catalyzed by P450c. POD activity was not inhibited by an antibody to P450d, but only by antibodies to P450c.

Phenacetin O-deethylase activity of human liver

A non-inhibitory MAb (3/4/2), was raised against rat P450c, which recognises PAH-inducible forms of cytochrome

P450 in other species including man (Sesardic et al. 1986). The binding site on rat P450c for this antibody has been identified (Edwards et al.1987) and on the basis of the published primary amino acid sequences for the human isozymes (Quattrochi et al. 1986, Jaiswal et al. 1987) it was predicted that MAb 3/4/2 would bind with equal affinity to the orthologue of both P450c and P450d in man.

This antibody was used to probe 41 different human liver samples by Western blotting. In all samples except one (Fig.2. lane 15) a single immunoreactive protein was detected, at Mr 54 KDa, comigrating with rat P450d. The concentration of this orthologue of rat P450d varied from < 0.53 to 33.5 pmol/mg of microsomal protein, a range of $>$ 60-fold. The content of the human orthologue to cytochrome P450c was below 0.5 pmol/mg, the limit of detection for the assay. This compares with the levels of P450c in livers of untreated rats (Sesardic et al. 1986). A Western blot for 16 of the samples is shown in Figure 2, with those at the extremes of the range in lanes 2 and 15.

Figure 2. Western blot with MAb 3/4/2 of microsomal proteins from samples of human liver.

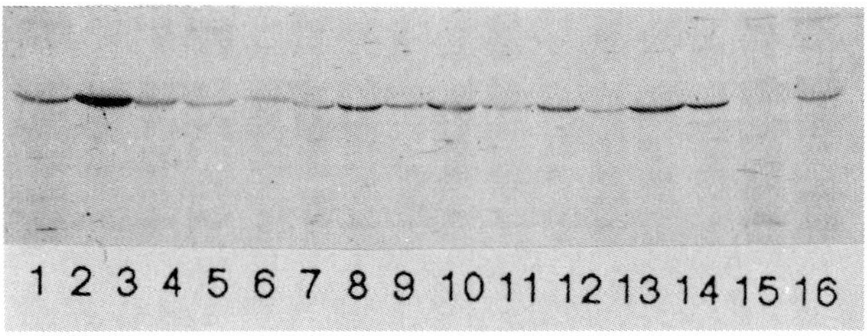

Thus, on the basis of its relative molecular weight, the relative amount of the isozyme and the fact that the epitope for MAb 3/4/2 on the two human isozymes is identical, it seems likely that the form of cytochrome P450 recognised by this antibody in human liver is the orthologue of rat form d and not form c. This would agree with data of both Wrighton et al. (1986) and Jaiswal et al. (1987) who could identify only the orthologue of rat form d in human livers.

Microsomal fractions from 41 different human liver samples were analysed for POD activity <u>in vitro</u>, using 4 μM phenacetin. There was a highly significant correlation between POD activity and the content of immunoreactive cytochrome P450, determined by Western blotting, $r_s = 0.874$ ($P < 0.001$) for the 28 transplant samples and $r_s = 0.846$ ($P < 0.001$) for the 13 biopsy samples. There was also a highly significant correlation when the samples were analysed together ($n=41$, $r_s = 0.706$, $P < 0.001$). Thus it seems likely that the form of cytochrome P450 orthologous to the rat P450d, also metabolises phenacetin in man.

Liver biopsy samples from known smokers (n=7) and non-smokers (n=8) were subjected to Western blotting, to quantify the immunoreactive cytochrome P450. The amount of this isoenzyme in liver samples from smokers was 3.5-fold higher ($P < 0.01$) than in those from non-smokers (Figure 3). High affinity POD activity was also determined in the same liver samples. This activity was also significantly higher (4.3-fold) in cigarette smokers ($P < 0.01$).

Figure. 3. Immunoqantification of the human orthologue of cytochrome P450d in hepatic biopsies from smokers (n=7) and non-smokers (n=8).

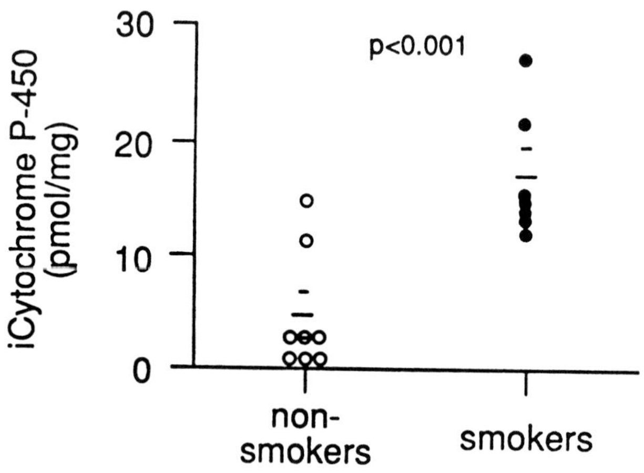

Thus it would appear that an isozyme of cytochrome P450 in man, which is orthologous to P450d in the rat, catalyses the O-deethylation of phenacetin and it is inducible by cigarette smoking. Based on this and additional data not shown here it is possible that this is the only isozyme catalyzing this reaction in man.

SUMMARY

Although both rat P450c and P450d are competent in supporting POD activity in reconstituted systems, in the liver of MC-treated rats, only P450d catalyses this reaction despite the presence of both isozymes. In extrahepatic tissues, where only P450c is induced by MC, this isozyme is responsible for POD activity. Thus, using phenacetin as a model substrate, it is apparent that there are serious limitations in extrapolating isozyme specificity obtained in reconstituted systems to microsomal metabolism. Nor is it possible to extrapolate isozyme specificity determined in one tissue to that of another, even in the same species.

Isozyme specificity must be defined with the species, tissue and substrate of interest, most conveniently achieved by the use of specific inhibitory antibodies.

ACKNOWLEDGEMENT
This work was supported by grants from the Medical Research Council. We are grateful to W. Levin and P.E. Thomas of Hoffmann La-Roche Inc., Nutley, New Jersey USA for supplying some of the antibodies used in this work.

REFERENCES

Boobis, A.R., Kahn, G.C., White, C., Brodie, M.J. & Davies, D.S., 1981, Biochemical Pharmacology, 30, 2451.
Boobis, A.R., Murray, S., Kahn, G.C., Robertz, G.-C. & Davies, D.S., 1983, Molecular Pharmacology, 23, 474.
Boobis, A.R., Hampden, C.E., Murray, S., Beaune, P. & Davies, D.S., 1985a, Biochemical Pharmacology, 19, 721.
Boobis, A.R., McQuade, J., Sesardic, D., Robson, R.T., Hayward, R.T., Lock, A.E., Elcombe, C.R., Rose, M.S. & Davies, D.S., 1985b, Biochemical Pharmacology, 34, 1671.
Edwards, R.J., Sesardic, D., Boobis, A.R. & Davies, D.S., 1987, Biochemical Society Transactions, 15, 455
Jaiswal, A.K., Nebert, D.W., McBride, O.W. & Gonzalez, F.J., 1987, Journal of Experimental Pathology, 3, 1.
Kahn, G.C., Rubenfield, M., Davies, D.S. & Boobis, A.R., 1987, Xenobiotica, 17, 179.
Murray, S. & Boobis, A.R., 1986, Biomedical and Environmental Mass Spectrometry, 1986, 91.
Quattrochi, L.A., Pendurthi, U.R., Okino, S.T., Potenza, C. & Tukey, R.H., 1986, Proceedings of National Academy of Science USA, 6731.
Reik, L.M., Levin, W., Ryan, D.E. & Thomas, P.E., 1982, Journal of Biological Chemistry, 257, 3950.
Sesardic, D., Boobis, A.R., McQuade, J., Baker, S., Lock, E.A., Elcombe, C.R., Robson, R.T., Hayward, C. & Davies, D.S., 1986, Biochemical Journal, 236, 567.
Thomas, P.E., Reik, L.M., Ryan, D.E. & Levin, W., 984, Journal of Biological Chemistry, 259, 3890.
Wrighton, S.A., Campanile, C., Thomas, P.E., Maines, S.A., Watkins, P.B., Parker, G., Mendez-Picon, G., Haniu, M., Shively, J.E., Levin, W. & Guzelian, P.S., 1986, Molecular Pharmacology, 29, 405.

UDP-GLUCURONOSYLTRANSFERASE

Purification and Characterization of Mammalian UDP-Glucuronosyltransferases

T.R. TEPHLY, B.L. COFFMAN, C.N. FALANY, M.D. GREEN, Y. IRSHAID, J.F. PUIG, S.A. KNAPP and J. BARON

Department of Pharmacology, The University of Iowa, Iowa City, IA 52242, USA

The formation of glucuronide conjugates represents an important process in mammalian organisms whereby numerous endogenous and exogenous compounds are converted to products of higher polarity than the parent substrate. The glucuronides formed are more readily excreted in the urine or bile than are the aglycone substrates. Glucuronidation occurs in many tissues and is catalyzed by a family of isozymes located in the endoplasmic reticulum and named collectively the UDP-glucuronosyltransferases (UDPGTs). Recently, a great deal of attention has been focused on the resolution, purification and characterization of UDPGTs from livers of laboratory animals and humans. Results of studies conducted by a number of laboratories have shown that marked heterogeneity exists with respect to the properties of these proteins in lower animals (Tukey & Tephly, 1981; Burchell, 1981; Bock et al. 1979; Falany & Tephly, 1983; Falany et al., 1986; Hochman & Zakim, 1983; Mackenzie et al., 1985; Roy Chowdhury et al., 1986) and in humans (Irshaid & Tephly, 1987).

Table 1 summarizes information regarding the minimum number of UDPGTs that are present in livers of animals or humans and which appear to have clearly discernible substrate specificities, kinetic properties, physical properties and/or inducibilities by various xenobiotics. In rat liver at least eight separate and distinct UDPGTs can be discerned. p-Nitrophenol UDPGT of rat liver is induced by polycyclic aromatic hydrocarbons such as 3-methylcholanthrene and reacts with a number of xenobiotics (Table 2). As yet, no endogenous substrate has been discovered for this enzyme. On the other hand, two other UDPGTs, 17β- and 3α-hydroxysteroid UDPGTs, which have been isolated to

Table 1. Hepatic UDP-Glucuronosyltransferases identified in various species

Rabbit Liver	
	p-nitrophenol UDPGT
	estrone UDPGT
	morphine UDPGT
Rat Liver	
	p-nitrophenol UDPGT
	17β-hydroxysteroid UDPGT
	3α-hydroxysteroid UDPGT
	morphine UDPGT
	digitoxigenin monodigitoxoside UDPGT
	estrone UDPGT
	bilirubin UDPGT
	4-hydroxybiphenyl UDPGT
Mouse Liver	
	pI 8.5 UDPGT
	pI 6.7 UDPGT
Pig Liver	
	p-nitrophenol UDPGT (2 forms) (estrone)
Human Liver	
	pI 7.4 UDPGT
	pI 6.2 UDPGT
	bilirubin UDPGT
	morphine UDPGT
	6-hydroxy bile acid UDPGT

homogeneity from rat liver react with both xenobiotics and steroid substrates. These UDPGTs are not inducible by 3-methylcholanthrene and do not react with 4-methylumbelliferone (Falany & Tephly, 1983; Green & Tephly, 1987; Kirkpatrick et al., 1984). The specific activity observed for the xenobiotic substrates is usually higher than that of endogenous substrates but turnover numbers favor endogenous substrates. Also, with steroid substrates, a high degree of glucuronidation stereo-specificity has been demonstrated, i.e., either a 17β- or 3α-glucuronidation of the hydroxyl moiety occurs when that group is present (Kirkpatrick et al., 1984; Falany et al., 1985). Estrone,

Table 2. Substrate specificity of rat liver p-nitrophenol UDPGT

Substrate	Relative Activity	Vm/Km[2]
p-nitrophenol	100	14,300
4-methylumbelliferone	60	-
1-naphthol	76	-
α-naphthylamine	10	1,050
β-naphthylamine	0.5	77
3-hydroxybenzo(a)pyrene	nd[1]	-
benzo(a)pyrene-3,6-quinol	26	-
benzo(a)pyrene-3,6-quinol monoglucoronide	3	-

[1]nd = not determined
[2]$nmol/\cdot min^{-1} \cdot mg^{-1}/mM$

Table 3. Substrate specificity of rat liver 17β- and 3α-hydroxysteroid UDPGTs

17β-Hydroxysteroid UDPGT	Relative Activity	Vm/Km[1]
testosterone	100	19,000
estradiol-17β	34	6,400
p-nitrophenol	850	825
1-naphthol	600	-
α-naphthylamine	900	2,440
β-naphthylamine	42	160
3α-Hydroxysteroid UDPGT	**Relative Activity**	**Vm/Km**
androsterone	100	56,000
etiocholanolone	150	284,000
3α-monohydroxylated C-24 bile acids	15-37	1,700-4,000
3α-dihydroxylated C-24 bile acids	2-10	82-250
α-naphthylamine	130	1,140
β-naphthylamine	16	83
4-aminobiphenyl	25	205

[1]$nmol/\cdot min^{-1} \cdot mg^{-1}/mM$

interestingly, which has a saturated steroidal A ring, does not react with either protein.

There are three UDPGTs that have been isolated and purified from rat liver microsomes which have quite narrow substrate specificities; namely morphine UDPGT, digitoxigenin monodigitoxiside UDPGT and bilirubin UDPGT (Puig & Tephly, 1987; Von Meyerinck et al., 1985; Burchell & Blanckaert, 1984). Other UDPGTs known to exist in rat liver microsomes but not fully characterized at this time are estrone, acetaminophen and 4-hydroxybiphenyl UDPGTs (Roy Chowdhury et al., 1986; Coffman et al., 1987; Puig & Tephly, 1987).

Many compounds have been shown to be glucuronidated by human liver microsomes (Bock et al., 1984). Recently, two human liver UDPGTs (Irshaid & Tephly, 1987) have been purified to homogeneity using chromatofocusing and affinity chromatography procedures. These proteins, referred to as pI 7.4 and pI 6.2 UDPGTs, have broad and overlapping substrate specificities (Table 4). The pI 7.4 isozyme does not react with either 4-aminobiphenyl or C-24 bile acids but reacts with estriol. The pI 6.2 isozyme does not react with estriol but reacts with 4-aminobiphenyl and C-24 bile acids.

Table 4. Substrate specificities of two human liver UDPGTs

pI 7.4 UDPGT	Relative Activity
p-nitrophenol	100
4-methylumbelliferone	98
α-naphthylamine	15
estriol	3
pI 6.2 UDPGT	**Relative Activity**
p-nitrophenol	100
4-methylumbelliferone	80
α-naphthylamine	9
4-aminobiphenyl	4
C-24 bile acids	2-3

Experimental evidence suggests that the glucuronidation of estriol by purified pI 7.4 UDPGT occurs at the 16-hydroxyl position. It is interesting to point out that neither of these forms reacts with estrone, 17β-estradiol, testosterone or androsterone (Irshaid & Tephly, 1987). Recent studies also show that neither protein reacts with morphine or 6-hydroxy bile acids (Radominska-Pyrek et al., 1986).

Localization of UDPGTs in Rat Liver.

Antibodies have been raised in rabbits against electroeluted, monomeric subunits of rat hepatic microsomal p-nitrophenol, 17β-hydroxysteroid and 3α-hydroxysteroid UDPGTs. Their antigen specificities were examined using purified enzymes as well as solubilized hepatic microsomal preparations obtained from untreated Sprague-Dawley and Wistar rats. Immunoblotting experiments showed that the anti-PNP-UDPGT recognizes only one polypeptide band in solubilized untreated hepatic microsomes corresponding to the PNP-UDPGT. This antibody does not recognize purified 17β- or 3α-hydroxysteroid UDPGTs. Antibodies to 17β- and 3α-hydroxysteroid UDPGTs do not recognize purified PNP-UDPGT or morphine UDPGT but do cross-react with the purified hydroxysteroid UDPGTs. In solubilized hepatic microsomal preparations, these antibodies also recognize more than one polypeptide band.

Utilizing immunohistochemical (i.e., avidin-biotin-peroxidase) staining techniques, we have investigated the intrahepatic localizations and intralobular distributions of p-nitrophenol, 17β-hydroxysteroid and 3α-hydroxysteroid UDPGTs in livers of untreated rats (Figure 1). Wistar rats were studied since this strain exhibits genetic variability in hepatic 3α-hydroxysteroid UDPGT activity: approximately 50% of Wistar rats possess high (HA) and 50% low (LA) glucuronidation activity towards androsterone and certain aromatic amines (Matsui & Hakozaki, 1979; Matsui & Watanabe, 1982; Green & Tephly, 1987) as a consequence of differences in the hepatic microsomal content of this UDPGT.

Immunoperoxidase staining for each UDPGT was readily detected within hepatocytes of both HA and LA Wistar rats. None of the UDPGTs appeared to be present within Kupffer or sinusoidal cells. Immunohistochemical observations indicate that antibodies raised against the three UDPGTs do not cross-react significantly in tissue sections and, further, that a number of marked differences exist among the three UDPGTs. (1) Centrilobular hepatocytes of HA and LA Wistar rats are stained much more intensely for p-nitrophenol UDPGT than are midzonal and periportal cells (Fig. 1A and D), whereas hepatocytes across the lobule are stained quite uniformly for 17β- and 3α-hydroxysteroid

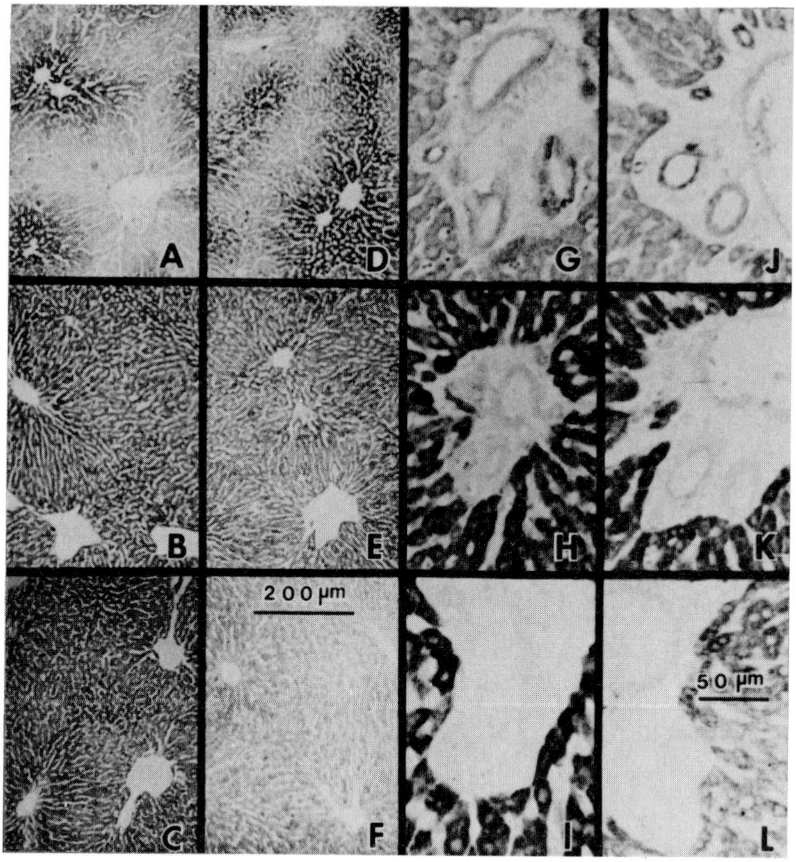

Figure 1. Immunohistochemical localization of p-nitrophenol [A, D, G and J] and 17β- [B, E, H and K] and 3α- [C, F, I and L] hydroxysteroid UDPGTs in livers of untreated HA [A-C and G-I] and LA [D-F and J-L] Wistar rats.

UDPGTs (Fig. 1 B, E, C and F). (2) Staining for p-nitro phenol UDPGT is also detected within bile duct epithelial cells and endothelial cells of the hepatic artery and portal vein (Fig. 1G and J). In contrast, staining for the 17β- and 3α-hydroxysteroid UDPGTs appears to be restricted to hepatocytes (Fig. 1H, K, I and L). (3) Consistent with results of biochemical analyses of hepatic microsomal

UDPGT activities, markedly less intense staining for 3α-hydroxysteroid UDPGT occurs in livers of LA Wistar rats in comparison with that seen in livers of HA Wistar rats (Fig. 1C and F), while qualitatively and quantitatively similar staining for the p-nitrophenol (Fig. 1A and D) and 17β-hydroxysteroid (Fig. 1B and E) UDPGTs is produced in livers of both HA and LA Wistar rats.

In summary, this paper has reviewed some of our knowledge on mammalian UDPGTs that have been isolated, purified and characterized. It is clear that numerous isozymes exist in the hepatic endoplasmic reticulum in all species studied so far. To understand how a given species glucuronidates chemicals one must consider the species under study and the characteristics of the specific UDPGTs involved. It has not been possible to extrapolate information from lower animals to humans with respect to the UDPGT proteins which have been isolated so far. Thus, in order to understand glucuronidation processes in humans, more work is needed on the study of human UDPGTs.

The immunohistochemical localization of UDPGTs is just beginning but, as antibodies are raised against purified UDPGTs, we will learn a great deal about their cellular levels and intercellular distributions in many organs where it is otherwise difficult to study these proteins.

(Supported by GM 26221 and GM 33253).

REFERENCES

Bock, K.W., Lilienblum, W. & Von Bahr, C., 1984, Drug Metabolism and Disposition, 12, 93.

Bock, K.W. Josting, D., Lilienblum, W. & Pfeil, H., 1979, European Journal of Biochemistry, 98, 19.

Burchell, B., 1981, Identification and purification of multiple forms of UDP-glucuronsyltransferase. In Reviews in Biochemical Toxicology, Vol. 3, edited by E. Hodgson, J.R. Bend & R.M. Philpot (Elsevier), p.l.

Burchell, B. & Blanckaert, N., 1984, Biochemical Journal, 223, 461.

Coffman, B.L., Green, M.D. & Tephly, T.R., 1987, The Pharmacologist, 29, 207.

Falany, C.N., Kirkpatrick, R.B. & Tephly, T.R., 1985, Comparison of rat and rabbit liver UDP-glucuronosyltransferase activities. In Advances in Glucuronide Conjugation, edited by S. Matern, K.W. Bock & W. Geork (Falk Symposium 40) (MTP Press), p. 41.

Falany, C.N. & Tephly, T.R., 1983, Archives of Biochemistry and Biophysics, 227, 248.

Falany, C.N., Green, M.D., Swain, E. & Tephly, T.R., 1986, Biochemical Journal, 238, 65.
Green, M.D. & Tephly, T.R., 1987, Cancer Research, 47, 2028.
Hochman, Y. & Zakim, D., 1983, Journal of Biological Chemistry, 256, 4143.
Irshaid, Y.M. & Tephly, T.R., 1987, Molecular Pharmacology, 31, 27.
Kirkpatrick, R.B., Falany, C.N. & Tephly, T.R., 1984, Journal of Biological Chemistry, 259, 6176.
Mackenzie, P.I., Joffe, M.M., Munson, P.J. & Owens, I.S., 1985, Biochemical Pharmacology, 34, 737.
Matsui, M. & Hakozaki, M., 1979, Biochemical Pharmacology, 28, 411.
Matsui, M. & Watanabe, H.K., 1982, Biochemical Journal, 202, 171.
Puig, J.F. & Tephly, T.R., 1986, Molecular Pharmacology, 30, 558.
Radominska-Pyrek, A., Zimniak, P., Irshaid, Y.M., Tephly, T.R. & Lester, R., 1986, Hepatology, 6, 1171.
Roy Chowdhury, J., Roy Chowdhury, N., Falany, C.N., Tephly, T.R. & Arias, I.M., 1986, Biochemical Journal, 233, 827.
Tukey, R.H. & Tephly, T.R., 1981, Archives of Biochemistry and Biophysics, 208, 565.
Von Meyerinck, L., Coffman, B.L., Green, M.D., Kirkpatrick, R.B., Schmoldt, A. & Tephly, T.R., 1985, Drug Metabolism and Disposition 13, 700.

Multiplicity and Structure of UDP-Glucuronosyltransferases as Revealed by Gene Cloning

PETER I. MACKENZIE and SAIKH J. HAQUE

aboratory of Developmental Pharmacology, National Institute of Child Health and Human Development, National Institutes of Health, Bethesda, MD 20892, USA

INTRODUCTION

The glucuronidation of thousands of lipophilic endogenous and foreign chemicals to hydrophilic conjugates is catalyzed by a family of UDP glucuronosyltransferases (UDPGT). The total number of isozymes in this family is unknown but evidence from chromatofocusing, purification and immunochemical studies suggests a minimum of eleven forms (Table 1). However, a definitive characterization of these forms in terms of their primary structures and substrate specificities has proved elusive due to the problems inherent in purifying catalytically-active, membrane-bound enzymes. We have therefore sought to circumvent problems of purity and optimalization of activities by the cloning of cDNAs to individual forms and their transfection into cultured mammalian cells, which provide a native membrane environment for expression of the cDNA-encoded enzyme. Data from these investigations will be summarized in this report.

CHARACTERIZATION OF UDPGT cDNAs

Our strategy was to synthesize and isolate cDNAs encoding several forms of UDPGT and to characterize them in terms of their amino acid sequences and substrate specificities. The methods to accomplish this goal have been described (Mackenzie 1986a,b, 1987). In brief, cDNA was synthesized from UDPGT-enriched polysomal poly(A) RNA and cloned into the SV40 expression vector pcD (Okayama & Berg 1983). Clone banks were prepared and UDPGT clones were selected by differential colony hybridization with cDNA prepared from UDPGT-enriched and unenriched poly(A) RNA.

Table 1. Forms of rat liver UDP glucuronosyltransferase

Form	MW	Inducer	Substrate	Glyco-protein	Clone name
GT_1A	56K	3MC	4-Nitrophenol 1-Naphthol	Yes	4NPGT
GT_1B	54K	3MC	1-Naphthol	-	-
GT_1C	61K	3MC	-	-	-
GT_2A	53K	Ø	Testosterone 4OH biphenyl	Yes	$UDPGT_r$-2
GT_2B	56K	Ø	Morphine	-	-
GT_3A	50	-	Testosterone	No	$UDPGT_r$-3 RLUG38
GT_3B	50	-	Testosterone	-	$UDPGT_r$-5
GT_4	53	-	Androsterone Bile acids	Yes	$UDPGT_r$-4 RLUG23
GT_5	--	Arochlor	Estrone	-	-
GT_6	53	Clofibrate	Bilirubin	-	-
GT_7	--	PCN	Digitoxigenin Monodigitoxide	-	-

Source: Bock *et al.* 1979; Matern *et al.* 1982; Falany & Tephly 1983; Burchell & Blanckaert 1984; Jackson & Burchell 1986; Matsui & Nagai 1986; Puig & Tephly 1986; Roy Chowdhury *et al.* 1986a,b, 1987; Coffman *et al.* 1987; Harding *et al.* 1987; Siest *et al.* 1987; von Meyerinck *et al.* 1985.

Plasmids with inserts greater than 1.6 Kbp were purified and transfected into COS cells, where the presence of T-antigen permitted extensive replication of the transfected plasmid. After 70 h, the cells were homogenized and assayed with a battery of aglycone substrates or nascent UDPGT was labelled with ^{35}S-L-methionine, immunoadsorbed and analyzed by SDS-PAGE. The results are shown in Fig. 1.

Fig. 1. Expression of cDNA-encoded UDP glucuronosyltransferases in cell culture. UDPGT-encoding expression plasmids containing Xho I inserts larger than 1.6 Kbp were transfected into COS cells and nascent UDPGT was labelled with S^{35}-L-methionine, 70 h later. The labelled enzyme was immunoadsorbed and analyzed on 7.5% SDS-PAGE gels. The M_rs of labelled polypeptides were 53K (●), 50K (O), 45K (▲) and 33K (Δ).

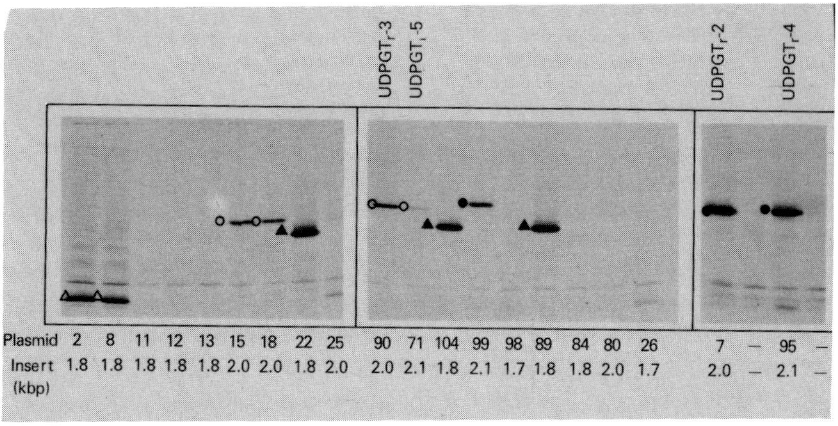

Several plasmids directed the synthesis of polypeptides that were recognized by UDPGT antibodies. As purified isozymes have molecular weights greater than 50 K daltons (see Table 1), only those cDNAs which expressed the larger polypeptides in transfect cells were characterized further.

UDPGT$_r$-2

The rat liver cDNA designated UDPGT$_r$-2, is 1961 nucleotides in length and encodes a protein of 529 residues. The protein contains a carboxy-terminal transmembrane segment and a leader sequence which is cleaved when synthesized in vitro in the presence of dog pancreatic microsomes (Mackenzie 1986a). The protein contains endoglycosidase H-sensitive glycosyl moieties, which is consistent with the presence of two potential N-linked glycosylation sites (Asn-x-$^{Ser}/_{Thr}$). The mRNA counterpart of UDPGT$_r$-2 is 2.3 Kb in length and is increased 5-fold by phenobarbital. The level of this mRNA is not significantly affected by 3-methylcholanthrene (3MC) or such inducers of drug-metabolizing enzymes as pyrazole, methylpyrazole, imidazole, rifampicin, dexamethasone or pregnenolone-16α-carbonitrile (unpublished data). The mRNA is elevated just after birth and is found predominantly in the liver.

$UDPGT_r$-2 when synthesized in COS cells glucuronidated the exogenous substrates, 4-hydroxybiphenyl, 4-methylumbelliferone and chloramphenicol. Of the endogenous substrates tested, $UDPGT_r$-2 was most active towards testosterone, but also displayed some activity towards dihydrotestosterone, β-estradiol and etiocholanolone (Fig. 2;

Fig. 2. Activities of $UDPGT_r$-2 expressed in transfected cells. $UDPGT_r$-2 cDNA was transfected into COS cells and assayed with ^{14}C-UDPGlcUA and the indicated aglycones 70 h later. The same cell homogenate was used for all assays. The radioactivities in the glucuronide-containing areas of the plates, corrected for those of the unknown glucuronide (arrowed) are graphed above the corresponding autoradiograph.

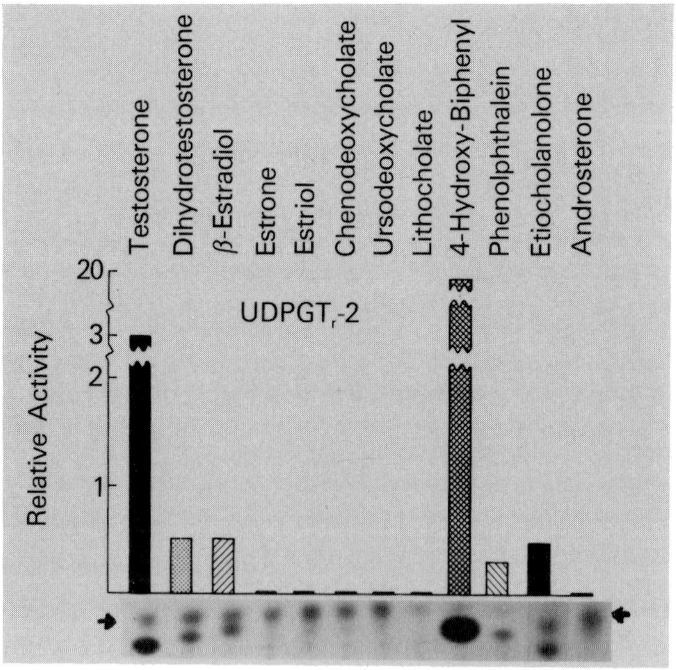

Mackenzie 1986a, 1987). It was also active toward an unknown, hydrophobic acceptor which was present in the cell homogenates (Fig. 2, arrowed) and in rat liver microsomal preparations (Mackenzie 1986a, 1987).

UDPGT$_r$-3

The cDNA designated UDPGT$_r$-3 is 1818 bp in length and encodes a 530 residue UDPGT which is 68% similar in amino acid sequence to UDPGT$_r$-2. In common with the latter protein, UDPGT$_r$-3 contains a cleavable leader peptide and a carboxy-terminal transmembrane region, but there are no potential asparagine-linked glycosylation sites (Mackenzie 1987). Dramatic alterations in the levels of its 2.3 Kb mRNA counterpart were not observed in rat livers treated with phenobarbital, 3MC or other foreign chemicals. The expression of UDPGT$_r$-3 is liver specific and is increased after birth (Mackenzie *et al*. 1984; Mackenzie 1987, unpublished data).

UDPGT$_r$-3 is active in the glucuronidation of testosterone, dihydrotestosterone and β-estradiol but, in contrast to UDPGT$_r$-2, has relatively little activity towards foreign chemicals (Fig. 3; Mackenzie 1987b).

Fig. 3. Activities of UDPGT$_r$-3 expressed in transfected cells. See Fig. 2 caption for details.

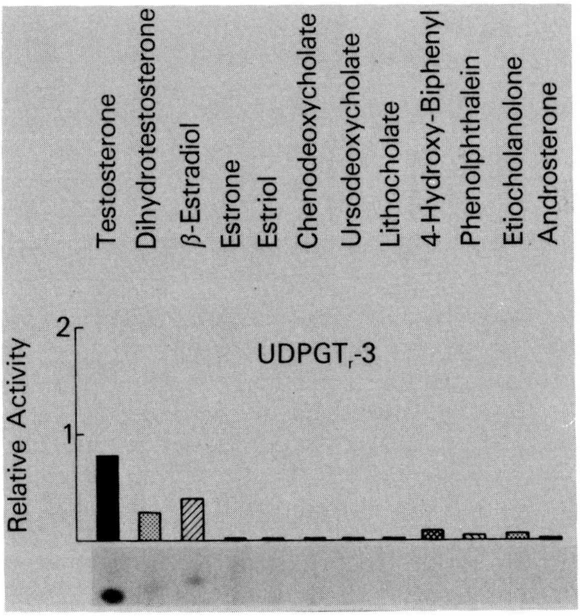

UDPGT$_r$-4

The cDNA of UDPGT$_r$-4 is 1970 bp long and encodes a protein of 530 amino acids which is 68% similar to UDPGT$_r$-2 and 85% similar to UDPGT$_r$-3 (Mackenzie 1986b). The encoded protein has a leader peptide and carboxy-terminal hydrophobic segment characteristic of other UDPGTs. This form contains one N-linked glycosylation site which is glycosylated when the enzyme is synthesized in COS cells. Similar to UDPGT$_r$-3, its mRNA counterpart is 2.3 Kb in length, is not significantly affected by 3MC or phenobarbital, is highest in the liver and appears postnatally (Mackenzie et al 1984; unpublished data).

In contrast to UDPGT$_r$-2 and 3, this enzyme has highest activity towards the endogenous substrates, etiocholanolone and androsterone. Activity towards the bile acids, lithocholate and ursodeoxycholate and the steroids, testosterone, dihydrotestosterone and β-estradiol is also detectable. The observation that this androsterone UDPGT also glucuronidates testosterone, is in agreement with the work of Matsui & Nagai (1986). The most preferred exogenous substrate of those tested is phenolphthalein (Fig. 4; Mackenzie 1986b), which explains why the glucuronidation of this aglycone is only detected in liver (Borghoff & Birnbaum 1985).

UDPGT$_r$-5

The UDPGT$_r$-5 cDNA is 1859 bp long and encodes a 530 residue protein which is 93% similar to UDPGT$_r$-3. The temporal and tissue-specific expression of this UDPGT and its regulation by foreign chemicals appears similar to that of UDPGT$_r$-3. It has slight activity towards testosterone and dihydrotestosterone and may therefore be a low activity variant of UDPGT$_r$-3 (unpublished data). Alternatively, the preferred substrate of this enzyme may not have been discovered.

CONCLUSION

The cDNAs encoding four forms of UDPGT have been cloned, sequenced and expressed in cultured mammalian cells. All forms glucuronidate steroids, are found predominantly in the liver and are expressed postnatally in rats. The mRNA level of the least homologous form, UDPGT$_r$-2, is elevated by phenobarbital, whereas this prototypic inducer has no effect on the mRNA levels of the more closely related UDPGT$_r$-3, 4 and 5. 3-Methylcholanthrene and some other foreign chemicals are without significant effect on all four forms. The possible identity of

Fig. 4. Activities of $UDPGT_r$-4 expressed in transfected cells. See Fig. 2 caption for details.

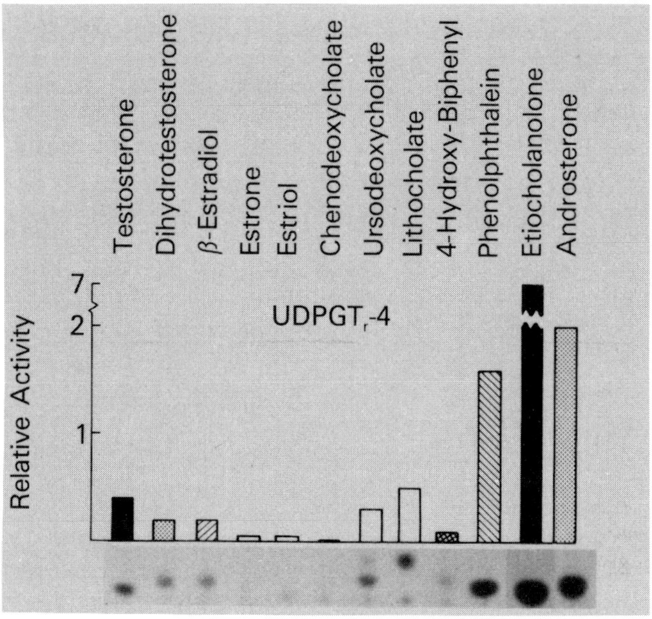

the forms encoded in these cDNAs to those purified or cloned in other laboratories is indicated in Table 1. These data demonstrate the utility of cDNA cloning in understanding the structural and catalytic relationships between various forms of UDPGT.

REFERENCES

Bock, K.W., Josting, D., Lilienblum, W. & Pfeil, H., 1979, European Journal of Biochemistry, 98, 19.

Borghoff, S.J. & Birnbaum, L.S., 1985, Drug Metabolism and Disposition, 13, 62.

Burchell, B. & Blanckaert, N., 1984, Biochemistry Journal, 223, 461.

Coffman, B.L., Green, M.D., Irshaid, Y.M. & Tephly, T.R., 1987, Federation Proceedings, 46, 863.

Falany, C.N. & Tephly, T.R., 1983, Archives of Biochemistry and Biophysics, 227, 248.

Harding, D., Wilson, S.M., Jackson, M.R., Burchell, B., Green, M.D. & Tephly, T.R., 1987, Nucleic Acids Research, 15, 3936.
Jackson, M.R. & Burchell, B., 1986, Nucleic Acids Research, 14, 779.
Mackenzie, P.I., 1986a, The Journal of Biological Chemistry, 261, 6119.
Mackenzie, P.I., 1986b, The Journal of Biological, Chemistry, 261, 14112.
Mackenzie, P.I., 1987, The Journal of Biological, Chemistry, 262, 9744.
Matern, H., Matern, S. & Gerok, W., 1982, The Journal of Biological Chemistry, 257, 7422.
Matsui, M. & Nagai, F., 1986, Biochemistry Journal, 234, 139.
Okayama, H. & Berg, P., 1983, Molecular and Cellular Biology, 3, 280.
Puig, J.F. & Tephly, T.R., 1986, Molecular Pharmacology, 30, 558.
Roy Chowdhury, N., Arias, I.M., Lederstein, M. & Roy Chowdhury, J., 1986, Hepatology, 6, 123.
Roy Chowdhury, N., Gross, F., Moscioni, A.D., Kram, M., Arias, I.M. & Roy Chowdhury, J., 1987, Journal of Clinical Investigation, 79, 327.
Roy Chowdhury, J., Roy Chowdhury, N., Falany, C.N., Tephly, T.R. & Arias, I.M., 1986, Biochemistry Journal, 233, 827.
Siest, G., Antoine, B., Fournel, S., Magdalou, J. & Thomassin, J., 1987, Biochemical Pharmacology, 36, 983.
von Meyerinck, L., Coffman, B.L., Green, M.D., Kirkpatrick, R.B., Schmoldt, A. & Tephly, T.R., 1985, Drug Metabolism and Disposition, 13, 700.

In Vitro Assessment of UDP-Glucuronosyltransferase Multiplicity and Substrate Specificity in Human Liver

J.O. MINERS, K.J. LILLYWHITE, A.P. MATTHEWS
and D.J. BIRKETT

Department of Clinical Pharmacology, Flinders Medical Centre,
Adelaide, South Australia 5042, Australia

INTRODUCTION

UDP-Glucuronosyltransferase (UDPGT) is responsible for the metabolism of a variety of drugs, drug metabolites, environmental chemicals and endogenous compounds, such as bilirubin, bile salts and steroid hormones. It is now well established that rat liver UDPGT exists as a family of differentially inducible isozymes which exhibit different but overlapping substrate specificities. Accumulating evidence from protein purification studies (Bock, et al., 1979; Falany et al., 1986; Puig & Tephly, 1986; Roy Chowdhury et al., 1986; Tephly et al., 1988) and from the isolation of cloned cDNAs to individual forms (Mackenzie & Haque, 1988) suggests the existence of at least eleven rat liver UDPGT isozymes. Despite the advances made in characterising multiple forms of UDPGT in laboratory animals, comparatively little is known about the multiplicity of this enzyme and the function of individual isozymes in humans. There is some in vitro evidence suggesting hetereogeneity of UDPGT in human liver (Bock et al., 1978; Bock et al., 1984; Dragacci et al., 1987; Pacifici et al., 1985) and two forms of UDPGT have recently been isolated from human hepatic tissue (Irshaid & Tephly, 1987). Studies in this laboratory have investigated whether substrates for specific rat liver UDPGT isozymes may be used to differentiate multiple forms of the enzyme in human liver. This communication describes the results of human liver microsomal kinetic and inhibitor studies with 1-naphthol (1NP) and 4-methylumbelliferone (4MU), substrates for the 3-methylcholanthrene inducible UDPGT isozymes(s) from rat liver, and with morphine, which is a substrate for the phenobarbitone UDPGT(s) from rat liver.

METHODS

Human liver samples were obtained from renal transplant donors; relevant details of the donors have been published elsewhere (Miners et al., 1987; Robson et al., 1987). Microsomal incubation conditions for 1NP, 4MU and morphine and procedures for the analysis of the glucuronides of these substrates have been published previously (Miners et al., 1987 & 1988). Kinetic parameters were obtained by computer-fitting data to the Michalis-Menten equation for a one- or two-enzyme system (Miners et al., 1988a & b). All results are expressed as mean ±S.D.

RESULTS AND DISCUSSION

1-Naphthol and 4-methylumbelliferone glucuronidation by human liver microsomes

Figure 1. Representative Eadie-Hofstee plots for 4MU glucuronidation by human liver microsomes (F6, activated) in the absence (open circles) and presence of 0.25 uM (open squares), 1.0 uM (closed circles) and 5.0 uM (closed squares) 1NP. Points are experimentally determined values while solid solid lines are the computer-generated curves of best fit.

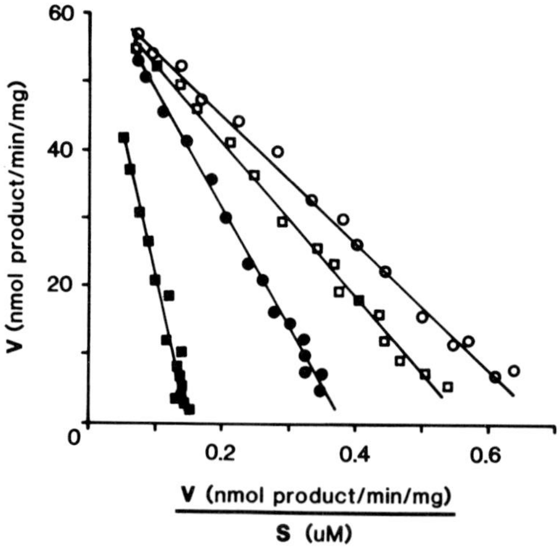

4MU glucuronidation followed Michaelis-Menten kinetics in both activated and native microsomes from all human livers studies (Fig.1) (Miners et al., 1987). The mean apparent K_m values in native and activated microsomes were 36.0 ± 9.5 uM and 91.8 ± 7.5 uM, respectively, while mean V_{max} values in native and activated microsomes were 14.9 ± 9.1 nmol/min/mg and 59.9 ± 24.4 nmol/min/mg, respectively.

In contrast to 4MU, 1NP exhibited biphasic glucuronidation kinetics (Fig.2) in both activated and native microsomes from all livers studied. Data for 1NP were best fitted to the Michaelis-Menten equation for a two enzyme system (Miners et al., 1987). For native microsomes the mean apparent K_m for the high affinity component of 1NP glucuronidation was 0.7 ± 0.3 uM while that for the low affinity component was 126.9 ± 21.5 uM. Detergent activation increased the apparent K_m of the high affinity component to 1.8 ± 0.1 uM but decreased the apparent K_m for the low affinity component to 87.2 ± 19.0 uM. Activation increased the V_{max} values for 1NP glucuronidation 3- to 4-fold; from 3.3 ± 1.8 to 12.8 ± 4.5 nmol/min/mg for the high affinity component and from 3.4 ± 1.6 to 10.2 ± 4.9 nmol/min/mg for the low affinity component.

Figure 2. Representative Eadie-Hofstee plots for 1NP glucuronidation by human liver microsomes (F6, activated) in the absence (panel A) and presence of 1,000 uM 4MU (panel B).

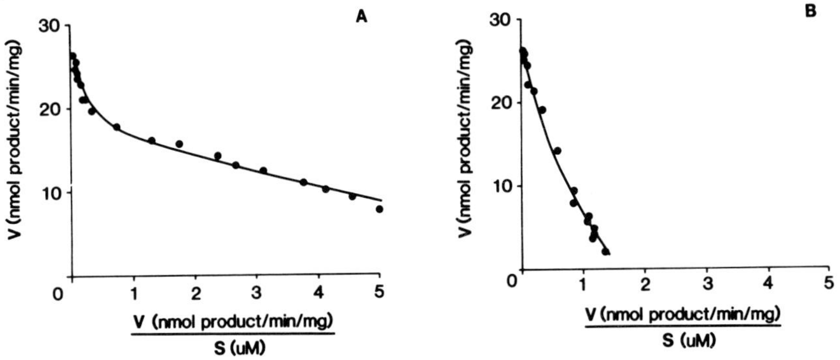

To determine the relationship between 4MU glucuronidation and the two phases of 1NP glucuronidation, cross inhibition studies with these substrates were performed (Miners et al., 1987). 1NP inhibited 4MU glucuronidation in a competitive manner (Fig. 1). 4MU selectively inhibited the high affinity component of 1NP glucuronidation but had no effect on the low affinity component (Fig. 2). Moreover, there was good agreement between estimates for the apparent K_m values for 4MU and the high affinity component of 1NP glucuronidation and their repective apparent K_i values determined in the cross-inhibition studies (Miners et al., 1987). These data suggest that the 4MU-UDPGT activity and the high affinity 1NP-UDPGT activity are the same enzyme form(s). The purification of two UDPGT isozymes from human liver has recently been reported (Irshaid & Tephly, 1987; Tephly et al., 1988). Both enzymes had similar K_m values for 4MU (viz. 170 uM and 230 uM) and would not have been differentiated in the microsomal kinetic studies described here. It is likely, but not certain, that the 4MU/high affinity 1NP-UDPGT activity characterised here is a composite of the two forms purified by Irshaid & Tephly. The low affinity 1NP-UDPGT activity characterised in the present work indicates the existence of at least one additional form of human liver UDPGT which glucuronidates 1NP.

A number of compounds known to be specific substrates for rat liver UDPGT isozymes were screened for their inhibitory effects on 4MU glucuronidation (Miners et al., 1987). Bilirubin, chloramphenicol, digitoxigenin monodigitoxoside, morphine, oestrone and testosterone were found not to inhibit 4MU glucuronidation, indicating that these substrates are metabolised by forms of human liver GT other than those involved in 4MU glucuronidation. 4-Nitrophenol and 2-aminophenol were competitive inhibitors of 4MU glucuronidation, suggesting a similar affinity of each of these compounds for the two putative isozymes involved in 4MU glucuronidation. Androsterone inhibited 4MU with atypical kinetics, perhaps indicating an allosteric effect. Overall, the substrate specificity of the 4MU/high affinity 1NP-UDPGT activity appears to be broadly comparable to that of the 3-methylcholanthrene inducible isozyme from rat liver, although some differences are apparent (Tephly et al., 1988).

A range of clinically used drugs known to undergo glucuronidation were also screened for inhibitory effects on 4MU glucuronidation in human liver microsomes. Results are summarised in Table 1 and indicate that the 4MU/high affinity 1NP-UDPGT activity may be associated with the metabolism of a limited number of drugs in man.

Table 1. Inhibitory effects of clinically used drugs on 4MU glucuronidation in human liver microsomes.

Inhibitory Drugs[a]	Non-inhibitory drugs
Salicylic acid	Paracetamol
Oxazepam	Salicylamide
Temazepam	Valproic acid
Lorazepam	Clofibric acid
Phenylbutazone	Probenecid
Sulphinpyrazone	Morphine
(+)/(-)-Fenoprofen	Chloramphenicol
	(+)/(-)-Naproxen
	(+)/(-)-Ibuprofen
	(±)-Ketoprofen
	(±)-Flurbiprofen

[a]4MU concentration 50 uM, inhibitor concentration 250 uM.

Morphine glucuronidation by human liver microsomes

Morphine glucuronidation exhibited biphasic glucuronidation kinetics in both native and detergent activated microsomes from human liver (Miners et al., 1988) and data were best fitted to the Michaelis-Menten equation for a two enzyme system. For native microsomes the mean appararent K_m of the high affinity component was 2.5 ± 0.6 uM and that of the low affinity component was 798 ± 123 uM. Detergent activation increased the apparent K_m values of both the low- and high-affinity components to 5.3 ± 1.6 uM and 1203 ± 289 uM, respectively. Similarly activation increased the V_{max} values for morphine glucuronidation approximately 3-fold; from 0.09 ± 0.02 to 0.30 ± 1.42 nmol/min/mg and from 3.63 ± 1.42 to 10.86 ± 1.64 nmol/min/mg for the low affinity component. Using mean kinetic data from native microsomes, the calculated contribution of the two morphine-UDPGT activities to morphine glucuronidation as a function of morphine concentration is shown in Fig. 3. For morphine concentrations below 5 uM the high affinity component accounts for 75% - 85% of morphine glucuronidation. Since plasma morphine concentrations rarely exceed 1.0 umol/l, even in cancer patients receiving high doses of the drug, the low affinity morphine-UDPGT activity would be expected to be of minimal importance in morphine glucuronidation _in vivo_.

Figure 3. Calculated contribution of the high- and low-affinity morphine-UDPGT activities towards morphine glucuronidation in native human liver microsomes as a function of morphine concentration.

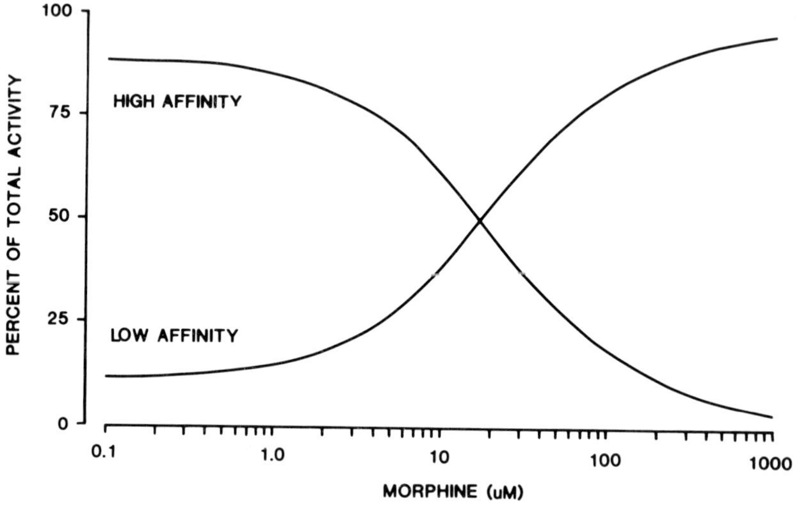

4MU, 1NP, 4-nitrophenol, chloramphenicol and 4-hydroxybiphenyl were screened for inhibitory effects on both the low and high affinity components of morphine glucuronidation. Each of these compounds were inhibitors of the low afinity component of morphine glucuronidation, with 4-hydroxybiphenyl being the most potent inhibitor (Miners et al., 1988). However, only 1NP and chloramphenicol were competitive inhibitors, the remaining compounds exhibiting mixed-type inhibition. 4MU, 1NP, 4-nitrophenol, chloramphenicol and 4-hydroxybiphenyl did not significantly inhibit the high affinity morphine-UDPGT activity.

Thus, both the high- and low-affinity morphine UDPGT-activities appear to be distinct to the 4MU/high affinity 1NP-UDPGT activity of human liver microsomes. Since the 4MU/high affinity 1NP-UDPGT activity apparently comprises two isozymes (Irshaid & Tephly, 1987), these data suggest the existence of at least four forms of UDPGT in human liver microsomes. The low affinity 1NP-UDPGT activity is also apparently distinct to the high affinity morphine UDPGT-activity but, since 1NP was a competitive inhibitor of the low affinity component of morphine glucuronidation, it is possible that the low affinity 1NP-UDPGT activity and the low affinity morphine-UDPGT activity represent the same enzyme form(s). The relationship between these two activities should become clearer as studies with purified human liver UDPGT isozymes progress.

ACKNOWLEDGEMENT

This work was supported by a grant from the National Health and Medical Research Council of Australia.

REFERENCES

Bock, K.W., Brunner, G., Hoensch, H., Huber, E. & Josting, D., 1978, European Journal of Clinical Pharmacology, 14, 367.

Bock, K.W., Josting, D., Lilienblum, W. & Pfeil, H., 1979, European Journal of Biochemistry, 98, 19.

Bock, K.W., Lilienblum, W. & von Bahr, C., 1984, Drug Metabolism and Disposition, 12, 93.

Dragacci, S., Thomassin, J., Souhali El Amri, H., Boissel, P. & Siest, G., 1987, European Journal of Clinical Pharmacology, 32, 485.

Falany, C.N., Green, M.D., Swain, E. & Tephly, T.R., 1986, Biochemical Journal, 238, 65.

Irshaid, Y.M. & Tephly, T.R., 1987, Molecular Pharmacology, 31, 27.

Mackenzie, P.I. & Haque, S.J., 1988, this volume.

Miners, J.O., Lillywhite, K.J., Matthews, A.P., Jones, M.E., Birkett, D.J., 1987, Biochemical Pharmacology, (in the press).

Miners, J.O., Lillywhite, K.J. & Birkett, D.J., 1988, Biochemical Pharmacology, (in the press).

Pacifici, G.M., Colizzi, C., Giuliani, L. & Rane, A., 1985. In Advances in Glucuronide Conjugation, edited by Matern, S., Bock, K.W. & Gerok, W., (MTP Press), p. 341.

Puig, J.F. & Tephly, T.R., 1986, Molecular Pharmacology, 30, 558.

Robson, R.A., Matthews, A.P., Miners, J.O., McManus, M.E., Meyer, U.A. & Birkett, D.J., 1987, British Journal of Clinical Pharmacology, 24, 293.

Roy Chowdhury, J., Roy Chowdhury, N., Falany, C.N., Tephly, T.R. & Arias, I.M., 1986, Biochemical Journal, 233, 1986.

Sawe, J., Pacifici, G.M., Kager, L., von Bahr, C. & Rane, A., 1982, Acta Anaesthesiologica Scandinavica, suppl. 74, 47.

Tephly, T.R., Coffman, B.L., Falany, C.N., Green, M.D., Irshaid, Y., Puig, J.F., Knapp, S.A. & Baron, J., 1988, this volume.

UDP-Glucuronosyltransferase: Adaptive Responses and Permanent Alterations in Preneoplastic Liver

K.W. BOCK, G. SCHIRMER and L.C. ERIKSON [1]

Department of Pharmacology and Toxicology, University of Gottingen, D-3400 Gottingen, FRG and [1] Department of Pathology, Huddinge Hospital, S-14186 Huddinge, Sweden

ABSTRACT

(1) Glucuronidation, catalyzed by an isozyme family of UDP-glucuronosyltransferases (GT), is a major pathway in the inactivation and elimination of bilirubin, steroids, drugs and carcinogens. A 3-methylcholanthrene(MC)-inducible phenol-GT (GT_{MC}) of rat liver is emphasized. (2) GT_{MC} with Mr 55 kDa mainly conjugates planar phenols and polyphenols, particularly of polycyclic aromatic hydrocarbons (PAHs). It appears to be co-induced with isozymes of P-450, probably as an adaptive program evolved to detoxify PAHs. (3) In contrast to P-450, GT_{MC} is permanently increased after initiation of hepatocarcinogenesis by a variety of carcinogens in early rat liver foci, liver nodules and differentiated hepatocellular carcinomas.

INTRODUCTION

Glucuronide formation is a major pathway in phase II of drug metabolism and an important detoxication reaction (Dutton, 1980). It is catalyzed by microsomal UDP-glucuronosyltransferase (GT, EC 2.4.1.17) which consists of a family of isozymes with different substrate specificities and monomeric molecular weights (Falany and Tephly, 1983), amino acid sequences (Jackson and Burchell, 1986; Mackenzie, 1986; Iyanagi et al., 1986) and different inducibilities by xenobiotics and hormones (Bock et al., 1979; Falany and Tephly, 1983; Schuetz et al., 1986). Some isozymes of rat liver appear to be more selective for bilirubin or steroids (Roy Chowdhury et al., 1986) while for others no typical endogenous substrates have yet been found. However, there is considerable overlap in substrate specificity between different isozymes. For example, testosterone-GT or 17ß-hydroxysteroid-GT efficiently conjugates planar phenols such as 4-nitrophenol and 1-naphthol (Falany and Tephly, 1983). This isozyme may be mainly responsible for conjugation of these phenols in untreated rat liver. In 3-methylcholanthrene(MC)-treated rats, on the other hand, phenol glucuronidation may be mainly carried out by GT_{MC} or GT_1, an isozyme which is scarce in untreated liver. In the following identification of this (these) inducible isozyme(s) by functional and molecular probes and its (their) permanent increase in preneoplastic rat liver is discussed.

METHODS

Rat liver GT_{MC} was purified as described by Bock et al. (1979) or by Falany and Tephly (1983) and antibodies were produced (Pfeil and Bock, 1982). Liver nodules and carcinomas were produced by intermittently feeding a diet containing 2-acetylaminofluorene (AAF) (Epstein et al., 1967) or by feeding N-nitrosomorpholine (160 mg/l drinking water) for 7 weeks (Moore et al., 1982). Benzo(a)pyrene-3,6-quinol glucuronidation was assayed by the method of Lilienblum et al. (1985).

RESULTS AND DISCUSSION

Adaptive increase of GT_{MC}

Stimulation of GT by MC is known since the report of Inscoe and Axelrod (1960). Differential induction of glucuronidation reactions by either phenobarbital or MC was found later and used as a major argument in favour of multiple isozymes of GT (Bock et al., 1973; Wishart, 1978). Phenobarbital- and MC-inducible GTs have been separated by anionic exchange chromatography and purified (Bock et al., 1979). GT_{MC} has also been isolated by chromatofocusing (Falany and Tephly, 1983). Exchange of GT_{MC} preparations and anti-GT_{MC} antibodies between laboratories revealed that the isolated isozymes with monomeric Mr of 55 kDa were very similar (Bock, Schirmer, Green, Tephly, unpublished). Recently cDNA clones coding for GT_{MC} have been obtained, allowing elucidation of the complete amino acid sequence (Iyanagi et al., 1986).

GT_{MC} accepts planar phenols as substrates such as 4-methylumbelliferone, 1-naphthol and 4-nitrophenol. However, the last two substrates are also efficiently conjugated by testosterone-GT. Due to this overlap in substrate specificity it is obviously difficult to identify GT_{MC}. Large PAH phenols appear to be more selectively metabolized Furthermore, regioselectivity of GT_{MC} for its substrates has also been noted. GT_{MC} recognizes the size and shape of PAH phenols. It accepts PAH phenols shorter than 1.3 nm when viewed from the hydroxyl group to the plane of the PAH molecule (Lilienblum et al., 1987). 'Longer' PAH phenols are mainly conjugated by phenobarbital-inducible isozymes. A particular substrate of GT_{MC} appears to be BP-3,6-quinol. Conjugation of this quinol proceeds in two steps in which the mono- and diglucuronides are formed. Formation of the two conjugates is strongly inducible by MC-treatment (10- and 40-fold; respectively; Lilienblum et al., 1985; Segura-Aquilar et al., 1986). Similar results are obtained at the cellular level. The marked difference in the inducibility of the two reactions suggests the existence of more than one MC-inducible GT. Conjugation of the quinol is of toxicological interest since this reaction prevents the formation of toxic semiquinones and of quinone/quinol redox-cycles with the production of reactive oxygen species (Lorentzen and Ts'o, 1977). GT_{MC} is probably co-induced with other drug metabolizing enzymes, for example isozymes of P-450, probably as an adaptive program evolved to detoxify PAHs.

Permanent increase of GT_{MC}

Increased GT_{MC} activities have been noted in rat liver nodules produced by a variety of carcinogens (Bock et al., 1982; Yin et al., 1982). Increased GT could also be demonstrated immunohistochemically in early and late liver foci (Fischer et al., 1985). A more detailed investigation was carried out recently with 4 liver nodules and 10 hepatocellular carcinomas using

Fig. 1. GT activities in rat liver nodules and hepatocellular carcinomas. On the abscissa identification numbers are listed. Enzyme activities were measured in microsomal fractions. Data represent the means of 2 to 4 determinations.

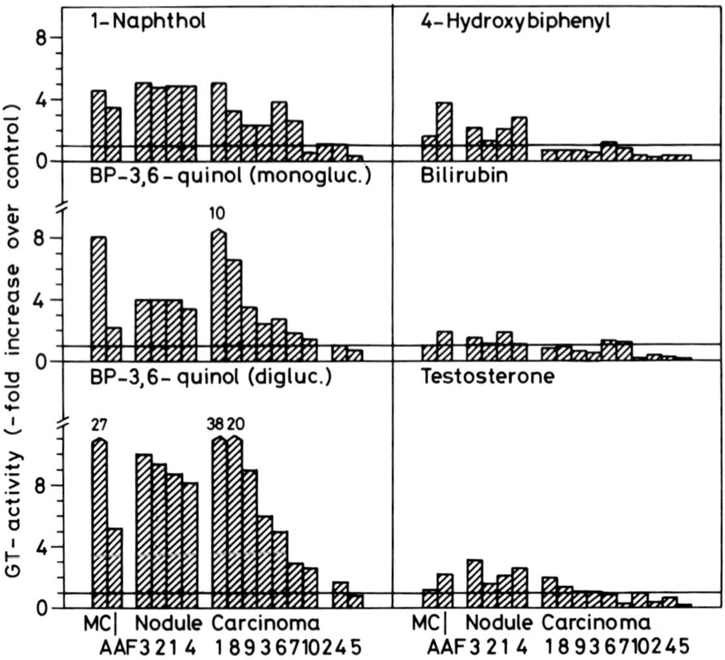

BP-3,6-quinol glucuronidation as a functional and immunoblot analysis as a molecular probe of GT_{MC}.

GT activities toward 1-naphthol and BP-3,6-quinol were markedly increased in liver nodules and some carcinomas (Fig. 1). In contrast, GT activities toward 4-hydroxybiphenyl, bilirubin and testosterone (substrates of other GT isozymes) were not elevated. As expected GT_{MC} activities in carcinomas were heterogeneous. They were highest in carcinoma 1 (classified histologically as well differentiated) and low in the poorly differentiated carcinomas 2 to 5 and 10. AAF-treatment for 7 days led to a moderate increase of GT_{MC} activities, but also of GT activity toward 4-hydroxybiphenyl (a substrate of phenobarbital-inducible GTs). It appears therefore that AAF may be a mixed-type inducer.

Fig 2. Immunoblot analysis of GT polypeptides of rat liver nodules and hepatocellular carcinomas using anti-GT_{MC} antibodies of broad specificity (A) and selective antibodies (B). Standard procedures were used for SDS-PAGE and blotting. GT polypeptides were visualized by sequential reactions with rabbit anti-GT antibodies and peroxidase conjugated porcine anti-rabbit antibodies. Tracks contained liver microsomes from untreated controls (C), MC- and AAF-treated rats, pooled nodules and selected carcinomas (Ca, 10 µg microsomal protein).

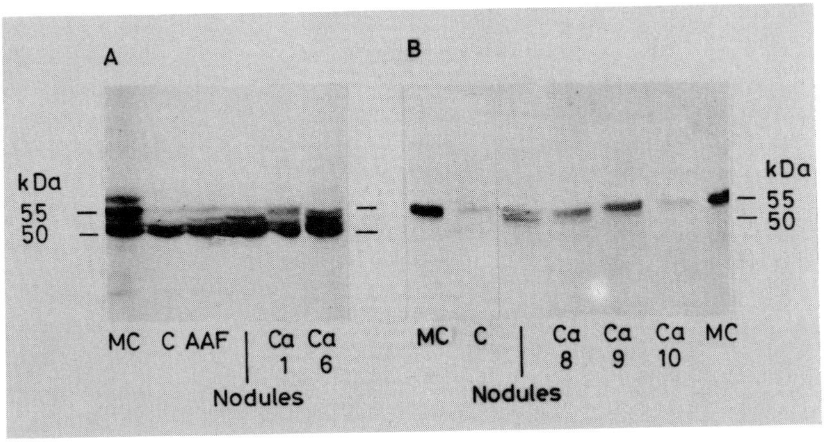

Production of selective antibodies to GT isozymes has proven to be very difficult. Progress has recently been made by using as antigens electroeluted proteins of electrophoretically separated GT polypeptides (Falany et al., 1986). Therefore in Fig. 2 immunoblots with two kinds of anti-GT_{MC} antibodies were compared: namely our anti-GT_{MC} antibodies of broader specificity (Fig. 2, A) and more selective antibodies, kindly provided by T.R. Tephly (Fig. 2, B). Only results with pooled nodules and some selected carcinomas are shown. The GT polypeptide at 55 kDa was clearly MC-inducible. It was moderately increased after short-term AAF-treatment and in liver nodules, and markedly increased in carcinomas 1, 8 and 9. Our antibodies recognized, in addition to GT_{MC}, testosterone-GT at 50 kDa and androsterone-GT at 52 kDa and an unknown 56 kDa polypeptide. These additional polypeptides were not recognized by the selective antibody. Interestingly, a moderately MC-inducible 53 kDa polypeptide was increased in liver after AAF-treatment, in nodules and in carcinoma 6. The 53 kDa polypeptide in nodules was also recognized by the more selective antibody and may be closely related to GT_{MC}. In fact it may be a second GT_{MC}, but its nature still has to be further explored.

Permanently increased GT_{MC} activity in liver nodules and carcinomas has to be distinguished from adaptive increases of this isozyme after treatment with MC-type inducers. The permanently increased GT activities can be further and reversibly enhanced in liver after treatment of nodule-bearing rats with inducers (Fischer et al., 1985). Hence adaptive and permanent changes can be superimposed. Increased GT_{MC} should lead to enhanced inactivation of many toxins. Together with permanent alterations of other drug metabolizing enzymes (for example decreased P-450) increased GT_{MC} often leads to enhanced toxin-resistance of initiated hepatocytes (Farber et al., 1984). This phenomenon may facilitate selective growth of initiated hepatocytes and may enhance the probability of multiple hits in their genome.

Future directions

Among many interesting points arising the following are stressed: (1) Isozymes similar to GT_{MC} are present in a number of extrahepatic tissues in the rat (Bock et al., 1980; Coughtrie et al., 1987). However, MC-inducibility and BP-3,6-quinol diglucuronide formation in relation to monoglucuronide formation is lower in kidney and small intestine than in liver (Koster et al., 1987). (2) MC-inducible phenol GT activities are present in other species such as mouse (Owens, 1977) and man (Bock et al., 1987). However, GT-inducibility appears to be much lower than in rat liver (Bock & Schirmer, 1987). Furthermore, mouse GT activities appear to be permanently decreased in liver foci and nodules (Stout & Becker, 1986; Fischer & Bock, unpublished). (3) The mechanism underlying the permanent increase of rat GT_{MC} in liver nodules and carcinomas remains to be elucidated. In the case of quinone reductase or DT diaphorase its DNA has been shown to be hypomethylated in rat liver nodules (Williams et al., 1986). Availability of cDNA clones for GT_{MC} (Iyanagi et al., 1986) may help to further our understanding.

ACKNOWLEDGEMENTS

The authors wish to thank Dr. Thomas R. Tephly, and Mitchell D. Green, University of Iowa, for providing antibodies to GT_{MC} and the Deutsche Forschungsgemeinschaft for financial support.

REFERENCES

Bock, K.W., Fröhling, W., Remmer, H. & Rexer, B., 1973, Biochimica Biophysica Acta, 327, 46.
Bock, K.W., Josting, D., Lilienblum, W. & Pfeil, H., 1979, European Journal of Biochemistry, 98, 19.
Bock, K.W., Clausbruch von, U.C., Kaufmann, R., Lilienblum, W., Oesch, F., Pfeil, H. & Platt, K.L., 1980, Biochemical Pharmacology, 29, 495.
Bock, K.W., Lilienblum, W., Pfeil, H. & Eriksson, L.C., 1982, Cancer Research, 42, 3747.
Bock, K.W., & Schirmer, G., 1987, Archives of Toxicology, Suppl. 10, 125.
Bock, K.W., Wiltfang, J., Blume, R., Ullrich, D. & Bircher, J., 1987, European Journal of Clinical Pharmacology, 31, 677.
Coughtrie, M.W.H., Burchell, B. & Bend, J.R., 1987, Biochemical Pharmacology, 36, 245.
Dutton, G.J., 1980, Glucuronidation of Drugs and other Compounds (CRC Press, Boca Raton, Florida.
Epstein, S., Ito, N., Markow, L. & Farber, E., 1967, Cancer Research, 27, 1702.
Falany, C.N. & Tephly, T.R., 1983, Archives of Biochemistry and Biophysics, 227, 248.
Falany, C.N., Green, M.D., Swain, E. & Tephly, T.R., 1986, Biochemical Journal, 238, 65.
Farber, E., 1984, Cancer Research, 44, 5463.
Fischer, G., Ullrich, D. & Bock, K.W., 1985, Carcinogenesis, 6, 605.
Inscoe, J.K. & Axelrod, J., 1960, Journal of Pharmacology and Experimental Therapeutics, 129, 128.

Iyanagi, T., Hanin, M., Sogawa, K., Fujii-Kuriyama, Y., Watanabe, S., Shively, J.E. & Anan, K.F., 1986, Journal of Biological Chemistry, 261, 15607.

Jackson, M.R. & Burchell, B., 1986, Nucleic Acids Research, 14, 779.

Koster, A. Sj., Schirmer, G. & Bock, K.W., 1986, Biochemical Pharmacology, 35, 3971.

Lilienblum, W., Bock-Hennig, B.S. & Bock, K.W., 1985, Molecular Pharmacology, 27, 451.

Lilienblum, W., Platt, K.L., Schirmer, G., Oesch, F. & Bock, K.W., 1987, Molecular Pharmacology, in the press.

Lorentzen, R.J. & Ts'o, P.O.P., 1977, Biochemistry, 16, 1467.

Mackenzie, P.I., 1986, Journal of Biological Chemistry, 261, 14112.

Moore, M.A., Mayer, D. & Bannasch, P., 1982, Carcinogenesis, 3, 1429.

Owens, I.S., 1977, Journal of Biological Chemistry, 252, 2827.

Pfeil, H. & Bock, K.W., 1983, European Journal of Biochemistry, 131, 619.

Roy Chowdhury, J., Roy Chowdhury, N., Falany, C.N., Tephly, T.R. & Arias, I.M., 1986, Biochemical Journal, 233, 827.

Schuetz, E.G., Hazelton, G.A., Hall, J., Watkins, P.B., Klaassen, C.D. & Guzelian, P.S., 1986, Journal of Biological Chemistry, 261, 8270.

Stout, D.L. & Becker, F.F., 1986, Cancer Research, 46, 2693.

Segura-Aquilar, J.W., Barreiro, V. & Lind, C., 1986 Archives of Biochemistry and Biophysics, 251, 266.

Williams, J.B., Lu, A.Y.H., Cameron, R.G. & Pickett, C.B., 1986, Journal of Biological Chemistry, 261, 5524.

Wishart, G.J., 1978, Biochemical Journal, 174,671.

Yin, Z., Sato, K., Tsuda, H. & Ito, N., 1982, Gann, 73, 239.

METABOLIC ACTIVATION : CELL TOXICITY AND CARCINOGENESIS

Functional Groups Activated via Flavin-Containing Monooxygenases

DANIEL M. ZIEGLER

Clayton Foundation Biochemical Institute
and The Department of Chemistry,
The University of Texas at Austin, TX 78712, USA

INTRODUCTION

The flavin-containing monooxygenase purified to homogeneity from porcine liver microsomes (Ziegler and Mitchell, 1972) catalyzes NADPH- and oxygen-dependent oxidation of a variety of functional groups bearing nucleophilic sulfur, selenium, nitrogen or phosphorus (Ziegler 1980, Tynes and Hodgson, 1985). Enzymes similar in composition and catalytic properties have also been isolated from mouse (Sabourin et al., 1984) and rat (Kimura et al., 1983) liver microsomes. More recently, a monooxygenase, similar in composition and mechanism to the liver enzyme, but with different immunologic and catalytic properties, has been isolated from rabbit lung microsomes (Williams et al., 1985, Tynes et al. 1985) and distinct tissue specific forms have also been detected in mice (Tynes and Hodgson, 1985). While oxidation of most functional groups catalyzed by these flavoproteins produces less reactive metabolites, some yield more reactive intermediates and the following sections will focus on the latter reactions that are catalyzed primarily by this enzyme in liver microsomes.

FUNCTIONAL GROUPS BEARING SULFUR

Thiols and Disulfides

As illustrated by representative examples in Table 1, alkyl or aryl thiols free from ionic groups, or ones containing only a single cationic group, are oxidized to disulfides through intermediate sulfenic acids.

While aminothiols are oxidized only to disulfides, thiols free from ionic groups are further oxidized to

TABLE 1

Thiols: Reaction Sequence and Kinetic Constants

REACTION

$$RSH \xrightarrow[O_2]{NADPH} RSOH \xrightarrow{RSH} RSSR + H_2O$$

EXAMPLES

Alkyl- and arylthiols	Km(μM)
Ethanethiol	58
Heptanethiol	15
Thiophenol	33
1,4-Butanedithiol	3
Dithiothreitol	470
Aminothiols	
Cysteamine	110
Piperazinylethanethiol	580
Guanidinoethanethiol	630

Km values based on kinetic studies with the purified porcine liver monooxygenase (Ziegler et al., 1980).

thiosulfenates. Although the latter reaction can be demonstrated in vitro, it is likely that in tissues, the intermediate sulfenic acid would react primarily with glutathione or protein thiols yielding the corresponding mixed disulfides (fig.1).

Extensive oxidation of tissue protein thiols in animals treated with cysteamine and other thiol radio-

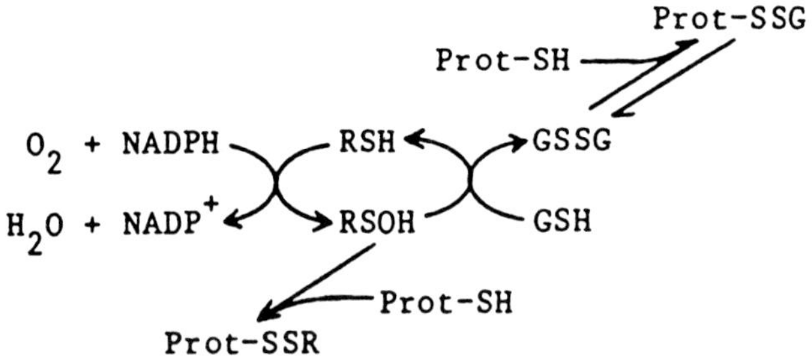

Figure 1. Oxidation of GSH and protein thiols by metabolic redox cycling of xenobiotic thiol substrates for the flavin-containing monooxygenase.

protective agents has been known for some time (Eldjarn and Nygaard, 1954). In addition, Lauterberg and Mitchell (1981) have shown that biliary efflux of GSSG increases almost 300 fold in rats treated with cysteamine. The oxidation of cellular thiols observed in vivo is apparently due to redox cycling of the nonphysiological thiols as illustrated in fig. 1. Since this futile redox cycle can be initiated by a variety of xenobiotic thiols, it is evident that cellular thiols will decrease, not increase, in animals treated with thiol substrates for the monooxygenase. Oxidation of cellular thiols by this mechanism undoubtedly contributes to the toxicity of foreign thiols.

Thiocarbamides and Related Compounds

The purified hog liver monooxygenase catalyzes the oxidation of thiocarbamides and mercaptoimidazoles to sulfinic acids through intermediate sulfenic acids as illustrated in Table 2. Both oxygenation reactions are fully enzymic and within the limits of detection the oxidation of thiocarbamides is catalyzed exclusively by the flavin-containing monooxygenase in liver microsomes (Poulsen

TABLE 2

Thiocarbamides and Mercaptoimidazoles: Reaction Sequence and Kinetic Constants

REACTION

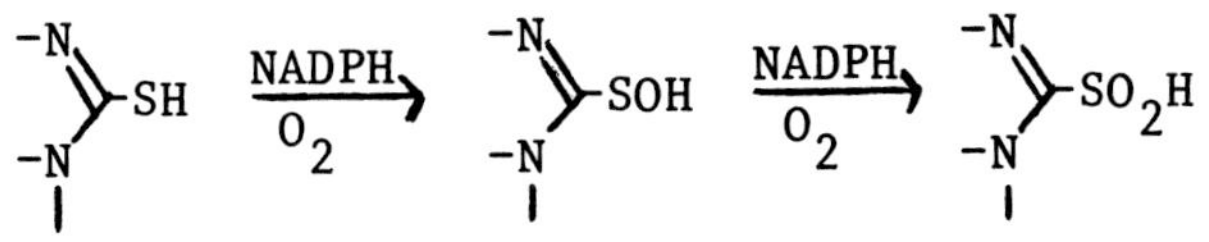

EXAMPLES

	Km'(μM)	Km"(μM)
Thiourea	22	140
Ethylenethiourea	31	51
Phenylthiourea	5	30
Thiocarbanilide	6	7
Methimazole	6	22
2-Mercaptobenzimidazole	11	23

Kinetic constants based on studies with the purified hog liver monooxygenase (Poulsen et al., 1979). Km' and Km" are Michealis constants for the first and second steps respectively.

et al., 1979).

Both formamidine sulfenates and sulfinates are potentially quite toxic, but they differ considerably in chemical properties. Formamidine sulfenates, like alkyl sulfenates, oxidize cellular thiols and the oxidation of GSH to GSSG by redox cycling similar to that of thiols (fig. 1) has been demonstrated in liver perfused with thiourea and phenylthiourea (Krieter et al., 1984). The thiocarbamide-induced biliary efflux of GSSG is completely insensitive to inhibitors selective for P-450 which demonstrates that the latter monooxygenase does not contribute significantly to the oxidation of these compounds in the intact organ.

Rapid reduction of the formamidine sulfenates by GSH would limit their further oxidation to formamidine sulfinates, but the rate of formation of the latter would increase as GSH is depleted by oxidation (fig. 1). While formamidine sulfinates do not oxidize thiols they are capable of alkylating amine groups (Ziegler, 1982) as illustrated in equation 1.

$$-N{=}\overset{\displaystyle SO_2^-}{\overset{|}{C}}-\underset{|}{N}- + \text{Prot-NH}_2 \longrightarrow -N{=}\overset{\displaystyle \text{Prot-NH}}{\overset{|}{C}}-\underset{|}{N}- + HSO_2^- \qquad (1)$$

Formation of stable covalent adducts with protein and/or nucleic acids by metabolically generated formamidine sulfinates may also contribute to the toxicity of thiocarbamides.

Mercaptoimidazoles are structurally similar to thiocarbamides and are oxidized to sulfinic acids by the same mechanism. The first oxidation product (imidazole sulfenates) also oxidizes GSH and protein thiols, and rapid methimazole-dependent biliary efflux of GSSG in perfused liver and live animals has been demonstrated (Krieter et al., 1984). On the other hand the dioxygenated products (imidazole sulfinates) are quite different in chemical properties and stability from the formamidine sulfinates (Ziegler, 1982). These differences in reactivity of their sulfinates may, in part, account for the differences in biological properties of mercaptoimidazoles and thiocarbamides.

Thioacids

Studies on substrate activity of thioacids with the purified monooxygenase have been quite limited, and all examples listed in Table 3 are taken from a single report (Taylor and Ziegler, 1987).

Products formed upon enzymic oxidation of thioacids have not been determined, but two moles of NADPH and oxygen are consumed per mole dithio acid oxidized. While several

TABLE 3

Thioacids: Reaction and Kinetic Constants

REACTION

$$R\text{-}\overset{S}{\overset{\|}{C}}\text{-}S^{-} \longrightarrow R\text{-}\overset{S}{\overset{\|}{C}}\text{-}SO^{-}$$

EXAMPLES

	Km(μM)
Thioacetate	200
Dithiobenzoate	3
Dithiosalicylate	3
N,N-Dimethyl-4-aminodithiobenzoate	14

Km values based on measurements with the porcine liver flavin-containing monooxygenase (Taylor and Ziegler, 1987).

reactions could account for the second oxidative reaction, the first step must involve oxidation of the dithioacid to the dithioperoxyacid. This metabolite is extremely reactive and readily oxidizes thiols and other cellular components. Although toxicity of dithioacids is largely unknown, their facile oxidation to this extremely reactive product suggests that they may be among the more toxic sulfur containing xenobiotics.

FUNCTIONAL GROUPS BEARING NITROGEN

Both the liver and lung flavin-containing monooxygenases catalyze oxidation of a variety of organic nitrogen compounds. Most funtional groups bearing this element are detoxified by N-oxidation, but secondary arylamines and 1,1-disubstituted hydrazines yield biologically more reactive metabolites and the following discussion is limited to these two functional groups. However, it should be pointed out that the oxidation of primary alkylamines, catalyzed exclusively by the pulmonary monooxygenase (Williams et al., 1985, Tynes et al., 1985) may also be a potential route for activation since, in general, hydroxylamines are chemically more reactive than amines.

Amines

As indicated by examples in Table 4, the porcine liver enzyme catalyzes N-oxidation of both dialkyl- and alkylaryl-amines to nitrones. While the effect on toxicity of the former is not known, N-oxidation is generally considered an early obligative event in the bioactivation of N-methylaryl-amines to carcinogenic metabolites. For example Kadlubar et al. (1976) have shown that N-oxidation of N-methyl-4-aminoazobenzene, catalyzed almost exclusively by the flavin-containing monooxygenase, produces the reactive nitrone which upon hydrolysis (equation 2) yields the primary hydroxylamine.

$$\mathrm{Ar\overset{O}{N}{=}CH_2 \xrightarrow{H_2O} ArNHOH + CH_2O} \qquad (2)$$

TABLE 4

Secondary Amines: Reaction Sequence and Kinetic Constants

REACTION

$$\mathrm{R{-}NHCH_2R \xrightarrow[O_2]{NADPH} R{-}\overset{OH}{N}CH_2R \xrightarrow[O_2]{NADPH} R{-}\overset{O}{N}{=}CHR}$$

EXAMPLES

	Km(μM)
N-Benzylamphetamine	120
Desmethyltrifluoperazine	9
Desipramine	250
N-Methylaniline	320
N-Methyl-4-Aminoazobenzene	200
Nortryptyline	500

Km's listed are for the first step. Although the second oxidation is also enzymic (Poulsen et al., 1974) the inherent instability of most secondary hydroxylamines precludes accurate determination of the second constant.

The role of this enzyme in the bioactivation of other arylamines has not been explored in depth. However, the loose substrate specificity for N-methylarylamines suggests a potential role for activation of N-methylarylamines administered directly or ones formed metabolically. The latter is a distinct possibility since Ansher and Jakoby (1986) have recently isolated two hepatic amine N-methyltransferases that catalyze N-methylation of a variety of amines including primary arylamines. Any N-methylarylamines formed by this route would be susceptible to activation by the mechanism illustrated in Table 5. Therefore, the role of these enzymes in the bioactivation of primary arylamines may be more significant than generally recognized.

Hydrazines

In a recent review Prough and Moloney (1985) state that the oxidation of 1,1-disubstituted hydrazines is catalyzed exclusively by the flavin-containing monooxygenase in liver microsomes. The studies of Prough et al. (1981) suggest that 1,1-disubstituted acyclic hydrazines are first oxidized to N-oxides. Dehydration of the hydrazine oxide followed by rearrangement of the intermediate diazene produces the hydrazone which upon hydrolysis yields an aldehyde and the monosubstituted hydrazine (Table 5).

While neither product is chemically more reactive than the disubstituted hydrazine, further oxidation of monosubstituted hydrazines formed by this route produces more reactive metabolites (Prough and Moloney, 1985).

TABLE 5

Hydrazines: Reaction Sequence and Kinetic Constants

REACTION

$$(R)_2N\text{-}NH_2 \xrightarrow[O_2]{NADPH} (R)_2\overset{O^-}{\overset{+}{N}}\text{-}NH_2 \xrightarrow{H_2O} (R)_2\overset{+}{N}{=}NH \longrightarrow RNHN{=}R$$

EXAMPLES

	Km(μM)
1,1-Dimethylhydrazine	430
1-Methyl-1-phenylhydrazine	80
N-Aminomorpholine	610
N-Aminopiperdine	30
N-Aminopyrrolidine	100

Km values taken from Prough et al. (1981).

REFERENCES

Ansher, S.S. and Jakoby, W.B., 1986, The Journal of Biological Chemistry, 261, 3996.
Eldjarn, L. and Nygaard, O., 1954, Archives of International Physiology, 62, 476.
Kadlubar, F.F., Miller, J.A. and Miller, E.C., 1976, Cancer Research, 36, 1196.
Kimura, T., Kodama, M. and Nagata, C., 1983, Biochemical and Biophysical Research Communications, 110, 640.
Krieter, P.A., Ziegler, D.M., Hill, K.A., and Burk, R.F., 1984, Molecular Pharmacology, 26, 122.
Lauterburg, B. and Mitchell, J.R., 1981, Hepatology, 1, 523.
Poulsen, L.L., Hyslop, R.M. and Ziegler, D.M., 1979, Archives of Biochemistry and Biophysics, 198, 78.
Poulsen, L.L., Kadlubar, F.F., and Ziegler, D.M., 1974, Archives of Biochemistry and Biophysics, 164, 774.
Prough, R.A., Freeman, P.C. and Hines, R.N., 1981, The Journal of Biological Chemistry, 256, 4178.
Prough, R.A. and Moloney, S.J., 1985, in Bioactivation of Foreign Compounds, edited by M.W. Anders (Academic Press, Inc., New York) Chapter 15, p. 433.
Sabourin, P.J., Smyser, B.P. and Hodgson, E., 1984, International Journal of Biochemistry, 16, 713.
Taylor, K.L. and Ziegler, D.M., 1987, Biochemical Pharmacology, 36, 141.
Tynes, R.E. and Hodgson, E., 1985, Archives of Biochemistry and Biophysics, 240, 77.
Tynes, R.E., Sabourin, P.J. and Hodgson, E., 1985, Biochemical and Biophysical Research Communiations, 126, 1069.
Williams, D.E., Hale, S.E., Meurhoff, A.S. and Masters, B.B.S., 1985, Molecular Pharmacology, 28, 381.
Ziegler, D.M., 1980, in Enzymatic Basis of Detoxication edited by W.B. Jakoby (Academic Press Inc., New York) Chapter 9, p. 201.
Ziegler D.M., 1982, in Metabolic Basis of Detoxication edited by W.B. Jakoby, J.R. Bend and J. Caldwell (Academic Press Inc., New York) Chapter 9, p. 171.
Ziegler, D.M. and Mitchell, C.H., 1972, Archives of Biochemistry and Biophysics, 150, 116.
Ziegler, D.M., Poulsen, L.L. and Duffel, M.W., 1980, in Microsomes, Drug Oxidations and Chemical Carcinogenesis, Vol. II, edited by M.J. Coon, A.H. Conney, R.W. Estabrook, H.V. Gelboin, J.R. Gillette and P.J. O'Brien (Academic Press Inc., New York) p. 637.

The Tissue Distribution of Glutathione Transferases

BRIAN KETTERER, DAVID J. MEYER, BRIAN COLES
and JOHN B. TAYLOR

Cancer Research Campaign Molecular Toxicology Research Group, Biochemistry Department, University College and Middlesex School of Medicine, Windeyer Building, Cleveland Street, London W1P 6 BD, UK

INTRODUCTION

Glutathione (GSH) transferases are dimeric enzymes of approximately 50,000 molecular weight. Wherever they have been sought in the animal and plant worlds they have been detected, but they have been most studied in mammals, in particular rat and man, where they have been shown to occur in most, if not all, tissues. (Mannervik, 1985a; Ketterer *et al*., 1986)

GSH Transferase Isoenzymes and Their Genetic Relationships

In rat and man the evidence is now clear that the soluble GSH transferases belong to a supergene family composed of at least three multigene families. In man, where the nomenclature is from the Greek alphabet and refers to dimers, the three multigene families are as follows, GSH transferases α-ε (the alpha family), GSH transferases μ and ψ (the mu family) and GSH transferases π, λ, ρ (the pi family) (see Mannervik *et al*., 1985; Hayes *et al*., 1987). In the rat, where a numerical nomenclature based on subunits is used (Jakoby *et al*, 1983), the alpha, mu and pi multigene families are represented by subunits 1, 2 and "10"; 3, 4, 6 and "9"; and 7, respectively. There is close homology between human and rat GSH transferases from the same family (Board *et al*., 1987; Rhoads *et al*., 1987; Mannervik *et al*., 1985 and Cowell *et al*, 1987). The mouse, it appears, also has the same three multigene families (Mannervik *et al*., 1985), indicating that the same type of supergene family may be widespread, at least in mammals. Evidence is emerging for yet another multigene family in the rat containing subunit 5: it is presumed that this family will also be found in other mammalian species.

The Enzymic Activities of GSH Transferases

GSH transferases are best known for their ability to catalyse the conjugation of lipophilic electrophiles with GSH, bringing about their detoxication and their eventual excretion. They will also catalyse a number of other GSH dependent reactions including the reduction of organic hydroperoxides (Se-independent GSH peroxidase activity) and the isomerisation of certain prostaglandins (Ketterer et al., 1986; Ketterer et al., 1987; Meyer & Ketterer, 1987).

TABLE Activity of GSH transferases towards a selection of substrates

substrate	μmol min^{-1} mg^{-1}						
	1-1	2-2	3-3	4-4	5-5	6-6	7-7
NABQI	24	48	6	3	nd	nd	60
BPDE	0.10	0.08	0.03	0.36	nd	nd	0.33
NPO	0.01	0.03	0.30	0.30	nd	nd	0.02
AFO	0.001	0.001	nil	nil	nd	nd	nil
HNE	2.6	0.7	2.7	6.9	nd	nd	nd
CHE	0.14	nd	nil	nil	nil	nd	nd
LinOOH	2.8	1.8	0.2	0.2	5.3	0.06	1.5
HOOMU	0.13	1.01	0.5	0.3	nd	1.3	1.3
DNAOOH	nil	nil	0.02	0.03	0.50	0.010	0.005

NABQI = N-acetylbenzoquinone imine; BPDE = (±)anti-benzo(a)pyrene-7,8-diol-9,10-oxide; NPO = 1-nitropyrene-4,5-oxide; AFO = aflatoxin B1-8,9-oxide; HNE = 4-hydroxy-non-2-enal (Mannervik et al); CHE = cholesterol-5α,6α-oxide; LinOOH = linoleic acid hydroperoxide; HOOMU = 5-hydroperoxymethyl uracil; DNAOOH = peroxidized DNA

The Table illustrates the activity of rat GSH transferase homodimers towards a number of substrates of biological importance, including electrophilic metabolites of a drug (paracetamol) and several carcinogens (benzo(a)pyrene, aflatoxin B1 and nitropyrene), a major electrophilic decomposition product of peroxidized lipids, (4-hydroxy-nonenal), a lipid hydroperoxide (linoleic acid hydroperoxide), a pyrimidine hydroperoxide (5-hydroperoxymethyl-uracil) and peroxidized DNA (DNA exposed to X-irradiation radiation). This Table gives an indication of how selective GSH transferase isoenzymes can be towards their substrates.

In the present paper we describe the application of a new method for analysing GSH transferases by reverse phase h.p.l.c. to a range of tissues, describe its advantages and discuss the significance of the results obtained.

METHODS

Estimates of the isoenzyme content of a number of tissues have been obtained by several methods including the use of specific substrates and inhibitors (Mannervik, 1985b), protein fractionation procedures (Ketterer et al., 1985), immunochemistry (Hayes & Mantle, 1986) and recombinant DNA techniques (Taylor et al, 1987). However, in our view the preferred technique is that of Ostlund Farrants et al. (1987). It involves first, the selection of the total GSH transferase fraction from a tissue extract by GSH Sepharose affinity chromatography and second, their separation by h.p.l.c. on a C18 reverse phase column using a 0.1% trifluoracetic acid in water-acetonitrile gradient (see Figs 1-4 for examples). This chromatographic system separates GSH transferases as their constituent subunits. In the rat at least 8 fractions are obtained including several forms of subunit 1. In two fractions where more than one subunit has the same retention time, further separation can be brought about by SDS-PAGE. Thus, in one peak which may contain subunits 4 and "9" (a variant of subunit 6), "9" electrophoreses faster than 4 and in the other, which may contain subunits 2, 5 and "10" (a subunit of the alpha class found only in young animals), relative rates of electrophoresis are "10" > 5 > 2. This method of GSH transferase analysis is relatively rapid and simple and, since there are so few steps losses are easily accounted for and quantitation is easy.

This technique can also fractionate human GSH transferase subunits. Alpha class enzymes can be separated into two subunit classes referred to as α_x and α_y and shown to exist as both homodimers and heterodimers. On the other hand μ and π emerge as single peaks. A good separation of α_x, α_y, μ and π is obtained.

RESULTS AND DISCUSSION

The distribution of GSH transferase isoenzymes of the soluble cytoplasm varies from tissue to tissue. When any one tissue is studied in detail, its GSH transferase content may change with development, sexual maturation, induction, tumour promotion and progression; and its transfer to *in vitro* culture.

The liver

Fig. 1 shows the analysis of the mature male rat liver. It is seen that subunits 1, 2, 3 and 4 are all abundant, but that subunit 7 is absent. At birth the picture is quite different. Only small amounts of GSH transferases are present and these are accounted for by subunits 2 and "10". The adult pattern begins to develop during the early weeks and at the approach of sexual maturity the male and female patterns diverge somewhat, the male having less of the alpha family and more of the mu family than the female. Inducers of drug metabolizing enzymes may increase the levels of one rather than all members of a family; for example phenobarbitone induces subunit 1 from the alpha family and subunit 3 from the mu family. During chemical hepatocarcinogenesis, preneoplastic foci appear first, followed by nodules, which, from the very beginning, show a specific differentiation pattern for a number of drug metabolizing enzymes including GSH transferases (Kitahara et al., 1984). Subunit 3 is much enhanced and subunit 7, which is normally restricted to bile duct cells, is expressed de novo in preneoplastic hepatocytes.

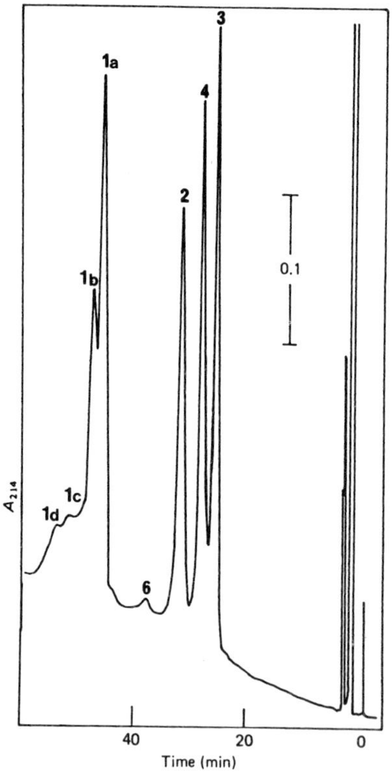

Fig 1. H.p.l.c analysis of GSH transferase subunits of rat liver

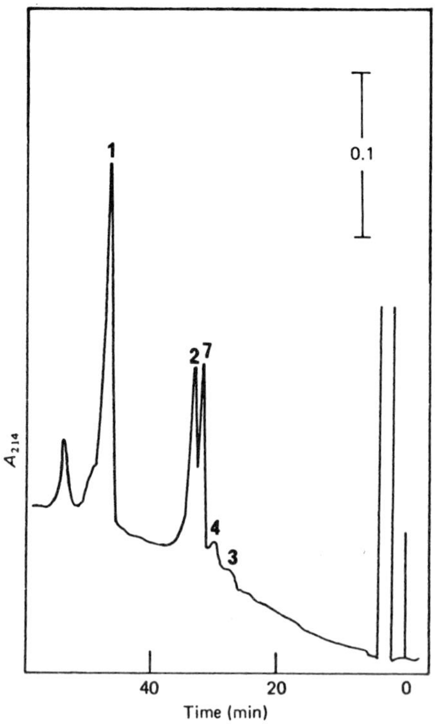

Fig 2. H.p.l.c. analysis of GSH transferase subunits of rat kidney

So dramatic and early is the appearance of subunit 7 in hepatocarcinogenesis, that it is regarded as the best marker for preneoplasia in the rat liver. Many foci and nodules eventually redifferentiate into normal liver, but some progress to primary hepatomas and these retain their preneoplastic pattern of GSH transferase expression (Meyer et al., 1985). Compared with other tissues, the liver contains the highest concentration of GSH transferases, in keeping with its function as the major organ of detoxication; yet it is surprising that it does not contain subunit 7, which is so effective in detoxication (see Table). By gaining subunit 7, preneoplastic cells presumably acquire superior powers of detoxication compared with surrounding normal liver and this may well assist in tumour promotion and progression.

Of particular interest is the observation that, when hepatocytes are put into cell culture, striking changes in their content of GSH transferases occur. Subunits 1 and 2 (i.e. the alpha family) are much reduced, while, in the case of the mu family, levels of subunit 4 are maintained and subunit 3 is increased. Also, as in hepatocarcinogenesis, subunit 7 is expressed de novo. The isolated hepatocyte may be a useful system in which to study factors which determine GSH transferase gene expression. It has already been shown that nicotinamide, which promotes the differentiated state, when added to the medium tends to restore levels of subunit 1 but not subunit 2 (Vandenberghe, Y., Glaise, D., Meyer, D.J., Guillouzo, A. & Ketterer, B., unpublished information)

The kidney

Fig 2. shows an analysis of the kidney. It differs from that of the liver in that the expression of the alpha family is much greater than that of the mu family and that subunit 7 is present. It is noteworthy that subunit 1 is the major component achieving such abundance only in the liver and the kidney.

The testis

After the liver, the organ which is richest in GSH transferase activity is the testis. It has two components namely the spermatogenic tubules and the interstitial cells each having a distinctive GSH transferase content (see Figs 3a and 3b). Spermatogenic tissue is noteworthy for its relatively high content of subunits 6 and "9". In contrast with liver and kidney, subunit 1 is absent. Not shown in these analyses is subunit 5 which is also present. The presence of both subunits 5 and 6, which are active in detoxifying both DNA-bound and free thymine peroxide is relevant to the function of the testis which is actively involved in the synthesis of DNA and the formation of sperm capable of carrying genetic information intact from from one generation to another. In this respect the protection against gentoxic electrophiles afforded by a spectrum of GSH transferase subunits is also important. It will be noted in Fig 3a that the spermatogenic tissue has a component with a retention time a little less than subunit 1a, which is as yet uncharacterized.

The adrenal and the mammary gland

These two organs are grouped together because they give a very similar analysis, subunit 2 being the predominant GSH transferase in both tissues. The adrenal gland shows sexual differentiation, the male containing more GSH transferases than the female (see Fig 4).

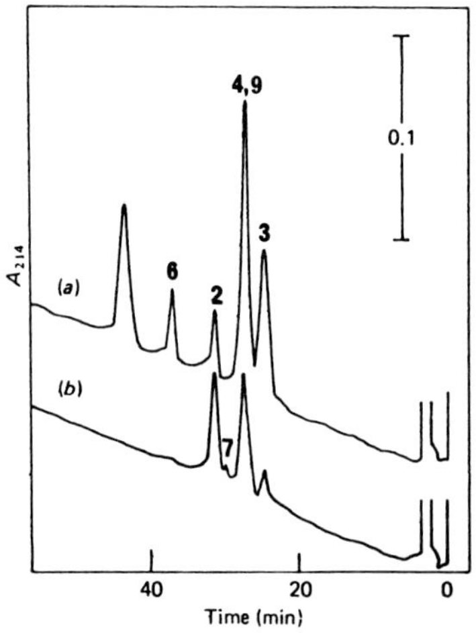

Fig 3. H.p.l.c. analysis of GSH transferase subunits of a) spermatogenic tissue and b) interstitial cells of the rat testis

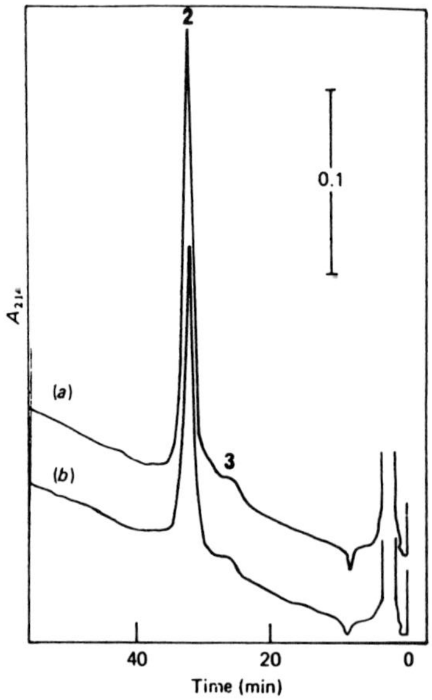

Fig 4. H.p.l.c. analysis of GSH transferase subunits of a) the male adrenal gland and b) the female adrenal gland of the rat

The mammary gland shows an effect of differentiation in GSH transferase expression in that the activity in the resting gland is only 5% that of lactating mammary gland.

Human Tissues

We have applied reverse-phase h.p.l.c. to human livers,

where in contrast to inbred rat populations, there is considerable inter-individual variation. Relative amounts of subunits α_x and α_y vary considerably, the subunit for μ is often absent (an observation first made by Warholm *et al*., 1981) and the subunit π is seen in the liver only occasionally (Johnson, P., Meyer, D.J., Lalor, E., Guengerich, F. & Ketterer, B., unpublished information). Reference to the Table shows that N-acetyl benzoquinone imine which is a substrate for the alpha class should be metabolized by all livers, but that benzo(a)pyrene-7,8-diol-9,10-oxide and 1-nitropyrene-4,5-oxide which are good substrates for the mu class, should be poorly detoxified in those individuals deficient in μ.

CONCLUSIONS

Since we last wrote on this topic (Ketterer *et al*., 1985) we have identified more GSH transferases, increased the number of substrates studied and investigated more tissues. We can make what seem to be reasonable suggestions concerning the distribution of GSH transferases in liver, the chemically induced hepatoma and the testis, however, it is stil difficult to offer a rationale for their specific distribution in some tissues and this raises the question as to whether or not, as yet, we fully understand the function of these proteins.

Acnowledgements

We would like to thank the Cancer Research Campaign and the Frances and Augustus Newman Trust for generous support throughout this research. We would also like to thank Miss Lucia Christodoulides for assistance in preparing this manuscript.

Illustrations are reprinted by permission from Biochemical Journal, 245, pp.423-428 copywrite (c) 19 The Biochemical Society, London.

REFERENCES

Board, P.G. & Webb, G.C., 1987, *Proceedings of the National Academy of Science, U.S.A.*, 84, 2377.

Cowell, I.G., Dixon, K.H., Pemble, S.E., Ketterer, B & Taylor, J.B., 1987, *Biochemical Society Transactions*, in press.

Hayes, J.D. & Mantle, T.J., 1986, *Biochemical Journal*, 233, 779.

Hayes, J.D., McLellan, L.I., Stockman, P.K., Howie, A.F., Hussey, A.J. & Beckett, G.J., 1987 human glutathione S-transferases; a polymorphic group of detoxication enzymes. In Glutathione S-Transferases and Carcinogenesis, edited by T.J. Mantle, C.B. Pickett and J.D. Hayes (Taylor and Francis, London) p.3.

Jakoby, W.B., Ketterer, B. & Mannervik, B., 1983, Biochemical Pharmacology 33, 2539.

Ketterer, B., Meyer, D.J., Coles, B. & Taylor, J.B. 1985 glutathione transferases: their multiplicity and tissue distribution. In Microsomes and Drug Oxidation, edited by A.R. Boobis, J. Caldwell, F. De Matteis & C.R. Elcombe (Taylor and Francis, London) p.167.

Ketterer, B., Meyer, D.J., Coles, B., Taylor, J.B. & Pemble, S., 1986 glutathione transferases and carcinogenesis. In Antimutagenesis and Anticarcinogenesis Mechanisms, edited by D.M. Shankel, P.E. Hartman, T. Kada & A. Hollaender (Plenum Press, New York) p. 103.

Ketterer, B., Tan, K.H., Meyer, D.J. & Coles, B., 1987 glutathione transferases: a possible role in the detoxication of DNA and lipid hydroperoxides. In Glutathione S-Transferases and Carcinogenesis, edited by T. Mantle, C.B. Pickett & J.D. Hayes, (Taylor and Francis, London) p. 149.

Kitahara, A., Satoh, K., Nishimura, K., Ishikawa, T., Ruiki, K., Sato, K., Tsuda, H. & Ito, N., 1984, Cancer Research, 44, 2698.

Mannervik, B., Alin, P., Guthenberg, C., Jensson, H., Tahir, M.K., Warholm, M. & Jornvall, H., 1985, Proceedings of the National Academy of Sciences, U.S.A., 82, 7202.

Mannervik, B., 1985, Advances in Enzymology and Related Areas of Molecular Biology, 57, 357.

Meyer, D.J., Beale, D., Tan, K.H., Coles, B. & Ketterer, B., 1985, FEBS Letters, 184. 139.

Meyer, D.J. & Ketterer, B., 1987 prostaglandin isomerase activity of purified rat glutathione (GSH) transferases. In Glutathione S-Transferases and Carcinogenesis, edited by T. Mantle, C.B. Pickett & J.D. Hayes, (Taylor and Francis, London) p.57

Ostlund Farrants, A.-K., Meyer, D.J., Coles, B. Southan, C., Aitken, A., Johnson, P.J. & Ketterer, B., 1987, Biochemical Journal, 245, 423.

Rhoads, D.M., Zarlego, R.P., & Tu, C.-P.D., 1987, Biochemical and Biophysical Research Communications, 145, 474.

Taylor, J.B., Pemble, S.E., Cowell, I.G., Dixon, K.H. & Ketterer, B., 1987, Biochemical Society Transactions, 15, 578.

Warholm, M., Guthenberg, C., Mannervik, B. & von Bahr, C., 1981, Biochemical and Biophysical Research Communications, 98, 512.

Control of Glutathione S-Transferase Gene Expression by Xenobiotics

CECIL B. PICKETT, CLAUDIA A. TELAKOWSKI-HOPKINS, VICTOR D.-H. DING and RON G. KING

Department of Molecular Pharmacology & Biochemistry, Merck Sharp & Dohme Research Laboratories, Rahway, NJ 07065, USA

INTRODUCTION

The glutathione S-transferases are a family of proteins that conjugate glutathione to various electrophilic ligands (Litwack et al. 1971; Arias et al. 1976; Jakoby & Habig 1980; Mannervik & Jensson 1982; Homma & Listowsky 1985). In addition to the conjugation activity, the transferases bind a number of hydrophobic compounds such as heme, bilirubin, polycyclic aromatic hydrocarbons and dexamethasone (Litwack et al. 1971; Arias et al. 1976; Jakoby & Habig 1980; Homma & Listowsky 1985)). To date there are structural data from DNA sequence analysis of cDNA clones as well as amino acid sequence analysis of purified subunits that indicate the existence of at least four glutathione S-transferase gene families (Lai et al. 1984; Pickett et al. 1984; Tu et al. 1984; Telakowski-Hopkins et al. 1985; Ding et al. 1985; Ding et al. 1986; Lai et al. 1986; Lai & Tu 1986; Alin et al. 1986). From Southern blot analysis of genomic DNA, we have estimated that approximately 5-7 genes constitute the glutathione S-transferase gene family encoding the basic subunits, Ya and Yc (Rothkopf et al. 1986).

In this article, we would like to summarize data from our laboratory that indicate the glutathione S-transferase Ya genes are transcriptionally activated by planar aromatic compounds such as 3-methylcholanthrene and β-naphthoflavone. The transcriptional activation leads to an elevation in the steady state level of Ya mRNA. We will also present direct evidence indicating that a Ya structural gene contains cis-acting regulatory elements in the 5' flanking region required for maximum basal promoter activity as well as responsiveness to β-naphthoflavone.

MATERIALS AND METHODS

RNA Blots and Nuclear Run-Off Assays

Poly(A^+)-RNA was isolated from total rat liver by oligo(dT)-cellulose affinity chromatography (Pickett et al. 1984). Total rat liver RNA was isolated by the procedure of Chirgwin et al. (1979). Various concentrations of poly(A^+)-RNA or total RNA were spotted onto nitrocellulose sheets using a minifold II slot blot apparatus (Schleicher & Schuell). Prehybridization and hybridization of the radiolabelled cDNA probes were done as described by Thomas (1980). Autoradiographs of the x-ray films were scanned with a densitometer.

Nuclear run-off assays using isolated rat liver nuclei were performed as described by McKnight and Palmiter (1979).

Construction of Glutathione S-Transferase Chloramphenicol Acetyltransferase Fusion Genes

A subclone of λGTB45-15 was constructed by ligating a 5.5 Kb fragment, which contained the 5' flanking region of the Ya gene, into the Eco R1 site of pBR325. This subclone, designated λpGTB45-15-5.5, was used in the construction of all transferase-CAT promoter fusion genes.

Transfection Assays and Measurement of Chloramphenicol Acetyltransferase Activity

Transfections were done by a modified calcium phosphate method (Gorman et al. 1982). For the measurement of chloramphenicol acetyltransferase activity transfected cells were washed and resuspended in 100 ul of 0.25 M tris-Hcl pH 7.8. The resuspended cells were lysed by freezing and thawing. Chloramphenicol acetyltransferase activity was measured as described by Gorman et al. (1982).

RESULTS AND DISCUSSION

Effect of 3-Methylcholanthrene on Glutathione S-Transferase Ya mRNA Level and Transcriptional Rates

Using *in vitro* translation systems and immunoprecipitation with specific antibodies raised against the purified glutathione S-transferase Ya Yc heterodimer, we have observed an increase in the translational activity of the Ya mRNA in response to 3-methylcholanthrene administration (Pickett et al. 1982). RNA blot analysis using a cDNA clone complementary to the Ya mRNA indicated that the Ya mRNA is elevated 4-fold at 12 hrs and 7-fold at 24 hrs after 3-methylcholanthrene administration (Pickett et al. 1984 and Table 1). The elevation in the steady state level of the Ya mRNA is due to an increase in the transcriptional rate of the Ya gene (Ding & Pickett 1985 and Table 1). The transcriptional rate of the Ya gene was maximally elevated 8-fold at 16 hrs after 3-methylcholanthrene administration (Table 1).

Effect of β-Naphthoflavone on Ya mRNA Level in Rat H5-6 Cells and Identification of Regulatory Elements in the 5' Flanking Region of the Ya Gene Responsible for Basal Level and Inducible Expression

In order to examine the function of the 5' flanking region of the glutathione S-transferase Ya structural gene, we constructed glutathione S-transferase promoter-CAT fusion genes. The glutathione S-transferase Ya structural gene contains seven exons separated by six introns and spans 11 Kb. The largest Eco R1 fragment (5.5 Kb), which contains exon 1, was used for the constructions (Telakowski-Hopkins et al. 1986). All the constructs contain 23 bp of intron 1, all of exon 1 (43 bp) and various lengths of the 5' flanking region of the Ya gene. These fragments were ligated into the Hind III site of pSVo CAT, a plasmid that contains the CAT structural gene. The pGTB .7 CAT construct contains nucleotides -1 to -650 and the pGTB 1.6 CAT construct contains nucleotides -1 to -1550 relative to the transcription start site. These constructs along with pSV_2 CAT were used in all subsequent transfection experiments. The plasmid, pSV_2 CAT, contains the CAT structural gene under the control of the SV_{40} enhancer and promoter (Gorman et al. 1982). When pGTB .7 CAT was transfected into the rat hepatoma cell line, CAT activity was significantly lower than the CAT activity observed with pSV_2 CAT (Fig. 1). The larger construct pGTB 1.6 CAT produced a CAT activity similar to pSV_2 CAT. When the rat hepatoma cells were treated with 50 uM β-naphthoflavone, we found only the CAT activity directed by pGTB 1.6 CAT was inducible 3-fold by β-naphthoflavone. The extent of induction of CAT activity parallels the increase in Ya mRNA seen in the rat H5-6 cell line after β-naphthoflavone induction. An approximately 3.0-4.0 fold increase in Ya mRNA can be detected with the Ya cDNA clone, pGTB38. Taken together these studies suggest that between -650 and -1550, there are two regulatory elements. One element is responsible for inducible expression of the Ya gene by planar aromatic compounds such as β-naphthoflavone and one that is required for maximum basal level expression. Transfection of these constructs into heterologous cells such as human Hep G2 produced a result that is identical to that seen in the rat H5-6 cells. Maximum basal level expression and β-naphthoflavone responsiveness were observed only when the pGTB 1.6 CAT construct was used.

Figure 1. Chloramphenicol acetyltransferase assay using rat H5-6 cells with glutathione S-transferase-CAT chimeric plasmids. Transient assays in the presence (+) or absence (-) or 50 uM β-naphthoflavone were performed with rat H5-6 cells using pSV_2 CAT, pGTB .7 CAT and pGTB 1.6 CAT.

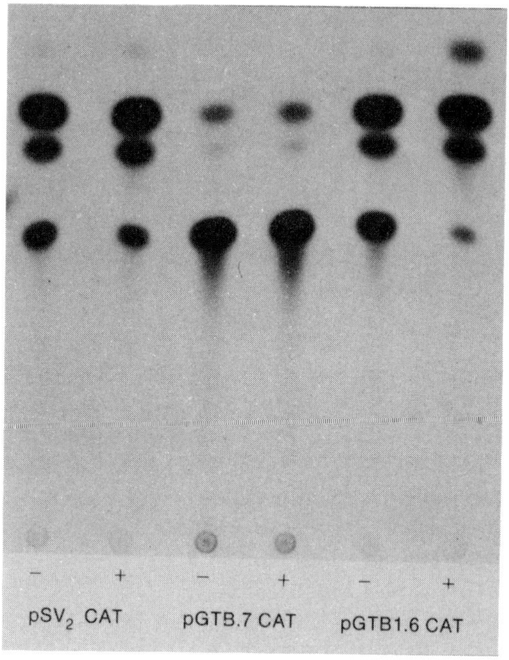

Role of the AH Receptor in the Transcriptional Regulation of the Ya Gene

In order to assess the role of the Ah receptor in the regulation of the glutathione S-transferase Ya gene, we transfected our constructs into mouse Hepa lclc7 cells and the mouse class I variant cells. The class I variants form few TCDD-receptor complexes; however those that form accumulate in the nucleus (Israel & Whitlock 1983; Miller et al. 1983; Whitlock & Galeazzi 1984). Transfection of pGTB 1.6 CAT construct into the wild type cell line, Hepa lclc7, resulted in a 4-fold induction in CAT activity in response to β-naphthoflavone (Fig. 2); however no induction was observed in the class I variants (Fig. 3). These studies suggest that a functional Ah receptor is required for induction. Similar results were seen with mouse class II variant cells, which express a normal number of receptors but the receptor-ligand complex is not translocated to the nucleus (Israel & Whitlock 1983; Miller et al. 1983; Whitlock & Galeazzi 1984).

Figure 2. Chloramphenicol acetyltransferase assay using mouse Hepa 1c1c7 wild type cells transfected with glutathione S-transferase-CAT chimeric plasmids. Transient assays in the presence (+) and absence (-) of 15 uM β-naphthoflavone were performed in mouse Hepa 1c1c7 cells using pGTB .7 CAT and pGTB 1.6 CAT.

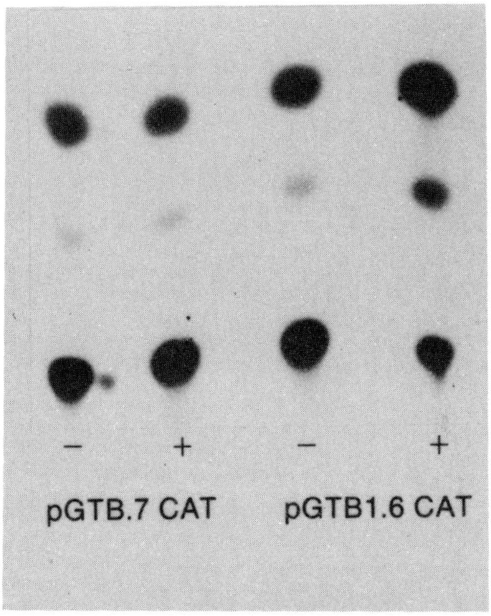

Figure 3. Chloramphenicol acetyltransferase assay using mouse class I variant cells transfected with glutathione S-transferase-CAT chimeric plasmids. Transient assays in the presence (+) and absence (-) of 15 uM β-naphthoflavone were performed in mouse class II variants using pSV$_2$ CAT, pGTB .7 CAT and pGTB 1.6 CAT.

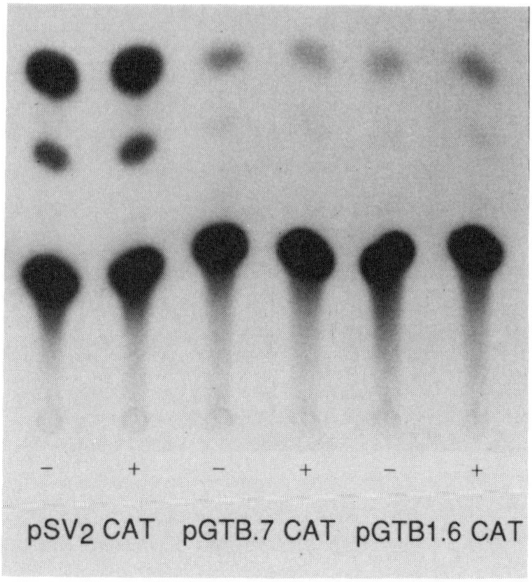

We do not know if the regulation of the glutathione S-transferase Ya gene by the Ah receptor is a direct or indirect mechanism. If the activation proceeds via an indirect mechanism, a second regulatory gene must be involved that encodes a trans-acting protein that interacts with the cis-acting regulatory element on the Ya gene. The transcriptional activation of this putative regulatory gene must be mediated by the Ah receptor.

In summary, we have provided evidence that at least two cis-acting regulatory elements control the expression of the glutathione S-transferase Ya gene. One element regulates basal level expression whereas the second element is required for β-naphthoflavone induction. The identification of these regulatory elements will allow for a detailed structural study to define the minimum DNA sequences that constitute these elements as well as provide an experimental system to define trans-acting proteins that function to control the expression of this gene.

ACKNOWLEDGEMENTS

We would like to thank Dr. Oliver Hankinson at the University of California, Los Angeles and Dr. J.P. Whitlock, Jr. at Stanford University for providing us with the cell lines. We would also like to thank Joan Kiliyanski for her assistance in the preparation of this manuscript.

Table 1. Effect of 3-methylcholanthrene on Ya mRNA level and transcriptional activity of the Ya gene.

Hours After 3-Methylcholanthrene Administration	Relative Ya mRNA Level[1]	Transcription Rate of Ya Genes (ppm)[2]
0	1.0	5
2	--	10
4	--	12
6	--	8
8	--	22
12	4.3	30
16	--	42
24	7.1	--

[1]Relative RNA levels are determined from RNA blot analysis followed by autoradiography and densiotimetric analysis of the autoradiograph.

[2]Parts per million hybridized was calculated based on the total cpm incubated with the nitrocellulose filter containing the Ya cDNA clone.

REFERENCES

Alin, P., Mannervik, B. & Jornvall, H., 1986, European Journal of Biochemistry, 156, 343-350.

Arias, I.M., Fleischner, G., Kirsch, R., Mishkin, S. & Gatmaitan, Z., 1976. In Glutathione: Metabolism and Function, edited by I.M. Arias & W.B. Jakoby (Raven Press, New York), p. 175-188.

Chirgwin, J.M., Przybyla, A.E., MacDonald, R.J. & Rutter, W.J., 1979, Biochemistry, 18, 5294-5299.

Ding, G.J.-F., Ding, V.D.-H., Rodkey, J.A., Bennett, C.D., Lu, A.Y.H. & Pickett, C.B., 1986, Journal of Biological Chemistry, 261, 7952-7957.

Ding, G.J.-F., Lu, A.Y.H. & Pickett, C.B., 1985, Journal of Biological Chemistry, 260, 13268-13271.

Ding, V.D.-H. & Pickett, C.B., 1985, Archives of Biochemistry and Biophysics, 240, 553-559.

Gorman, C.M., Mofat, L.F. & Howard, B.H., 1982, Molecular and Cellular Biology, 2, 1044-1051.

Homma, H. & Listowsky, I., 1985, Proceedings of the National Academy of Sciences USA, 82, 7165-7169.

Israel, D.I. & Whitlock, J.P., Jr., 1983, Journal of Biological Chemistry, 258, 10390-10394.

Jakoby, W.B. & Habig, W.H., 1980. In Enzymatic Basis of Detoxification, edited by W.B. Jakoby (Academic Press, New York), p. 63-94.

Lai, H.-C.J., Grove, G. & Tu, C.-P.D., 1986, Nucleic Acids Research, 14, 6101-6114.

Lai, H.-C.J., Li, N.-Q., Weiss, M.J., Reddy, C.C. & Tu, C.-P.D., 1984, Journal of Biological Chemistry, 259, 5536-5542.

Lai, H.-C.J., Tu, C.-P.D., 1986, Journal of Biological Chemistry, 261, 13793-13799.

Litwack, G., Ketterer, B. & Arias, I.M., 1971, Nature (London), 234, 466-467.

Mannervik, B. & Jensson, H., 1982, Journal of Biological Chemistry, 257, 9909-9912.

McKnight, G.S. & Palmiter, R.D., 1979, Journal of Biological Chemistry, 254, 9050-9058.

Miller, A.G., Israel, D. & Whitlock, J.P., Jr., 1983, Journal of Biological Chemistry, 258, 3523-3527.

Pickett, C.B., Telakowski-Hopkins, C.A., Ding, G.J.-F., Argenbright, L. & Lu, A.Y.H., 1984, Journal of Biological Chemistry, 259, 5182-5188.

Pickett, C.B., Telakowski-Hopkins, C.A., Donohue, A.M., Lu, A.Y.H. & Hales, B.F., 1982, Biochemical and Biophysical Research Communications, 104, 611-619.

Rothkopf, G.S., Telakowski-Hopkins, C.A., Stotish, R.L. & Pickett, C.B., 1986, Biochemistry, 25, 993-1002.

Telakowski-Hopkins, C.A., Rodkey, J.A., Bennett, C.D., Lu, A.Y.H. & Pickett, C.B., 1985, Journal of Biological Chemistry, 260, 5820-5825.

Telakowski-Hopkins, C.A., Rothkopf, G.S. & Pickett, C.B., 1986, Proceedings of the National Academy of Sciences USA, 83, 9393-9397.

Thomas, P., 1980, Proceedings of the National Academy of Sciences USA, 77, 5201-5205.

Tu, C.-P.D., Lai, H.-C.J., Li, N.-Q., Weiss, M.J. & Reddy, C.C., 1984, Journal of Biological Chemistry, 259, 9434-9439.

Whitlock, J.P., Jr. & Galeazzi, D.R., 1984, Journal of Biological Chemistry, 259, 980-985.

Role of Glutathione Transferases in Chemical Deactivation and Activation

BENGT MANNERVIK and U. HELENA DANIELSON

Department of Biochemistry, Arrhenius Laboratory, University of Stockholm, S-106 91, Sweden

INTRODUCTION

Glutathione transferases in several animal species investigated can be divided into three classes of cytosolic enzymes, Alpha, Mu, and Pi, and a membrane-associated, 'microsomal' enzyme (Mannervik et al. 1985a). The classification was originally based on specific activities with several substrates, inhibition characteristics with a set of inhibitors, and cross-reactions with antibodies (Mannervik et al. 1984a), but amino acid sequence determinations were required for a more complete understanding of the relationships among the various forms of the enzyme (Mannervik et al. 1985a). Available data suggest that the determination of the sequence of the first 10 to 20 N-terminal amino acids of an animal glutathione transferase would be sufficient for assigning the enzyme to one of the classes. A limitation is that the majority of the known class Alpha enzymes have blocked N-terminal amino acids. In a recent review (Mannervik & Danielson 1988) all the known primary structures were compiled and it was obvious that the recently determined amino acid sequence of a glutathione transferase from the helminth *Schistosoma japonicum* (Smith et al. 1986) was homologous to the mammalian sequences previously studied. The mammalian enzymes within a class show 60-80% sequence homology, whereas cytosolic enzymes of different classes, although clearly related, do not express more than 30% homology. The microsomal glutathione transferase does not display any obvious sequence homology to any of the cytosolic enzymes. The latter enzymes have obviously evolved by divergent evolution from a common ancestral protein whereas the microsomal enzyme may have a different origin.

The structural relationships among the multiple forms of cytosolic glutathione transferase raise the question about the distinction between isoenzymes and separate enzymes. It would appear that members (from the same biological species) within a class should be regarded as true isoenzymes. For example, the six binary combinations of the rat subunits 3, 4 and 6 are isoenzymes in class Mu. Rat transferase 7-7 (of class Pi) and rat transferase 3-3 are more distantly related than proper isoenzymes, but perhaps not different enough to be regarded as separate enzymes. A difficulty in the assignment is that the concept of isoenzymes is normally based on a common substrate specificity, whereas the physiological substrates of the glutathione transferases are not known and even the 'isoenzymes' within the same class show distinct substrate specificities with many of the nonphysiological compounds used to measure enzyme activity.

SUBSTRATES OF GLUTATHIONE TRANSFERASES

It is well established that a multitude of chemical compounds with various functional groups may serve as substrates for glutathione transferases (Jakoby 1978; Chasseaud, 1979). However, the majority of these substances have no biological occurrence and cannot have influenced the evolution of the transferases. It has consequently been proposed that the physiologically relevant substrates of the enzymes should be found among products of oxidative metabolism such as organic hydroperoxides, quinones, epoxides, and activated alkenes (Mannervik et al. 1985b; Mannervik 1985, 1986). Such activated metabolic intermediates are produced by processes catalyzed by monooxygenases and other enzymes. In most cases, reactions with glutathione cause a deactivation of reactive molecules, even though vic-dihaloalkanes and some related substances become chemically activated by glutathione conjugation (cf. Igwe 1986). Formation of leukotriene C_4 (LTC_4) from the naturally occurring epoxide LTA_4 may be considered as a physiological rather than a chemical activation, since LTC_4 is involved in the etiology and pathology of various diseases.

The finding that, under standard assay conditions, rat glutathione transferase 4-4 was selectively inhibited by the anti-inflammatory drug indomethacin (Nicholls & Ahokas, 1984; Ålin et al., 1985b) and the coincidence that of the conventional glutathione transferases this enzyme is the most active with LTA_4 as a substrate (Mannervik et al., 1984b) suggested a pharmacologically significant connection. However, the specific strong inhibition of transferase 4-4 was only observed with high (millimolar) concentrations of 1-chloro-2,4-dinitrobenzene and not with low concentrations of this substrate (Danielson & Mannervik 1987). Neither was indomethacin a more potent inhibitor for transferase 4-4 than for other forms of glutathione transferase when alternative substrates were used. Thus, the effect of indomethacin on the glutathione transferases has no obvious link to the metabolism of the leukotrienes.

A particularly important group of substrates of glutathione transferases are 4-hydroxyalkenals (Ålin et al. 1985a; Danielson et al. 1987), which may be formed from polyunsaturated fatty acids during lipid peroxidation. The most active enzyme with these substrates is rat glutathione transferase 8-8, a class Alpha enzyme (Jensson et al. 1986). A detailed study of a homologous series (C_5 to C_{15}) of 4-hydroxyalkenals was undertaken with 15 rat, mouse and human glutathione transferases. Catalytic efficiencies (k_{cat}/K_m values) of a series of homodimeric rat glutathione transferases are given in Figure 1 and the incremental Gibbs free energy of binding of substrate in the transition state was evaluated from the k_{cat}/K_m values (Danielson et al. 1987).

In general, increased hydrophobicity (i.e. increased length of the hydrocarbon chain) is accompanied by increased binding energy. Each CH_2 group of the substrate contributes approximately 2.9 kJ/mol until the increase in binding energy is counteracted by steric limitations of the active site.

CATALYTIC MECHANISM

The discovery that 4-hydroxyalkenals are particularly good substrates for the glutathione transferases suggests that the α,β-unsaturated carbonyl structure may be an important determinant for some of the biologically relevant substrates of the enzymes. Many natural products from e.g. plants contain this reactive structure. Several of the non-biological substrates commonly used to assay the enzymes, e.g. trans-4-phenyl-3-buten-2-one and ethacrynic acid also carry the same functional group. The glutathione conjugation of α,β-unsaturated carbonyl compounds involves nucleophilic attack by the sulfur atom of glutathione on the β-carbon. This addition reaction is facilitated by a base which deprotonates the sulfhydryl group as well as by

Figure 1. Catalytic efficiencies of rat glutathione transferases with a homologous series of 4-hydroxyalkenals. Note the differences in scaling of the y-axis in the two panels.

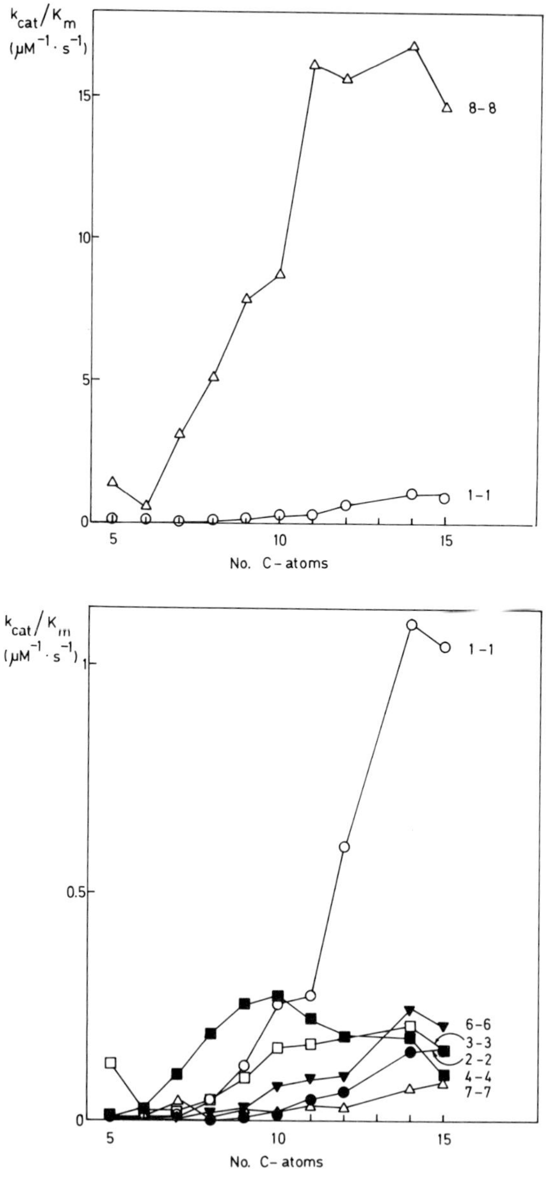

an acid protonating the oxygen of the carbonyl group. We therefore propose that the active site of a glutathione transferase active in this reaction should contain a base (e.g. an imidazol group) and an acid (e.g. a carboxyl group). A hypothetical active site capable of such acid-base catalysis is shown in Figure 2.

Figure 2. Hypothetical active site of a glutathione transferase subunit acting on an alkenal substrate.

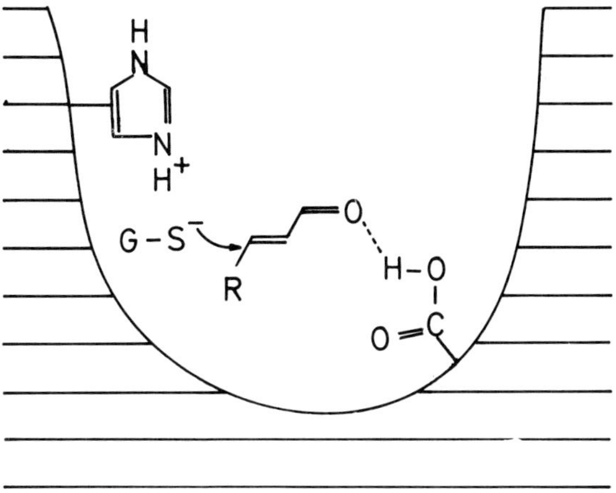

Evidence for an essential histidine residue in the active site of a human class Mu transferase was recently published (Awasthi et al. 1987). It is obvious that a corresponding base-assisted ionization of glutathione accompanied by a protonation of an oxygen in the functional group of the electrophilic substrate would be equally plausible for the transferase-catalyzed reactions involving epoxides and organic hydroperoxides. Further work is needed to corroborate the model in Figure 2.

GLUTATHIONE TRANSFERASES AND CANCER

Glutathione transferases catalyze the inactivation of many carcinogens. In particular, a number of epoxides have been studied. The class Mu transferases are significantly more active than the class Alpha and Pi transferases with most epoxides even though the class Pi enzymes are the most active ones with the ultimate carcinogenic metabolite of benzo(a)pyrene, (+)-anti-benzo(a)pyrene-7,8-diol-9,10-oxide (cf. Robertson et al. 1986a, b).

In view of the discovery that only approximately 60% of the human population express the class Mu transferase, it was proposed that the individuals in this fraction may have a better protection against certain toxic compounds (Warholm et al. 1983). The finding that the presence of this enzyme type can be measured in mononuclear leukocytes by use of trans-stilbene oxide as substrate (Seidegård et al. 1985, 1987) has allowed studies of smokers, demonstrating that lack of the class Mu transferase is associated with a significantly higher incidence of lung cancer (Seidegård et al. 1986).

Recent work from several laboratories have demonstrated elevated concentrations of glutathione and glutathione transferases in cancer cells (cf. Mantle et al. 1987). It appears as if the transferases are part of a general protection and resistance phenotype of neoplastic cells (cf. Farber 1984). Some of the chemotherapeutics used in the clinical treatment of cancer are inactivated by glutathione conjugation and the transferases may thus contribute to drug resistance. In a study of rat brain tumor cells it was found that the relative amount of glutathione transferase 4-4 was elevated in a cell line with increased resistence against 1,3-bis-(2-chloroethyl)-1-nitrosourea (BCNU) (Smith et al. 1987). Using purified glutathione transferases, it was demonstrated that transferase 4-4 was the most efficient of the rat enzymes in catalyzing the degradation of BCNU. Similar studies involving purified human enzymes likewise indicated that the class Mu transferase was the most efficient enzyme in the inactivation of BCNU (S. Uhlig & B. Mannervik, unpublished work). Considering the class Mu transferase polymorphism in humans (Warholm et al. 1983), it seems likely that patients expressing a class Mu gene may be more prone to develop resistance against BCNU. Other enzyme forms appear to be involved in the inactivation of alternative chemotherapeutics.

It would appear that glutathione transferases may play several roles in the development of cancer. First, they may prevent initiation by inactivating initiators such as epoxides. Secondly, they may afford protection to the cancer cells, once a neoplastic lesion has occurred. Consequently, glutathione transferases may have both a beneficial and a detrimental role in the organism depending on the particular circumstances under which its activity is expressed.

Acknowledgements

This work was supported by the Swedish Natural Science Research Council.

REFERENCES

Ålin, P., Danielson, U.H. & Mannervik, B., 1985a, FEBS Letters, 179, 267.

Ålin, P., Jensson, H., Guthenberg, C., Danielson, U.H., Tahir, M.K. & Mannervik, B., 1985b, Analytical Biochemistry, 146, 313.

Awasthi, Y.C., Bhatnagar, A. & Singh, S.V., 1987, Biochemical and Biophysical Research Communications, 143, 965.

Chasseaud, L.F., 1979, Advances in Cancer Research, 29, 175.

Danielson, U.H. & Mannervik, B., 1987, Biochemical Journal (in the press).

Danielson, U.H., Esterbauer, H. & Mannervik, B., 1987, Biochemical Journal, 247, 707.

Farber, E., 1984, Biochimica et Biophysica Acta, 738, 171.

Igwe, O.J., 1986, Biochemical Pharmacology, 35, 2987.

Jakoby, W.B., 1978, Advances in Enzymology, 46, 383.

Jensson, H., Guthenberg, C., Ålin, P. & Mannervik, B., 1986, FEBS Letters, 203, 207.

Mannervik, B., 1985, Advances in Enzymology, 57, 357.

Mannervik, B., 1986, Chemica Scripta, 26B, 281.

Mannervik, B. & Danielson, U.H., 1988, Glutathione transferases - structure and catalytic activity, CRC Critical Reviews in Biochemistry (in the press).

Mannervik, B., Guthenberg, C., Jensson, H., Tahir, M.K., Warholm, M. & Ålin, P., 1984a, In Proceedings of IUPHAR 9th International Congress of Pharmacology, Vol. 3, edited by W. Paton, J. Mitchell & P. Turner (Macmillan Press, Baringstoke), p. 225.

Mannervik, B., Jensson, H., Ålin, P., Örning, L. & Hammarström, S., 1984b, FEBS Letters, 175, 289.

Mannervik, B., Ålin, P., Guthenberg, C., Jensson, H., Tahir, M.K., Warholm, M. & Jörnvall, H., 1985a, Proceedings of the National Academy of Sciences USA, 82, 7202.

Mannervik, B., Ålin, P., Guthenberg, C., Jensson, H. & Warholm, M., 1985b, In Microsomes and Drug Oxidations, edited by A.R. Boobis, J. Caldwell, F. De Matteis & C.R. Elcombe (Taylor & Francis, London), p. 221.

Mantle, T.J., Pickett, C.B. & Hayes, J.D. (editors), 1987, Glutathione S-Transferases and Carcinogenesis (Taylor & Francis, London).
Nicholls, F.A. & Ahokas, J.T., 1984, Biochemical and Biophysical Research Communications, 119, 1034.
Robertson, I.G.C., Guthenberg, C., Mannervik, B. & Jernström, B., 1986a, Cancer Research, 46, 2220.
Robertson, I.G.C., Jensson, H., Mannervik, B. & Jernström, B., 1986b, Carcinogenesis, 7, 295.
Seidegård, J., DePierre, J.W. & Pero, R.W., 1985, Carcinogenesis, 6, 1211.
Seidegård, J., Pero, R.W., Miller, D.G. & Beattie, E.J., 1986, Carcinogenesis, 7, 751.
Seidegård, J., Guthenberg, C., Pero, R.W. & Mannervik, B., 1987, Biochemical Journal, 246, 783.
Smith, D.B., Davern, K.M., Board, P.G., Tiu, W.U., Garcia, E.G. & Mitchell, G.F., 1986, Proceedings of the National Academy of Sciences USA, 83, 8703.
Smith, M.T., Doane-Setzer, P., Evans, C.G., Castro, V., Tahir, M.K. & Mannervik, B., 1987, Proceedings of the American Association for Cancer Research, 28, 278.
Warholm, M., Guthenberg, C. & Mannervik, B., 1983, Biochemistry, 22, 3610.

Mechanisms of Cell Toxicity -- The Thiol/Calcium Hypothesis

S. ORRENIUS, H. THOR, D.J. McCONKEY,
P. NICOTERA and G. BELLOMO

Department of Toxicology, Karolinska Institutet,
Box 60400, S-104 01 Stockholm, Sweden

INTRODUCTION

Tissue necrosis has long been known to be associated with the accumulation of large amounts of calcium in the necrotic tissue, and it has been proposed that the calcium ion may play a critical role in the development of lethal alterations of cell structure and function (Campbell 1983). Although the generality of this hypothesis has been questioned (Starke et al. 1986), the involvement of Ca^{2+} in the development of lethal cell injury is now supported by a large number of observations. For example, Shanne and associates (1979) found that removal of Ca^{2+} from the medium exerted a protective effect against the toxicity of a number of agents in cultured hepatocytes. This and similar observations led to the proposal that an influx of extracellular Ca^{2+} could play a predominant role in the development of irreversible cell damage (Farber 1981). However, subsequent studies have shown that an influx of extracellular Ca^{2+} is not necessarily required for cytotoxicity, and that a disruption of intracellular Ca^{2+} homeostasis may also result in irreversible cell damage.

Studies in this and other laboratories have demonstrated that oxidative stress in hepatocytes is associated with the inhibition of Ca^{2+} translocases involved in the maintenance of intracellular Ca^{2+} homeostasis (Orrenius & Bellomo 1986). This may lead to a sustained elevation of cytosolic Ca^{2+} concentration which, in turn, can cause the activation of various Ca^{2+}-dependent degradative enzymes, including phospholipases, proteases, and endonucleases, whose activities may combine to cause irreversible damage (Figure 1). The following sections will be concerned with the mechanisms by which a perturbation of intracellular Ca^{2+} homeostasis can cause cytotoxicity and cell death.

Fig. 1. Schematic illustration of the thiol/calcium hypothesis of toxic cell injury.

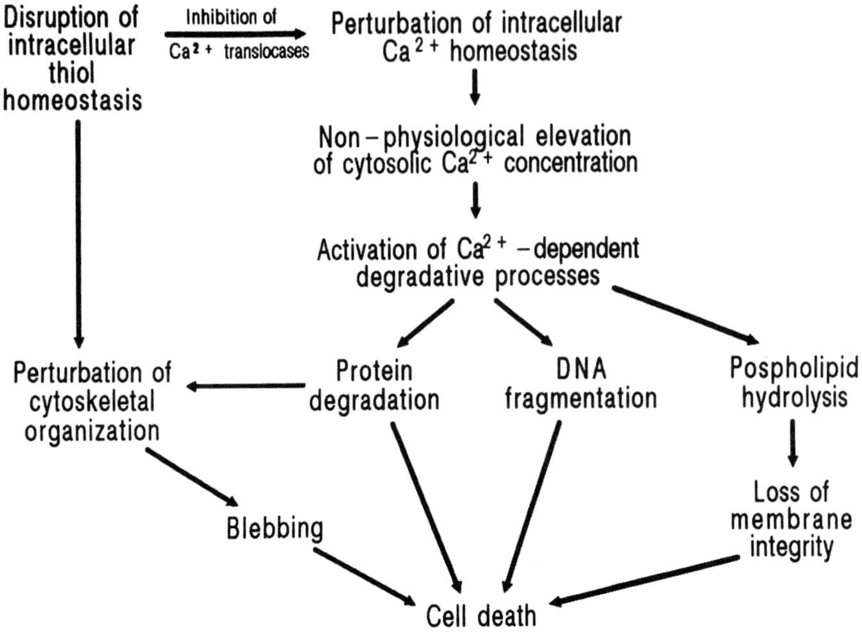

CALCIUM COMPARTMENTATION IN HEPATOCYTES

The low resting concentration of Ca^{2+} in the cytosol of hepatocytes (∿0,1 μM) is maintained by active compartmentation processes and by Ca^{2+} binding to specific proteins, including calmodulin. The mitochondria and the endoplasmic reticulum represent the predominant sites of intracellular Ca^{2+} sequestration which can exchange Ca^{2+} with the cytosolic compartment and thereby buffer the effects of normal Ca^{2+} utilization. The plasma membrane Ca^{2+} translocase then plays a major role in the maintenance of the concentration gradient existing between the extra- and intracellular environments by actively extruding Ca^{2+} from the cell.

Mitochondrial Ca^{2+} homeostasis is regulated by a cyclic mechanism, involving Ca^{2+} uptake by an energy-dependent pathway, and Ca^{2+} release which, in the liver, is probably mediated by a Ca^{2+}/H^+ antiporter. The latter appears to be regulated by the redox level of intramitochondrial pyridine nucleotides (Lehninger et al. 1978), although membrane-bound protein thiols may also be important in modulating mitochondrial Ca^{2+} fluxes. The active transport of calcium ions through the endoplasmic reticular and plasma membranes is mediated by Ca^{2+}-stimulated, Mg^{2+}-dependent ATPases which appear to be dependent on free sulfhydryl groups for their activity (Moore et al. 1975; Bellomo et al. 1983).

Thus, it seems that the major mechanisms responsible for Ca^{2+} movements across cellular membranes are all dependent on protein thiol groups. It is therefore conceivable that oxidation and/or arylation of these critical thiol groups may result in the inactivation of the translocases involved in the regulation of intracellular Ca^{2+} compartmentation, leading to an alteration of the level of cytosolic Ca^{2+}.

ROLE OF GLUTATHIONE AND PROTEIN THIOL MODIFICATION IN TOXICITY

Damage to biological systems caused by generation of active oxygen species is often referred to as "oxidative stress" (Sies 1986). Since aerobic cells possess antioxidant systems to trap and/or inactivate such reactive species, oxidative stress is the result of an imbalance between the generation of active oxygen species and their inactivation by the protective systems. Glutathione (GSH) plays a unique role in cellular defence against reactive intermediates, inasmuch as it functions as a reductant in the metabolism of hydrogen peroxide and organic hydroperoxides and can also bind electrophilic metabolites (Orrenius & Moldéus 1984). During the glutathione peroxidase-catalyzed metabolism of hydroperoxides, GSH serves as an electron donor and the glutathione disulfide (GSSG) formed in the reaction is subsequently reduced back to GSH by glutathione reductase at the expense of NADPH. Under conditions of oxidative stress, when the cell must cope with large amounts of H_2O_2 and/or organic hydroperoxides, the glutathione reductase is unable to match the rate of glutathione oxidation, and GSSG accumulates. In an apparent effort to avoid the detrimental effects of increased intracellular levels of GSSG, e.g. formation of mixed disulfides with protein thiols, the cell actively excretes the disulfide which can lead to a depletion of the intracellular glutathione pool.

Other thiol groups of toxicological significance are those present in cellular proteins. In general, sulfhydryl groups are highly reactive and can participate in several different types of reactions, such as arylation, oxidation, and thiol-disulfide exchange. All of these reactions may be involved in the modification of protein thiols as a consequence of their interaction with reactive intermediates formed during the metabolism of toxic chemicals. Such modifications of protein thiols may be particularly important during oxidative stress (Di Monte et al. 1984).

MENADIONE-INDUCED OXIDATIVE STRESS

During the past several years our laboratory has been actively engaged in the study of possible mechanisms of cell killing by oxidative stress in isolated hepatocytes. A model compound extensively used in our studies is menadione (2-methyl-1,4-naphthoquinone). In isolated liver cells menadione metabolism involves both one- and two-electron reduction pathways which differ in their contributions to cytotoxicity (Thor et al. 1982). One-electron reduction of quinones results in the formation of semiquinone radicals which can rapidly reduce dioxygen, forming the superoxide anion radical ($O_2^{\bar{\cdot}}$), and regenerating the parent quinone. Dismutation of $O_2^{\bar{\cdot}}$ to hydrogen peroxide, and the production of other highly reactive species, quickly lead to a condition of oxidative stress as redox cycling of the quinone continues.

Exposure of hepatocytes to toxic concentrations of menadione results in the extensive formation of $O_2^{\bar{\cdot}}$ and H_2O_2 and the oxidation of GSH and pyridine nucleotides. Following GSH depletion, a dose- and time-dependent loss of protein thiols precedes cell death in hepatocytes exposed to menadione (Di Monte et al. 1984). Although a general decrease in protein thiols has been observed in our experiments in which cells were exposed to cytotoxic levels of menadione, this does not imply that all protein thiols are equally important for cell survival. It appears that the fraction of critical thiols may be quite small. As mentioned above, the proper functioning of intracellular Ca^{2+} translocases is dependent on thiol groups in specific proteins, and the disruption of Ca^{2+} homeostasis resulting from their modification appears to be ultimately related to the onset of cytotoxicity. Other critical thiols affected by menadione metabolism are located on microfilament proteins, whose modification may result in the disruption of the structural integrity of the plasma membrane and the onset of widespread bleb formation.

APPEARANCE OF SURFACE BLEBS IN HEPATOCYTES EXPOSED TO TOXIC AGENTS

Incubation of hepatocytes with menadione results in an alteration of surface morphology characterized by a loss of microvilli and the appearance of multiple blebs on the surface of the hepatocytes (Thor et al. 1982). Many other toxic agents cause simular alterations in surface structure, indicating that plasma membrane blebbing is a common event in the progression of toxic injury (Jewell et al. 1982). Blebs usually appear before any sign of increased plasma membrane permeability is observed and seem to be initially reversible, i.e. they often disappear when the toxic agent is removed from the incubation.

Cell surface morphology is thought to be determined by the organization of cortical microfilaments associated with the plasma membrane. This assumption is supported by the finding that two classes of compounds, the cytochalasins and phalloidins which disrupt cortical microfilament structure, cause bleb formation on the surface of hepatocytes similar to that observed with other toxic agents.

As will be discussed below, during oxidative stress a perturbation of intracellular Ca^{2+} homeostasis is intimately related to a modification of the intracellular thiol status. Both disruption of intracellular thiol and Ca^{2+} homeostasis may result in alterations of cytoskeletal structures. Thiol oxidation can perturb microfilament organization through S-S-cross-linking of actin molecules (Mirabelli et al. 1987). A change in intracellular Ca^{2+} distribution can affect cytoskeletal structure because the calcium ion and its associated binding proteins play a pivotal role in regulating the cytoskeletal network (Cheung 1980). Alternatively, Ca^{2+}-activated catabolic enzymes, such as cytosolic proteases, can cleave cytoskeletally-associated proteins and thus cause alteration of the microfilament network and blebbing (Collier & Wang 1982; Nicotera et al. 1986a,b). It is therefore conceivable that both thiol modification and Ca^{2+}-dependent processes may combine to produce cytoskeletal alterations during oxidative stress.

DISRUPTION OF INTRACELLULAR CALCIUM HOMEOSTASIS DURING OXIDATIVE STRESS

The availability of non-invasive techniques to measure Ca^{2+} content in intracellular compartments has made it possible to monitor alterations in Ca^{2+} distribution during the development of toxicity in hepatocytes. Thus, in studies with isolated hepatocytes we have been able to demonstrate that menadione-induced oxidative stress causes the mobilization of Ca^{2+} from both mitochondria and the endoplasmic reticulum (Thor et al. 1982). Further studies with isolated subcellular fractions have shown that menadione metabolism impairs the ability of mitochondria to take up and retain Ca^{2+} by causing the oxidation of pyridine nucleotides (Moore et al. 1986). Menadione also inhibits Ca^{2+} uptake by rat liver microsomes through a mechanism which seems to involve oxidation/arylation of thiol group(s) necessary for Ca^{2+}-ATPase activity (Thor et al. 1985). The incubation of hepatocytes with menadione therefore causes the release into the cytosol of Ca^{2+} which cannot be resequestered. Normally, this would result in only a transient rise in the cytosolic Ca^{2+} concentration, because the plasma membrane Ca^{2+}-ATPase would remove this Ca^{2+} from the cell, and the Ca^{2+} concentration would return to its usual, very low level. However, recent studies in our laboratory have shown that the hepatic plasma membrane Ca^{2+}-ATPase is inhibited by agents, including menadione, which oxidize membrane protein thiols (Nicotera et al. 1985). The redox cycling of menadione can therefore inhibit the Ca^{2+} translocases present in the mitochondria, endoplasmic reticulum, and plasma membrane of hepatocytes, leading to a sustained rise in cytosolic Ca^{2+}.

Under physiological conditions the cytosolic Ca^{2+} concentration in hepatocytes is modulated by various hormones and other Ca^{2+}-mobilizing agents. To investigate the difference between physiological and toxicological fluctuations in cytosolic Ca^{2+} levels we have recently compared the effects of an alpha-adrenergic agonist, phenylephrine, with those of both non-toxic and toxic concentrations of menadione in Quin-2-loaded hepatocytes. Our study has shown that both the extent and the duration of the rise in cytosolic Ca^{2+} differ between non-toxic and toxic conditions. Toxic concentrations of menadione caused a sustained elevation of cytosolic Ca^{2+}, which reached micromolar concentrations and fully saturated the binding capacity of the entrapped dye. Furthermore, in view of the ability of reducing agents to prevent protein thiol modification and to protect hepatocytes from menadione toxicity, we investigated the effect of dithiothreitol on the increase in cytosolic Ca^{2+} caused by menadione. Treatment of hepatocytes with dithiothreitol prevented both the increase in cytosolic Ca^{2+} and the onset of cytotoxicity in hepatocytes exposed to menadione.

MECHANISMS OF CALCIUM-MEDIATED CYTOTOXICITY

The discovery of a relationship between a sustained increase in cytosolic Ca^{2+} concentration and the toxicity of menadione and several other agents in hepatocytes has led to a search for mechanism(s) by which the increased cytosolic Ca^{2+} could trigger cytotoxicity. It has been proposed that a sustained increase in Ca^{2+} concentration may cause abnormal stimulation of physiological functions, including various Ca^{2+}-dependent degradative processes.

Ca^{2+}-activated phospholipases are widely distributed in mammalian cells, and it has been suggested that an enhanced rate of phospholipid hydrolysis may result in irreversible cell damage. This assumption is supported by the observation that inhibitors of phospholipase activity can prevent ischemic cell death in liver and heart, and by the finding that phospholipid breakdown is enhanced during tissue injury (Chien et al. 1979). We have recently investigated whether phospholipase activation may contribute to the toxicity of the reactive disulfide cystamine in hepatocytes (Nicotera et al. 1986a). Indeed, exposure to cystamine did increase the rate of phospholipid hydrolysis, and this effect followed the elevation of cytosolic Ca^{2+} and preceded loss of cell viability. However, in our study, phospholipase inhibitors failed to prevent cystamine toxicity.

Intracellular protein degradation is also known to be stimulated by a rise in cytosolic Ca^{2+}, and a group of cytosolic, thiol-sensitive proteases, whose activity depends on micromolar concentrations of Ca^{2+}, seems to be involved in this process (Murachi 1983). These proteases function in the cleavage of membrane receptors, cytoskeletal proteins, and protein kinase C. Exposure of hepatocytes to toxic levels of cystamine resulted in a stimulation of Ca^{2+}-dependent, non-lysosomal proteases which appeared to be responsible for cystamine toxicity. Moreover, Ca^{2+}-activated proteases also seem to be involved in the toxicity of other agents, i.e. extracellular ATP and ionophore A23187, which cause a sustained increase in cytosolic Ca^{2+} by selectively interfering with the normal Ca^{2+} influx - efflux balance (Nicotera et al. 1986b). However, it remains to be established whether this mechanism of toxicity may apply also to other toxic compounds and to different cell systems.

A third catabolic process found to be sensitive to a non-physiological elevation of cytosolic Ca^{2+} concentration is that of an endogenous endonuclease, whose activity has been associated with "apoptosis", or programmed cell death (Wyllie 1980). Thus, both glucocorticoid hormones and the calcium ionophore A23187 stimulate apoptosis in isolated mouse thymocytes (Cohen & Duke 1984; Wyllie et al. 1984). Characteristic morphological changes accompany this process, including plasma membrane blebbing and extensive chromatin condensation.

As noted above, we have found that plasma membrane blebbing can often be linked to a disruption of Ca^{2+} homeostasis, and that the blebbing observed in hepatocytes treated with ionophore A23187 is associated with the activation of Ca^{2+}-dependent proteases (Nicotera et al. 1986b). The appearance of blebs on apoptotic cells could therefore be an indication that a perturbation of Ca^{2+} homeostasis accompanies this process. Since menadione metabolism creates such a disruption, we have recently studied its effects on endogenous endonuclease activity in hepatocytes (McConkey et al. 1987). In the presence of moderately toxic concentrations of menadione, endonucelase activation was apparent after about 2 hours; maximal DNA cleavage was seen with 100 µM menadione after 4 hours of incubation. Moreover, activation of the endonuclease preceded cell death by an appreciable period of time and was associated with the appearance of plasma membrane blebbing, characteristic of menadione-induced oxidative stress.

Interestingly, morphological alterations associated with menadione treatment of hepatocytes follow those observed with apoptotic thymocytes quite closely. The formation of surface blebs has already been mentioned, but chromatin condensation is also seen and a close similarity exists between DNA cleavage products isolated from apoptotic thymocytes and from menadione-treated hepatocytes. Gel electrophoresis of these products has shown that host chromatin is preferentially cleaved into oligonucleosome-length fragments by the endonuclease in each case (Wyllie 1980; McConkey et al. 1987). Thus, it appears that the activation of hepatocyte endonuclease activity is not only a critical step in the sequence of events involved in programmed cell death in the liver but can also occur during the development of oxidative cell injury.

CONCLUDING REMARKS

A series of potentially toxic mechanisms are activated when a cell is exposed to agents which cause oxidative stress. They include a perturbation of the normal redox balance, thiol depletion, and disruption of intracellular Ca^{2+} homeostasis. The pathways through which a perturbation of intracellular Ca^{2+} homeostasis may cause irreversible damage appear to involve the activation of degradative enzymes such as phospholipases, proteases, and endonucleases. Although additional mechanisms may contribute to cell killing, our findings suggest that a disruption of Ca^{2+} homeostasis is a critical event in the development of lethal cell injury.

REFERENCES

Bellomo, G., Mirabelli, F., Richelmi, P. & Orrenius S., 1983, FEBS Letters, 163, 136.

Campbell, A.K., 1983, Intracellular Calcium: Its Universal Role as Regulator, (Wiley and Sons).

Cheung, W.Y., 1980, Science 217, 1257.

Chien, K.R., Pfau, R.G. & Farber, J.L., 1979, American Journal of Pathology, 97, 505.

Cohen, J.J. & Duke, R.L., 1984, Journal of Immunology, 132, 38.

Collier, N.L. & Wang, K., 1982, Journal of Biological Chemistry, 257, 6937.

Di Monte, D., Bellomo, G., Thor, H., Nicotera, P. & Orrenius, S., 1984, Archives of Biochemistry and Biophysics, 235, 343.

Farber, J.L., 1981, Life Sciences, 29, 1289.
Jewell, S.A., Bellomo, G., Thor, H., Orrenius, S. & Smith, M.T., 1982, Science, 214, 1257.
Lehninger, A.L., Vercesi, A. & Bababunmi, E., 1978, Proceedings of the National Academy of Sciences USA, 75, 1690.
McConkey, D., Nicotera, P., Hartzell, P., Wyllie, A.H. & Orrenius, S., 1987, Journal of Biological Chemistry (submitted).
Mirabelli, F., Salis, A., Marinoni, V., Finardi, G., Bellomo, G., Thor, H. & Orrenius, S., 1987, Journal of Biological Chemistry (submitted).
Moore, L., Chen, T., Knapp, H.R. & Landon, E., 1975, Journal of Biological Chemistry, 250, 4562.
Moore, G.A., O'Brien, J.P. & Orrenius, S., 1986, Xenobiotica, 16, 873.
Murachi, T., 1983, In Calcium and Cell Function, Vol. IV, edited by W.Y. Cheung, (Academic Press) p. 376.
Nicotera, P., Hartzell, P., Baldi, C., Svensson, S.Å., Bellomo, G. & Orrenius, S., 1986a, Journal of Biological Chemistry, 261, 14628.
Nicotera, P., Hartzell, P., Davis, G. & Orrenius, S., 1986b, FEBS Letters, 209, 139.
Nicotera, P., Moore, M., Mirabelli, F., Bellomo, G. & Orrenius, S., 1985, FEBS Letters, 181, 149.
Orrenius, S. & Bellomo, G., 1986 In Calçium and Cell Function, Vol. VI, editied by W.Y. Cheung, (Academic Press) p. 185.
Orrenius, S. & Moldéus, P., 1984, Trends in Pharmacological Sciences, 5, 432.
Shanne, F.A., Kane, A.B., Young, E.E. & Farber, J.L., 1979, Science, 206, 700.
Sies, H., 1986, Angewandte Chemie International Editions English, 25, 1058.
Starke, P.E., Hoek, J.B. & Farber, J.L., 1986, Journal of Biological Chemistry, 261, 3006.
Thor, H., Hartzell, P., Svensson, S.Å., Orrenius, S., Mirabelli, F., Marinoni, V. & Bellomo, G., 1985, Biochemical Pharmacology, 34, 3717.
Thor, H., Smith, M.T., Hartzell, P., Bellomo, G., Jewell, S.A. & Orrenius, S., 1982, Journal of Biological Chemistry, 257, 12419.
Wyllie, A.H., 1980, Nature, 284, 555.
Wyllie, A.H., Morris, R.G., Smith, A.C. & Dunlop, D., 1984, Journal of Pathology, 14, 67.

Drug Metabolism and Toxicity in Hypoxic States

JAMES C. VELTMAN and M.W. ANDERS

Department of Pharmacology, University of Rochester,
601 Elmwood Avenue, Rochester, NY 14642, USA

Hypoxic states alter the metabolism and toxicity of drugs and chemicals. The in vivo toxicity of, for example, carbon tetrachloride (CCl_4) (Shen et al. 1979) is increased in animals maintained under hypoxic conditions. This hypoxic potentiation of toxicity is thought to be due to increased reductive bioactivation, which is favored by hypoxic conditions. Both the halogenated hydrocarbon and oxygen compete for electrons provided by cytochromes P-450 (Nastainczyk et al. 1978); because oxygen is more easily reduced than the halogenated compounds, the reductive bioactivation of halothane and CCl_4 is favored by decreased cellular oxygen concentrations.

The possibility that hypoxic states may alter the response of cells to chemical insult also needs to be considered. Indeed, hypoxic conditions alone produce centrilobular liver injury (Lemasters et al. 1981; Shingu et al. 1982), and hypoxia may increase the toxicity of compounds that undergo oxidative metabolism (Jones 1985). Mild hypoxia, the so-called neahypoxia, may produce a large number of changes in metabolic processes (Jones et al. 1985) and may labilize cells to the cytotoxic effects of chemicals.

In the experiments described here, we have investigated the role of glutathione (GSH) homeostasis in CCl_4-induced cytotoxicity and the effect of neahypoxia on the cytotoxicity of N,N-bis(2-chloroethyl)-N-nitrosourea (BCNU).

MATERIALS AND METHODS

Hepatocytes were isolated from male Long-Evans rats according to the procedure of Moldéus et al. (1978), except that all buffers were supplemented with 25 mM HEPES and 2% B.S.A. Cell suspensions with a viability greater than

90% by trypan blue exclusion, were diluted to a final concentration of 2 x 10^6 cells/ml. Hepatocytes were incubated with either 95% O_2:5% CO_2 or 5% O_2:90% N_2:5% CO_2 at 37°C. A 5% O_2 concentration yields an initial O_2 concentration of 52 μM. Presuming a respiration rate of 12 nmol $O_2/10^6$ cells/min, the O_2 concentration decreases to approximately 37, 20, and 6 μM after 1, 2, and 3 h, respectively (Jones, 1985). At the indicated times, cell viability was measured as were the concentrations of malondialdehyde (MDA) (Lindstrom et al. 1970), GSH (Reed et al. 1980), and ATP (Lamprecht and Trautschold, 1974).

RESULTS AND DISCUSSION

A comparison of CCl_4- and BCNU-induced cytotoxicity in normoxic (95% O_2) and hypoxic (5%) hepatocytes is shown in Table 1. Hepatocytes incubated under 5% O_2 were more susceptible to CCl_4- and BCNU-induced cytotoxicity than were similarly treated cells incubated under 95% O_2.

Table 1. Cytotoxicity of BCNU and CCl_4 in normoxic and hypoxic isolated rat hepatocytes.

Treatment	% Viability	
	95% O_2	5% O_2
Control	92	93
CCl_4 (2.5 mM)	77	49
CCl_4 (5.0 mM)	75	42
CCl_4 (7.5 mM)	74	40
BCNU (25 μM)	89	72
BCNU (50 μM)	88	60
BCNU (100 μM)	88	48

Isolated rat hepatocytes (2 x 10^6 cells/ml) were incubated with the indicated concentrations of CCl_4 or BCNU at 37°C. One hour later, the cell viability was determined by trypan blue exclusion. Each value is the mean of three experiments.

Lipid peroxidation was also potentiated by hypoxia. Normoxic hepatocytes incubated with 2.5 mM CCl_4 showed little increase in MDA production over control normoxic cells. CCl_4-exposed hypoxic cells produced 3-times the amount of MDA as normoxic cells. Hypoxia potentiates CCl_4-induced lipid peroxidation in microsomes (Noll & DeGroot 1984) and in hepatocytes (DeGroot & Noll 1986). The role of GSH in preventing lipid peroxidation and CCl_4-induced cytotoxicity in hypoxic cells, however, is not clear. Although hypoxia compromises some hepatic detoxification pathways, reactions with GSH are not affected (Jones 1985). In our studies, however, cellular GSH concentrations were depleted in hypoxic cells exposed to CCl_4; these cells had a GSH concentration of approximately 2-3 nmol/10^6 cells compared with the concentration of 50-55 nmol/10^6 cells observed in normoxic control hepatocytes. This finding suggests two questions: 1) Are hepatocyte GSH concentrations conserved in hypoxic cells treated with CCl_4 and an antioxidant? 2) Does GSH prevent CCl_4-induced cytoxicity by preventing lipid peroxidation?

The data in Table 2 provide answers to these questions. Hypoxic hepatocytes were incubated with vehicle, the antioxidant N,N'-dipheny-p-phenylenediamine (DPPD), or precursor amino acids (L-glu, L-met, gly) for GSH biosynthesis, in the presence or absence of CCl_4. After 3 h, CCl_4 reduced cell viability and GSH contents and increased MDA production. The addition of DPPD delayed, but did not prevent, CCl_4induced cytotoxicity. In cells incubated with CCl_4 alone or with DPPD and CCl_4, GSH concentrations were depleted and lipid peroxidation was prevented. In contrast to DPPD, precursor amino acids prevented CCl_4-induced cytotoxicity, but did not block fully CCl_4-induced lipid peroxidation. Hypoxic hepatocytes displayed a biphasic response to CCl_4 in that cellular GSH concentrations initially dropped but after 3 h increased to over 3 times the concentration in hypoxic control cells (Table 2). In other experiments, hepatocytes were incubated with BCNU (150 μM, 1 h) under 95% O_2 and later incubated under 5% O_2 in the presence of CCl_4. After 3 h, BCNU augmented CCl_4-induced cytotoxicity (42% and 20% in suspensions treated with CCl_4 or with BCNU and CCl_4, respectively). Lipid peroxidation was only slightly elevated.

These findings indicate that during reductive bioactivation of CCl_4 in hypoxic cells, GSH homeostasis becomes disrupted, and this effect is more closely associated with cytotoxicity than lipid peroxidation. This idea is consistent with studies that report a lack of correlation between the degree of lipid peroxidation and cytotoxicity (Waller et al. 1983) and with the hypothesis that the major event in CCl_4 metabolism is the covalent binding of free radicals to microsomal protein and lipid (Recknagel 1983). Moreover, earlier studies with rats (Shen et al 1982) showed a 2-3-fold increase in the amount of covalent binding of CCl_4 metabolites in hypoxic compared with normoxic animals. This raises speculation that hypoxia may either augment trichloromethyl radical formation (Wolf et al. 1980) or facilitate the binding of metabolites to macromolecules (Tribble et al. 1987).

Table 2. Effects of CCl_4 and antioxidants or precursor amino acids on cytotoxicity, lipid peroxidation, and cellular glutathione concentrations in hypoxic rat hepatocytes.

Treatment	Viability (%)	Malondialdehyde (ng/10^6 cells)	Glutathione (nmol/10^6 cells)
Control	83	0.17	24.4
CCl_4	16	0.67*	2.9*
DPPD	81	0.10	21.9
DPPD and CCl_4	19	0.13	5.7*
A.A.	81	0.18	60.8*
A.A. and CCl_4	75	0.39*	80.0*

Isolated hepatocytes (2.0 x 10^6 cells/ml) were incubated with either vehicle (ethanol, < 0.5%) or the listed compounds for 3 h at 37°C under 5% O_2. Afterwards, cell viability, lipid peroxidation, and cellular glutathione concentrations were determined. A.A. = precursor amino acids (5 mM L-glu, L-met, gly); DPPD = N,N-diphenyl-p-phenylenediamine (10 µM); CCl_4, 5.0 mM. Data represent the mean of three experiments. *$p < .05$ when compared with controls.

The data in Table 1 show a dose-dependent increase in cytotoxicity in hepatocytes incubated in 5% O_2 and BCNU. Because BCNU undergoes reductive inactivation to nontoxic metabolites (Potter & Reed 1982), an alternative mechanism must be invoked for hypoxic potentiation of BCNU-induced cytotoxicity.

Table 3 shows a comparison of the effects of hypoxia and BCNU on hepatocyte GSH and ATP concentrations. Incubation of hepatocytes in 5% O_2 yields a medium O_2 concentration of 36 μM after 1 h, which may produce a neahypoxic condition (Jones et al. 1985). The absence of any loss of cellular ATP content confirms this contention as do data from Table 1 showing no loss of viability accompanying hypoxia. Hypoxia alone lowered GSH concentrations. This lower GSH concentration corresponds to the reported heterogeneous GSH concentration within the liver lobule, which may be attributed to the different lobular O_2 concentrations (Slater et al. 1985). Addition of BCNU lowered hepatocyte GSH concentrations in both normoxic and hypoxic cells. BCNU did not increase intracellular glutathione disulfide concentrations in either normoxic or hypoxic cells and increased the extracellular glutathione disulfide content in only the normoxic cells. The degradation products of BCNU include 2-chloroethyl isocyanate, 2-chloroethylamine, and 2-chloroethanol, all of which may react with GSH (Reed 1985). Hence, the observed loss of GSH contents in hepatocytes incubated with BCNU may be the result of conjugation reactions of GSH and BCNU degradation products. These GSH conjugates may be formed under either oxygen concentrations, but the data in Table 3 indicate that GSH synthesis may be compromised during hypoxia and, thus, cannot offset the loss of GSH induced by BCNU.

Table 3. Effects of BCNU on ATP and glutathione concentrations in normoxic and hypoxic hepatocytes.

Treatment	Glutathione ($nmol/10^6$ cells)		ATP ($nmol/10^6$ cells)	
	95% O_2	5% O_2	95% O_2	5% O_2
Control	32.0	25.0	18.5	21.0
BCNU (100 μM)	25.0	13.5*	16.5	6.2*

Isolated hepatocytes were incubated under the described conditions at 37°C for 1 h. Values represent the mean of 3 experiments. *$p < .05$ when compared with control value.

BCNU-induced cytotoxicity in hypoxic cells is also associated with a loss of cellular ATP content (Table 3). If GSH S-conjugates are formed, their export by plasma membrane ATPases (Nicotera et al. 1985) may promote the hydrolysis of ATP. This assertion is supported by data showing that the loss of ATP in hepatocytes incubated with BCNU and 5% O_2 is accompanied by an increase in ADP concentrations.

The data indicate that the combined loss of cellular GSH and ATP concentrations is associated with BCNU-induced cytotoxicity in hypoxic cells. Table 4 shows a comparison of the effects of compounds that increase or decrease cellular ATP or GSH concentrations on cell viability in hepatocytes incubated with BCNU in 95% O_2 and 5% O_2. Ethionine, a known depletor of cellular ATP content, is not cytotoxic; however, when combined with BCNU, cell viability of normoxic hepatocytes is reduced by 50%. Moreover, the GSH concentration of these hepatocytes was also reduced to 16 nmol/10^6 cells. Both of these values were comparable to those observed in BCNU-exposed hypoxic hepatocytes. Likewise, treatment of normoxic cells with both diethyl maleate and BCNU increased cytotoxicity. Because diethyl maleate depletes cellular GSH by forming GSH S-conjugates (Högberg et al. 1975), the increased cytotoxicity may be the consequence of both reduced cellular GSH and greater rates of ATP hydrolysis in extruding GSH S-conjugates. As expected, ATP concentrations in normoxic hepatocytes incubated with both diethyl maleate and BCNU decreased from 20 to 3 nmol/10^6 cell, which is similar to the ATP content in hypoxic hepatocytes exposed to BCNU alone. Adenosine increases hepatocyte ATP concentrations (Lund et al. 1975). When adenosine was added to hepatocytes incubated with BCNU and 5% O_2, no cytotoxicity was observed, and ATP concentrations were 50% greater than in hypoxic control cells. This finding also indicates that hepatocytes in a neahypoxic state cannot resynthesize enough ATP to match the rate of energy consumption stimulated by these conjugates. If the loss of ATP is prevented, as observed after incubation with adenosine, BCNU-induced cytotoxicity was also prevented.

Although neahypoxic conditions may increase the reductive bioactivation of CCl_4, it is apparent from the data that hypoxia may also potentiate chemically induced cytotoxicity by rendering hepatocytes more susceptible to toxic injury and more prone to enter a state of bioenergetic hypoxia that compromises vital cellular functions (Aw & Jones 1985). This hypothesis is supported by data showing similar cytotoxic effects of BCNU and its degradation products, 2-chloroethanol, 2-chloroethyl isocyanate and 2-chloroethylamine, in hypoxic, but not normoxic, hepatocytes.

Table 4. Cytotoxicity of BCNU and ethionine, adenosine, or diethyl maleate in normoxic and hypoxic rat hepatocytes.

Treatment	Cytotoxicity (%) 95% O_2	5% O_2
Control	93	88
BCNU (100 μM)	87	57
BCNU and Ethionine (1 mM)	55	N.D.
BCNU and Adenosine (1 mM)	88	83
BCNU and Diethyl Maleate (300 μM)	41	N.D.

Hepatocytes were incubated with either 95% O_2 or 5% O_2 and the listed compounds at 37°C. One hour later cell viability was assessed by trypan blue exclusion. N.D. = not determined. Each value represents the mean of two experiments.

ACKNOWLEDGMENT

This research was supported by N.I.E.H.S. grant ES03126 to M. W. Anders and National Research Service Award HL07475 to J. C. Veltman. The authors thank Ms. Sandra Morgan for typing the manuscript.

REFERENCES

Aw, T. & Jones, D., 1985, American Journal of Physiology, 249, C385.

DeGroot, H. & Noll, T., 1986, Biochemical Pharmacology, 35, 15.

Högberg, J., Orrenius, S. & O'Brien, P., 1975, European Journal of Biochemistry, 59, 449.

Jones, D., 1985, The role of oxygen concentration in oxidative stress: Hypoxic and hyperoxic models. In Oxidative Stress, edited by H. Sies (Academic Press), p. 151.

Jones, D.P., Kennedy, F.G., Andersson, B.S., Aw. T.Y. & Wilson, E., 1985, Molecular Physiology, 8, 473.

Lamprecht, W. & Trautschold, J., 1974, Adenosine-5'-triphosphate determination with hexokinase and glucose-6-phosphate dehydrogenase. In Methods of Enzymatic Analysis, Vol. 4, edited by H. Bergmeyer (Verlag Chemie/Academic Press), p. 2101.

Lemasters, J.J., Ji, S. & Thurman, R.G., 1981, Science, 213, 662.
Lindstrom, T., Anders, M. & Remmer, Jr., 1978, Experimental and Molecular Pathology, 28, 48.
Lund, P., Cronell, N. & Krebs, H., 1975, Biochemical Journal, 152, 593.
McLain, G.E., Sipes, I.G. & Brown, B.B., Jr., 1979, Anesthesiology, 51, 321.
Moldéus, P., Högberg, Jr. & Orrenius, S., 1978, Methods in Enzymology, 51, 60.
Nastainczyk, W., Ullrich, V. & Sies, H., 1978, Biochemical Pharmacology, 27, 387.
Nicotera, P., Moore, M., Bellomo, G., Mirabelli, F. & Orrenius, S., 1985, Journal of Biological Chemistry, 260, 1999.
Noll, T. & DeGroot, H., 1984, Biochimica Biophysica Acts, 795, 356.
Potter, D. & Reed, D., 1982, Archives of Biochemistry and Biophysics, 216, 158.
Recknagel, R., 1983, Life Science, 33, 401.
Reed, D., 1985, Nitrosoureas. In Oxidative Stress, edited by H. Sies (Academic Press), p. 115.
Reed, D., Babson, J., Beatty, P., Brodie, A., Ellis, W. & Potter, D., 1980, Analytical Biochemistry, 106, 55.
Shen, E.S., Garry, V.F. & Anders, M.W., 1982, Biochemical Pharmacology, 31, 3787.
Shingu, K., Eger, E.I., II & Johnson, B.G., 1982, Anesthesia and Analgesia, 61, 820.
Slater, T., Cheeseman, K. & Ingold, K., 1985, Philosophical Transaction of the Royal Society of London, B311, 633.
Tribble, D., Aw, T., & Jones, D., 1987, Hepatology, 7, 377.
Waller, R., Glende, E., Recknagel, R., 1983, Biochemical Pharmacology, 32, 1613.
Wolf, C., Harrelson, W., Jr., Nastainczyk, W., Philpot, R., Kalyanaraman, B. & Mason, R., 1980, Molecular Pharmacology, 18, 553.

Rat Liver Cell Type Specific Patterns of Drug Metabolizing Enzymes and Consequences for the Control of Genotoxic Metabolites

F. OESCH, W.M. LAFRANCONI, M. ARAND and P. STEINBERG

Institute of Toxicology, University of Mainz,
Obere Zahlbacher Str. 67, D-6500 Mainz,
Federal Republic of Germany

Rat liver preparations, in most cases whole liver homogenates, have been widely used as an in vitro model system for studying the metabolism of xenobiotics. However, the liver is not homogeneous with respect to either metabolism or in the way it responds to insults from xenobiotics, e.g. when adult rats are exposed to vinyl chloride tumors develop from the sinusoidal lining cells without apparent damage to the parenchymal cells (Creech and Johnson, 1974; Maltoni and Lefermine, 1974). The basis of this heterogeneous response is unknown, but it may be due to a number of factors including differences between cell populations in repair after insult, the heterogeneous distribution of xenobiotics within the liver and differences in the activation of xenobiotics to toxic metabolites. It is this last point we will refer to in this paper.

The parenchymal cells are the most abundant cell population in the liver: they account for 90 to 95 per cent of the total cell mass and 65 per cent of the total cell number (Weibel et al., 1969). The nonparenchymal cells of the liver are primarily sinusoidal lining cells (endothelial, Kupffer, fat-storing and pit cells), but also hemopoietic, bile duct and blood vessel wall cells. When a xenobiotic enters the hepatic circulation the first population of cells it encounters are the sinusoidal lining cells. By the time it reaches the parenchymal cells, it is already partially altered by whatever enzyme activities may be present in the nonparenchymal cells. While much has been written about the metabolic potential of parenchymal cells, a very limited number of reports have focused on xenobiotic metabolizing enzymes in nonparenchymal cells (Cantrell and Bresnick, 1972; Morland and Olsen, 1977). Therefore, we have measured the activities of some important xenobiotic metabolizing enzymes in nonparenchymal cells and compared them with those determined in parenchymal cells in order to better characterize the enzyme distribution in the rat liver. Furthermore, we have studied the genotoxic effects of certain model carcinogens activated by parenchymal and nonparenchymal cells in the Salmonella/ mammalian microsome mutagenicity test (Ames test) to determine the toxicological significance of the enzyme activities measured.

Materials and Methods

All chemicals were purchased from commercial sources and were of the highest purity grade available. ± trans-7,8-Dihydroxy-7,8-dihydrobenzo(a)pyrene was prepared by Dr. Karl Platt at our Institute with methods described by McCaustland and Engel (1975) using the synthetic modification suggested by Fu and Harvey (1977). Benzphetamine was a generous gift from Dr. Anthony Y.H. Lu, Merck Sharpe and Dohme (Rahway, N.J.). The Salmonella typhimurium tester strain TA-100 was kindly supplied by Dr. Bruce Ames (University of California, Berkeley, CA).

Male Sprague-Dawley rats (150-200 g body weight) were purchased from Süddeutsche Versuchstierfarm (Tuttlingen, FRG), housed four to a cage, and allowed free access to both water and food until used. Aroclor 1254 in corn oil was administered as a single intraperitoneal dose (500 mg/kg body weight) 5 days before killing. Untreated rats received appropriate volumes of corn oil.

Cell isolation and characterization

Monodispersed preparations of hepatocytes were obtained using a two-step collagenase perfusion method described by Glatt et al. (1981). The heterogeneous suspension of liver cells was resolved into two major fractions by differential centrifugation. The suspension was centrifuged and washed three times at 50 g for 3 min and the supernatants saved. The final pellet contained > 95 % parenchymal cells. The supernatants, containing the nonparenchymal cells, were pooled and centrifuged at 300 g for 5 min. The cells in the pellet were resuspended and the contaminating parenchymal cells removed from the preparation by pronase E digestion (0.25 %) at 10 °C for 60 min. The cells were then freed of cellular debris and remaining erythrocytes by layering the cell suspension on a 1.090 g/cm^3 Ficoll cushion and centrifuging at 1400 g for 15 min.

Cells were identified based on: a) their morphology at the light and electron microscopic levels (Knook and Sleyster, 1976); b) the distribution of pyruvate kinase isoenzyme activities (van Berkel, 1972). Viability was assessed by counting the proportion of cells that excluded 0.25 % trypan blue.

Enzyme assays

Benzphetamine N-demethylase was determined colorimetrically by following the formation of formaldehyde (Lu et al., 1972). Ethoxyresorufin O-deethylase activity was measured fluorimetrically according to the method of Burke and Mayer (1974). Cytochrome P-450 content was determined as described by Omura and Sato (1964). Epoxide hydrolase activity was measured radiometrically with [^{3}H]Benzo(a)-pyrene 4,5-oxide (Schmassmann et al., 1976). Glutathione transferase and UDP- glucuronosyltransferase activities were measured spectrophotometrically according to the methods of Habig et al. (1974) and Bock et al. (1983), respectively, using 1 mM 1-chloro-2,4-dinitrobenzene to assay for glutathione transferase and 5 mM 1-naphthol as the substrate for UDP-glucuronosyltransferase. Proteins were measured by the method of Lowry et al. (1951) with bovine serum albumin as standard.

Cell-mediated bacterial mutagenicity test

Freshly prepared parenchymal cells (1 x 10^{6} cells) or nonparenchymal cells (10 x 10^{6} cells) were incubated for 1 hour at 37 °C with about 1.7 x 10^{8} Salmonella typhimurium (TA-100) and the indicated amount of test compound dissolved in 10 µl acetone as described by Glatt et al. (1981). The contents of each tube were then mixed with molten top agar, poured onto histidine-poor agar plates and incubated at 37 °C in the dark for 3 days, after which, the number of histidine-independent bacterial colonies was counted.

Results and Discussion

The cell isolation procedure chosen (pronase E digestion at 10 °C) provided nonparenchymal cell populations that were essentially (more than 99 %) free from contamination with whole parenchymal cells and parenchymal cell debris: pronase E selectively destroys parenchymal cells, while the low temperature prevents the phagocytosis of debris by the nonparenchymal cells. On the other hand, the parenchymal cell preparations contained less than 5 % nonparenchymal cells.

The parenchymal cells possessed greater enzyme activities than the nonparenchymal cells (Table 1). The differences were particularly large for benzphetamine N-demethylase (280-fold), ethoxyresorufin O-deethylase (14-fold) and the total amount of cytochrome P-450 (6-fold). The differences in the specific activities between the two cell populations were much smaller (2.0- to 2.5-fold) in the case of the postoxidative enzymes (cytosolic glutathione transferase, UDP-glucuronosyltransferase and microsomal epoxide hydrolase). This indicates that nonparenchymal cells might have a relatively lower ability to oxidize xenobiotics to reactive electrophiles and a greater ability to conjugate or hydrolyze those products that may be formed. Indeed, freshly isolated parenchymal cells were shown to metabolize vinyl chloride to reactive chemical species which bound to cellular protein (Ottenwälder et al., 1983). However, when nonparenchymal cells were incubated with vinyl chloride the protein binding of metabolites was only 40 % of that observed with parenchymal cells.

Treatment of the rats with Aroclor 1254, a commercial polychlorinated biphenyl mixure which simultaneously exhibits phenobarbital and 3-methylcholanthrene-type inducing properties, consistently led to an increase in all enzyme activities measured in both cell populations (Table 1). In parenchymal cells the most dramatic increase was seen with the ethoxyresorufin O-deethylase (150-fold), while the other activities measured and the content of cytochrome P-450 were increased 3- to 6-fold. In the nonparenchymal cells ethoxyresorufin O-deethylase activity was also remarkably increased (260-fold) by the Aroclor 1254 treatment (Note: the ratio between the means of induced and untreated is 261. However, because of the large error in the very low control activity, this represents only a rough estimation of the increase in activity by the Aroclor 1254 treatment). Likewise, benzphetamine N-demethylase activity in the nonparenchymal cells, only spuriously present in untreated animals, was also dramatically increased in Aroclor 1254-treated animals (the ratio between the means of induced and untreated was 180). Despite the much more dramatic changes in the cytochrome P-450 dependent enzyme activities in nonparenchymal cells, the concentration of cytochrome P-450 in nonparenchymal cells was increased by the Aroclor 1254 treatment to a similar extent (5.9-fold) to that observed in parenchymal cells (4.8-fold). This finding remains unexplained. It is possible that the nonparenchymal cells possess enzymes other than the cytochrome P-450 dependent ones, which can metabolize ethoxyresorufin and benzpheta-

Table 1. Drug metabolizing enzyme activities in rat liver parenchymal and nonparenchymal cells isolated from untreated and Aroclor 1254-pretreated rats[a]

Enzyme assay	Cell type	Untreated	Aroclor 1254-pretreated
Ethoxyresorufin O-deethylase[b]	Parenchymal	14 ± 5	2153 ± 456*
	Nonparenchymal	1 ± 1	261 ± 109*
Benzphetamine N-demethylase[b]	Parenchymal	830 ± 128	4175 ± 722*
	Nonparenchymal	3 ± 4	538 ± 174*
Cytochrome P-450[c]	Parenchymal	416 ± 53	2400 ± 590*
	Nonparenchymal	66 ± 11	454 ± 103*
Cytosolic glutathione transferase[d]	Parenchymal	656 ± 64	2289 ± 444*
	Nonparenchymal	305 ± 98	754 ± 196*
UDP-glucuronosyl transferase[b]	Parenchymal	3190 ± 696	13043 ± 1409*
	Nonparenchymal	1290 ± 196	4648 ± 668*
Microsomal epoxide hydrolase[b]	Parenchymal	2460 ± 470	8569 ± 1365*
	Nonparenchymal	1223 ± 897	6543 ± 1444*

[a] All enzyme activities were assayed in broken cell preparations. Values are means ± SD from at least six experiments per group.

[b] Activity is expressed as pmol product formed/min/mg whole homogenate protein.

[c] Assayed spectrophotometrically in microsomal preparations. Units are pmol cytochrome P-450/mg microsomal protein.

[d] Activity is expressed as nmol product formed/min/mg whole homogenate protein.

* Indicates significantly different from the corresponding control value ($p < 0.05$, Student's t-test).

mine and which are induced by Aroclor 1254 (e.g. the flavine containing mixed function oxidase). Another possibility is that the cytochrome P-450 found in nonparenchymal cells from untreated animals is almost exclusively involved in the metabolism of endogenous substrates; after treatment with Aroclor 1254 a larger proportion of the cytochrome P-450 would be then able to use benzphetamine and ethoxyresorufin as substrates.

The postoxidative enzyme activities were also increased in both cell populations after administration of Aroclor 1254 to the rats, but the increase was not as great as for the oxidative activities. As a result of the differential induction, the ratio of the measured cytochrome P-450 dependent activities to the postoxidative activities was drastically decreased in the nonparenchymal cells by the Aroclor 1254 treatment. Nevertheless, even after induction, this ratio was still consistently lower in the nonparenchymal than in the parenchymal cells.

The majority of chemical carcinogens require metabolic activation to reactive electrophilic intermediates, which interact with vital cellular macromolecules such as DNA. In order to assess the activation potential of the parenchymal and nonparenchymal cell populations, the ability of these cells to mediate mutagenicity in bacteria was monitored using the Ames test. Representative carcinogens were chosen from 3 major groups: the polycyclic aromatic hydrocarbons (benzo(a)pyrene and ± trans-7,8-dihydroxy-7,8-dihydrobenzo-(a)pyrene), the aflatoxins (aflatoxin B_1) and the aromatic amines (2-aminoanthracene). Table 2 shows that the nonparenchymal cells were consistently weaker mediators of bacterial mutagenicity than were the parenchymal cells. The one exception was ± trans-7,8-dihydroxy-7,8-dihydrobenzo(a)-pyrene (the immediate precursor of the major ultimate mutagen and carcinogen benzo(a)pyrene-7,8-dihydrodiol-9,10-epoxide), which was activated to a similar extent by the two cell populations.

In summary, xenobiotic metabolizing enzyme activities were found to be consistently lower in the nonparenchymal than in the parenchymal cells. Despite the normally low enzyme activities in nonparenchymal cells, these cells were capable of metabolizing some carcinogens to genotoxic species, which indicates that not only parenchymal cells but also nonparenchymal cells may contribute to the hepatic disposition and toxicity of xenobiotics.

Table 2. Activation of carcinogens to mutagens by rat liver parenchymal and nonparenchymal cells

Carcinogen	μg	Histidine revertants per plate Parenchymal cells	 Nonparenchymal cells
Benzo(a)pyrene	5	208	42
	10	231	48
	25	239	56
	50	481	73
	75	662	109
	100	875	143
± trans-7,8-Dihydroxy-7,8-dihydrobenzo(a)-pyrene	0.25	745	316
	0.50	1532	1380
	0.75	1918	2541
	1.00	1204	2569
Aflatoxin B_1	0.1	181	52
	0.5	876	373
	1.0	1695	724
	1.5	1190	612
2-Aminoanthracene	0.5	1282	271
	1.0	2163	614

One million parenchymal cells and 10^7 nonparenchymal cells (roughly equal protein concentration) from Aroclor 1254-pretreated male Sprague-Dawley rats were incubated for 1 hour with bacteria (Salmonella typhimurium TA-100) and the indicated amount of test compound. The number of histidine-independent colonies was counted after plating the bacteria and incubating for 3 days in the dark at 37 °C. Values are means from 3 experiments.

Acknowledgements

This work was supported by the Deutsche Forschungsgemeinschaft (SFB 302) and the Alexander von Humboldt Foundation (W.M.L., P.S.). We also thank Ms. S. Pollok for typing this manuscript.

References

Bock, K.W., Burchell, B., Dutton, G.J., Hanninen, O., Mulder, G.J., Owens, I.S., Siest, G. & Tephly, T.R., 1983, Biochemical Pharmacology, 32, 953.

Burke, M.D. & Mayer, R.T., 1974, Drug Metabolism and Disposition, 2, 583.

Cantrell, E. & Bresnick, E., 1972, Journal of Cell Biology, 52, 316.

Creech, J.L. Jr. & Johnson, M.N., 1974, Journal of Occupational Medicine, 16, 150.

Fu, P.P. & Harvey, R.G., 1977, Tetrahedron Letters, 2059.

Glatt, H.R., Billings, R., Platt, K.L. & Oesch, F., 1981, Cancer Research, 41, 270.

Habig, H., Pabst, M.J. and Jakoby, W.B., 1974, Journal of Biological Chemistry, 249, 7130.

Knook, D.L. & Sleyster, E.Ch., Experimental Cell Research, 99, 444.

Lowry, O.H., Rosebrough, J., Farr, A.L. & Randall, R.J., 1951, Journal of Biological Chemistry, 193, 265.

Lu, A.Y.H., Somogyi, A., West, S., Kuntzman, R. & Conney, H.H., 1972, Archives of Biochemistry and Biophysics, 152, 457.

Maltoni, C. & Lefermine, G., 1974, Environmental Research, 7, 387.

McCaustland, D.J. & Engel, J.F., 1975, Tetrahedron Letters, 2549.

Morland, J. & Olsen, H., 1977, Drug Metabolism and Disposition, 5, 511.

Omura, T. & Sato, R., 1964, Journal of Biological Chemistry, 239, 2370.

Ottenwälder, H., Kappus, H. & Bolt, H.M., 1983, Archives of Toxicology, Suppl. 6, 266.

Schmassmann, U., Glatt, H.R. & Oesch, F., 1976, Analytical Biochemistry, 74, 94.

Van Berkel, T.J.J., Koster, J.F. and Hulsmann, W.C., 1972, Biochimical et Biophysica Acta, 276, 425.

Weibel, E.R., Stäubli, W., Gnägi, R.H. and Hess, F.A., Journal of Cell Biology, 42, 68.

Binding of Metabolically Formed Bay-Region Diol Epoxides to DNA

DONALD M. JERINA, ALBERT M. CHEH, ANJU CHADHA, HARUHIKO YAGI and JANE M. SAYER

Laboratory of Bioorganic Chemistry, NIDDK, National Institute of Health, Bethesda, MD 20892, USA

INTRODUCTION

In 1976 we had proposed bay-region diol epoxides as the metabolites responsible for the tumorigenic activity of the alternant polycyclic aromatic hydrocarbons provided that the hydrocarbon is carcinogenic and that such metabolites are formed (Jerina & Daly 1976). The general pathway for the formation of these metabolites (Fig. 1) consists of i) oxidation of a benzo ring of the hydrocarbon by cytochrome P450 to form a non-bay-region arene oxide, ii) hydration of this arene oxide by microsomal epoxide hydrolase to form a trans dihydrodiol, and iii) epoxidation of the benzo-ring double bond by cytochrome P450 to form the bay-region diol epoxide. Since only trans but not cis dihydrodiols are formed by mammalian enzymes, only two diastereomeric diol epoxides are possible; diol epoxide-1 (DE-1) in which the benzylic hydroxyl group and epoxide oxygen are cis and diol epoxide-2 (DE-2) in which these two groups are trans. The situation is, however, complicated by the facts that both the (R,R)- and (S,S)-enantiomers of the dihydrodiol can be formed, and that each can be metabolized to both diastereomeric diol epoxides. Thus, the formation of four optically active bay-region diol epoxides is possible in mammals (Fig. 2).

Evidence is now available which indicates that bay-region diol epoxides are ultimate carcinogens for well over a dozen alternant polycyclic aromatic hydrocarbons including some azapolycyclics. As with other chemical carcinogens, the diol epoxides are thought to exert their effects via covalent interaction with DNA such that a transformed cell results upon replication. Although numerous studies have examined the binding of hydrocarbons to DNA, relatively few of these have been at a structural level (Jeffrey 1984)

<u>Figure 1</u>: General illustration of the formation of bay-region diol epoxides by the combined action of cytochrome P450 (P450) and epoxide hydrolase (EH). Absolute configuration is not implied.

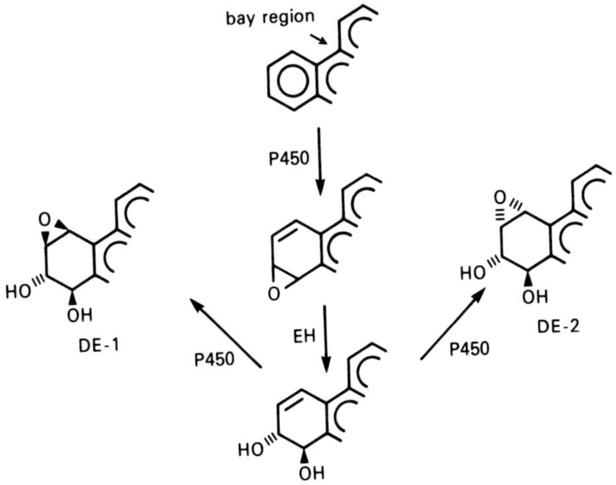

<u>Figure 2</u>: Four possible optically active bay-region diol epoxides from the (R,R)- and (S,S)-dihydrodiol enantiomers. The dotted line represents a mirror plane to illustrate enantiomeric relationships. Subsequent reference to these diol epoxides in the text will utilize the R/S notation going around the ring starting with the benzylic hydroxyl group. Thus the structure in the upper left is designated (R,S,R,S).

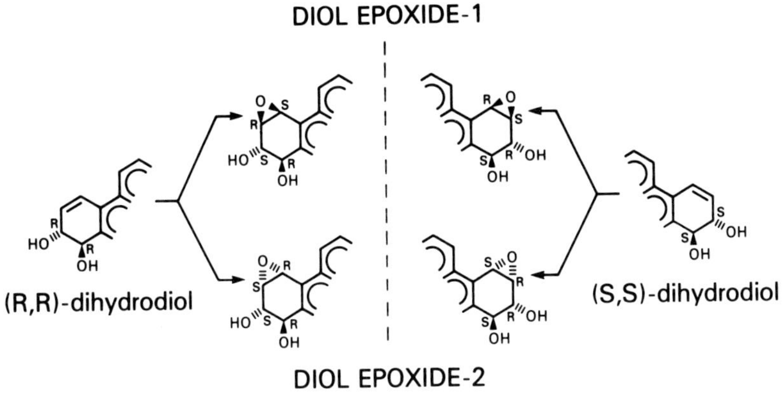

and none of these have been able to correlate directly a specific lesion with cellular transformation. The present chapter compares the <u>in vitro</u> specificity for the covalent binding to calf-thymus DNA of the sets of four optically active bay-region diol epoxides from benzo(a)pyrene, benzo(c)phenanthrene, dibenz(a,j)anthracene and benz(a)-anthracene (Fig. 3) in order to further our understanding of the initiation of the carcinogenic process.

SOLVOLYSIS OF DIOL EPOXIDES

The most thoroughly studied reaction of diol epoxides is their hydrolysis in aqueous media (Whalen et al. 1977; Sayer et al. 1981, 1984). This process follows the rate equation

$$k_{obs} = k_H a_{H+} + k_o$$

to produce biphasic pH rate profiles with a first order dependence on acid concentration at low pH (k_H) and a pH-independent region at higher pH (k_o). Products consist of tetraols resulting from cis and trans addition of water at the benzylic epoxide position as well as keto diols from rearrangement of the epoxide. Distribution of these products is dependent on the specific hydrocarbon, the diol epoxide diastereomer, and the pH. DE-2 diastereomers undergo mainly trans hydration in both pH regions whereas substantial cis hydration and formation of keto diols (k_o) occurs for DE-1 diastereomers. Since these hydrolyses are subject to general acid catalysis (Whalen et al. 1979), products characteristic of the k_H process can be observed at neutral pH in the presence of high buffer concentrations. The effects of 1 mg/ml of calf thymus DNA on the lifetimes of several diol epoxides are shown in Table 1.

Although the kinetics of reactions of diol epoxides from hydrocarbons other than BP have not been studied in detail, it is clear from these qualitative observations that there is a substantial acceleration of the rate of diol epoxide disappearance in the presence of DNA. Several effects are possible when DNA is added to the hydrolysis medium for a diol epoxide. For example, DNA could sequester the hydrocarbon while stabilizing a cationic transition state for epoxide ring opening by electrostatic interactions (much like an anionic detergent), it could function as a general acid catalyst for hydrolysis, and its many potential nucleophilic groups could compete with water and thus covalently bind to the diol epoxide. In the case of BP diol epoxides, the rate acceleration by DNA has been shown

<u>Figure 3</u>: Structure of the (R,S,S,R)-enantiomers of the diol epoxide-2 diastereomers derived from BP, BPh, DBA and BA. As illustrated in Figure 2, three other optically active isomers are possible for each hydrocarbon. The diol epoxides are oriented such that the substituents on the benzo-rings are superimposable.

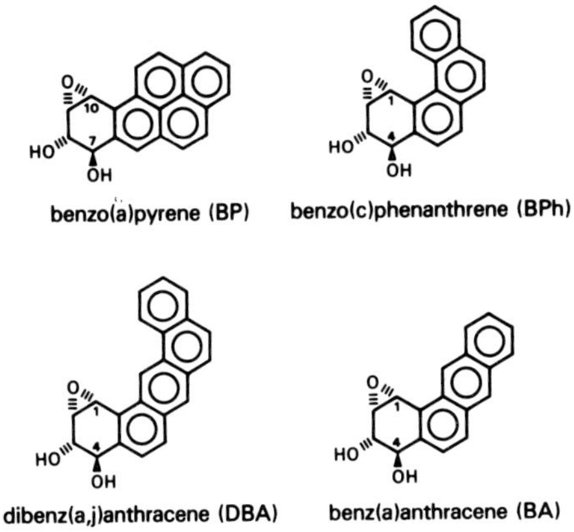

Table 1. Comparison of lifetimes of diol epoxides under solvolytic conditions and in the presence of calf thymus DNA (1 mg/ml).

Diol epoxide of	Isomer	$T_{1/2}$ (solvolysis)[a] minutes	$T_{1/2}$ (DNA)[b] minutes
BP	1	2.7	<0.3[c]
	2	43	<1
BA	1	36	~2[d]
	2	210	<2[d]
DBA	1	84	<5
	2	300	<8
BPh	1	924	~7
	2	480	4-7

[a]At 25 °C, pH 7.0, in 1:9 dioxane:water. [b]At 37 °C, pH 6.8-7.0, 1:10 acetonitrile:water or acetone:water unless otherwise noted. [b]Estimated from the data of Islam et al. 1987, at 25 °C in 4% dioxane in water. [d]At pH 7.4.

to involve pre-equilibrium formation of a non-covalent complex, followed by hydrolytic cleavage of the bound diol epoxide (Geacintov et al. 1982; Michaud et al. 1983; Islam et al. 1987). With these diol epoxides, the preponderance (>98%) of products formed results from DNA-catalyzed hydrolysis rather than covalent adduct formation. When physically bound to DNA, these diol epoxides undergo both hydronium-ion-dependent and pH-independent hydrolyses, with the latter process making a much more substantial contribution to the hydrolysis of DE-1 than DE-2. The hydronium-ion-dependent solvolyses of DNA-bound diol epoxides may involve either specific acid catalysis (direct reaction of hydronium ion with the epoxide) or general acid catalysis by the small fraction of DNA in the protonated phosphodiester form. An additional mechanism whereby DNA could accelerate the ring opening of diol epoxides involves electrostatic stabilization of a cationic transition state by the polyanionic phosphodiester "backbone" of DNA. The involvement of such electrostatic interactions is suggested by the large accelerations (ca. 50-fold) in the hydrolysis of DNA-bound BP diol epoxide-2 when the shielding effect of counterions is diminished by decreasing the concentration of added salts from 0.1 M to ca. 10^{-3} M.

ADDUCT FORMATION

Binding of the 7,8-diol-9,10-epoxides of BP (Baird & Pruess-Schwartz 1987) and the 3,4-diol-1,2-epoxides of BPh (Jerina et al. 1986; Agarwal et al. 1987; Dipple et al. 1987) has been described in detail elsewhere. The present report concentrates more on our preliminary finding for the binding of the optically active 3,4-diol-1,2-epoxides of DBA and BA to calf thymus DNA <u>in vitro</u>. The general procedure consists of reaction of the individual isomers with DNA and enzymatic hydrolysis to nucleosides and nucleoside adducts. UV spectra (Fig. 4) as well as chromatographic comparison with products obtained from individual deoxynucleotides allowed assignment of the peaks obtained from alkylated DNA. As with the adducts from BP and BPh, mass spectral, CD, and NMR evidence (data not presented) indicates that the DBA adducts also result from cis and trans opening of the oxirane ring at the benzylic position by the N^2-amino group of dGuo and the N^6-amino group of dAdo. Chemical characterization of the BA adducts is currently in progress. The results of these studies are summarized in Table 2.

The extent of binding versus hydrolysis shows a remarkable variation between the sixteen diol epoxides ranging from as little as 0.2% for BP (S,R,R,S)-DE-2 to as

Figure 4: Typical UV spectra of adducts from DBA and BA. The highly characteristic spectrum of the DBA tetraols allows ready identification of dGuo and dAdo adducts. Such discrimination is more difficult for adducts formed from the BA diol epoxides.

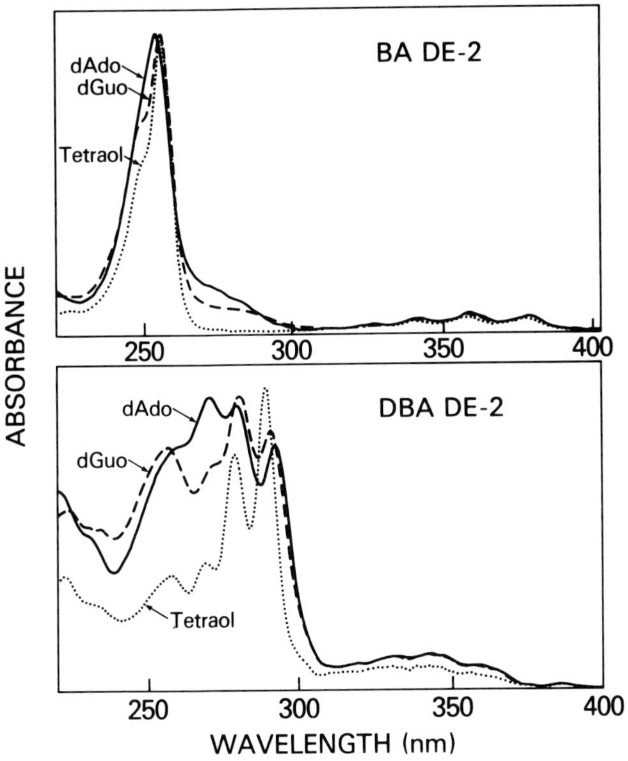

much as 75% for BPh (R,S,S,R)-DE-2. When diol epoxides from a given hydrocarbon are compared, the extent of binding varies to a much lesser extent. This variation is typically less than 3-fold. In general, the (R,S,S,R)-DE-2 isomers tend to bind better, but the efficiency of their binding relative to hydrolysis is not substantially enhanced compared to the other three isomers. As stated earlier, the exocyclic amino groups of dGuo and dAdo are the major sites of attack although minor amounts of binding to dCytd may also occur. Since our structural data are as yet incomplete, adducts which arise by cis and trans opening of the epoxide ring at the benzylic position are considered together. With the exception of the data for the diol epoxides from BPh, dGuo is typically a 2- to 20-fold better target than dAdo. In the case of BPh, adducts from dAdo

Table 2. Comparative extents of in vitro covalent binding to calf thymus DNA (1 mg/ml) and adduct distribution for the optically active isomers of selected polycyclic aromatic hydrocarbons.

Diol epoxide of	Enantiomer	Per cent bound to DNA	Adduct distribution (per cent)	
			dGuo	dAdo
BP[a]	(S,R,S,R)-DE-1	2-5	~95	~5
	(R,S,R,S)-DE-1	2-5	~95	~5
	(R,S,S,R)-DE-2	2-5	~94	~6
	(S,R,R,S)-DE-2	<0.2	~50	~50
BA[b]	(S,R,S,R)-DE-1	~10	74	26
	(R,S,R,S)-DE-1	~12	63	37
	(R,S,S,R)-DE-2	~24	78	22
	(S,R,R,S)-DE-2	~10	67	33
DBA[b]	(S,R,S,R)-DE-1	3-4	69	31
	(R,S,R,S)-DE-1	3-4	66	34
	(R,S,S,R)-DE-2	9-10	80	20
	(S,R,R,S)-DE-2	~3	76	23
BPh[c]	(S,R,S,R)-DE-1	~60	14	86
	(R,S,R,S)-DE-1	~55	45	55
	(R,S,S,R)-DE-2	~75	34	66
	(S,R,R,S)-DE-2	~60	42	58

[a]From Meehan and Straub, 1979 and Baird, unpublished. [b]Results from present study. [c]Jerina et al. 1987; Dipple et al. 1987.

predominate, as much as 6-fold in the case of (S,R,S,R) DE-1. Model building suggests the hydrocarbon adducts of dGuo might lie in the minor groove of B-DNA while adducts at the exocyclic amino group of dAdo might lie in the major groove. At present it is not known whether the differences in the shapes of BP and BPh account for the marked differences in their ability to bind to the exocyclic amino group of dAdo.

METABOLISM AND TUMORIGENICITY

A high degree of stereoselectivity is associated with the diol epoxide pathway of metabolism. Several hydrocarbons have been studied with liver enzymes with the finding that the (R,S,S,R) DE-2 isomer greatly predominates (Jerina et al. 1986). Much smaller amounts of (S,R,R,S) DE-1 and the remaining two isomers are formed. Studies of the binding of BP (Baird & Pruess-Schwartz 1987) and BPh (Pruess-Schwartz et al. 1987) to DNA reflect this metabolic trend. Studies of the metabolism induced binding of BA to DNA (MacNicoll et al. 1981) have indicated that adducts are formed from the bay-region 3,4-diol 1,2-epoxides as well as the non-bay-region 8,9-diol 10,11-epoxides. Structural studies of the adducts formed on binding of DBA are presently unavailable.

In terms of tumorigenic response, DE-2 diastereomers are generally much more active than DE-1 diastereomers (Jerina et al. 1986). Based on comparison of activity for racemic DE-2 diastereomers, the following order of activity emerges: BPh DE-2 > BP DE-2 > BA DE-2. As of yet, tumor data are unavailable for diol epoxides from DBA. Given the above ranking, no correlation between reactivity of DE-2 diastereomers of BPh, BP and BA with DNA (Table 2, percent bound) and tumorigenic response is evident, although there is some indication that the generally more active (R,S,S,R)-enantiomer binds better than the (S,R,R,S)-enantiomer or either DE-1 enantiomer. Comparison of total dGuo to dAdo adducts with tumorigenic response for the sixteen diol epoxides also shows no clear trend. Although gross binding studies of the present type are highly informative in terms of comparison of reactivity and determination of adduct structures, more refined structural studies as to damage at specific sites on DNA may be required before clear correlations with tumorigenicity emerge.

ACKNOWLEDGEMENT

The authors are pleased to thank W. Levin, A. W. Wood and A. H. Conney at Hoffmann-La Roche, A. Dipple at Frederick Cancer Research, W. Baird at Purdue University, and D. L. Whalen at the University of Maryland for their exceptional contributions to various aspects of the present studies.

REFERENCES

Agarwal, S.K., Sayer, J.M., Yeh, H.J.C., Pannell, L.K., Hilton, B.D., Pigott, M.A., Dipple, A., Yagi, H. & Jerina, D.M., 1987, Journal of the American Chemical Society, 109, 2497.

Baird, W.M. & Pruess-Schwartz, D., 1987, Polycyclic Aromatic Hydrocarbon-DNA Adducts and Their Analysis, In Polycyclic Aromatic Hydrocarbon Carcinogenesis, edited by S.K. Yang & D.B. Silverman (CRC Press) (in the press).

Dipple, A., Pigott, M.A., Agarwal, S.K., Yagi, H., Sayer, J.M. & Jerina, D.M., 1987, Nature, 327, 535.

Geacintov, N.E., Yoshida, H., Ibañez, V. & Harvey, R.G., 1982, Biochemistry, 21, 1864.

Islam, N.B., Whalen, D.L., Yagi, H. & Jerina, D.M., 1987, Journal of the American Chemical Society, 109, 2108.

Jeffrey, A.M., 1984, Polycyclic Aromatic Hydrocarbon-DNA Adducts. In Polycyclic Hydrocarbons and Carcinogenesis, edited by R.G. Harvey (ACS Symposium Series 283) (American Chemical Society), p. 187.

Jerina, D.M. & Daly, J.W., 1976, Oxidation At Carbon, In Drug Metabolism - From Microbe to Man, edited by D.V. Parke & R.L. Smith (Taylor & Francis Ltd.), p. 13.

Jerina, D.M., Sayer, J.M., Agarwal, S.K., Yagi, H., Levin, W., Wood, A.W., Conney, A.H., Pruess-Schwartz, D., Baird, W.M., Pigott, M.A. & Dipple, A., 1986, Reactivity and tumorigenicity of bay-region diol epoxides derived from polycyclic aromatic hydrocarbons. In Biological Reactive Intermediates III, edited by J.J. Kocsis, D.J. Jollow, C.M. Witmer, J.O. Nelson & R. Snyder (Plenum Publishing Corporation), p.11.

MacNicoll, A.D., Cooper, C.S., Ribeiro, O., Pal, K., Hewer, A., Grover, P.L. & Sims, P., 1981, Cancer Letters, 11, 243.

Meehan, T. & Straub, K., 1979, Nature, 277, 410.

Michaud, D.P., Gupta, S.C., Whalen, D.L., Sayer, J.M. & Jerina, D.M., 1983, Chemico-Biological Interactions, 44, 41.

Pruess-Schwartz, D., Baird, W.M., Yagi, H., Jerina, D.M., Pigott, M.A. & Dipple, A., 1987, Cancer Research, 47, 4032.

Sayer, J.M., Whalen, D.L., Friedman, S.L., Paik, A., Yagi, H., Vyas, K.P. & Jerina, D.M., 1984, Journal of the American Chemical Society, 106, 226.

Sayer, J.M., Yagi, H., Croisy-Delcey, M. & Jerina, D.M., 1981, Journal of the American Chemical Society, 103, 4970.

Whalen, D.L., Montemarano, J.A., Thakker, D.R., Yagi, H. & Jerina, D.M., 1977, Journal of the American Chemical Society, 99, 5522.

Whalen, D.L., Ross, A.M., Montemarano, J.A., Thakker, D.R., Yagi, H. & Jerina, D.M., 1979, Journal of the American Chemical Society, 101, 5086.

A Mechanism for Bisulfate Co-Carcinogenesis with Benzo(A)pyrene: Potential of Epoxidation by Peroxidases

T.E.ELING, J.F.CURTIS, M.F.HUGHES and R.P.MASON

Laboratory of Molecular Physics, NIEHS, PO Box 12233, Research Triangle Park, NC 27709, USA and Curriculum in Toxicology, University of North Carolina, Chapel Hill, NC 27514, USA

INTRODUCTION

Hydroperoxide-dependent oxidation of chemicals, especially when triggered by prostaglandin H synthase (PHS) have been extensively investigated in recent years (Marnett & Eling, 1983; Krauss & Eling, 1984). Interest in these reactions derives from the possibility that they could represent alternate pathways for the metabolic activation of carcinogens. Many classes of carcinogens, including polycyclic aromatic hydrocarbons and aromatic amines are metabolized by hydroperoxide-dependent oxidation to reactive electrophilic agents. Hydroperoxide-dependent reactions can be divided into three distinct mechanisms as shown in Figure 1.

Figure 1. Mechanisms responsible for hydroperoxide-dependent oxidations during arachidonic acid metabolism.

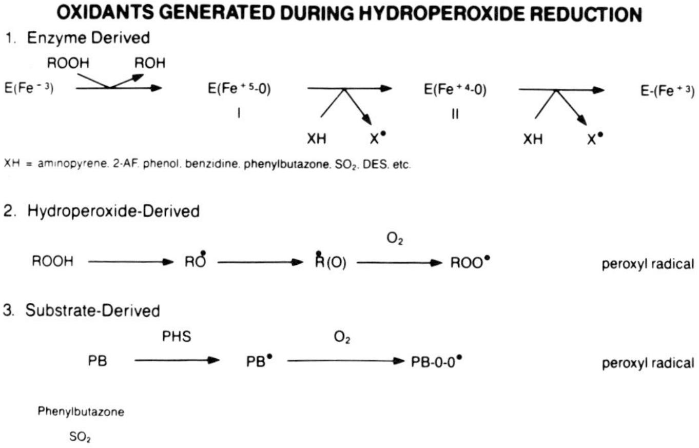

For example, aromatic amines, as well as other classes of carcinogens, undergo an enzyme-derived one electron oxidation (mechanism 1) to free radical metabolites that can react with nucleic acids and are mutagenic in the Ames assay (Krauss & Eling, 1984). The hydroperoxidase activity of PHS is similar mechanistically to horseradish peroxidase (HRP), but the enzymes differ in substrate specifity (Markey et al., 1987). The dihydrodiols of polycyclic aromatic hydrocarbons are not classical peroxidase substrates but undergo epoxidation mediated by peroxyl radicals (mechanism 2) generated during the oxidation of arachidonic acid (Dix et al., 1985). Epoxidation of dihydrodiols is of particular interest since it frequently represents the terminal enzymatic step in polycyclic aromatic hydrocarbon induced carcinogenesis. Interestingly, epoxidation by this mechanism of dihydrodiols by this mechanism is stereoselective in that the anti-diolepoxide is predominately formed. For the third mechanism of hydroperoxide-dependent oxidations, a primary substrate is oxidized by the peroxidase, the resulting radical then reacts with molecular oxygen yielding a peroxyl radical which is the ultimate epoxidizing agent. By this mechanism significant enhancement of diolepoxide formation from benzo(a)pyrene-7,8-dihydrodiol occurs as has been described for the anti-inflammatory agent, phenylbutazone (Reed et al., 1984).

Sulfur dioxide, a common environmental pollutant, exists as bisulfite and sulfite in aqueous solution at neutral pH (see Figure 2, line 1).

Figure 2. Bisulfite oxidation and the proposed epoxidation of BP-7,8-diol to diolepoxides.

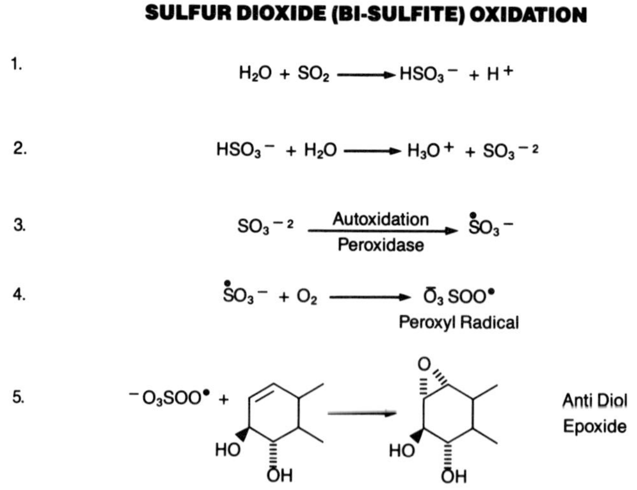

Metal catalyzed autooxidation or peroxidase catalyzed oxidation (Mottley et al., 1982) of bisulfite yields the sulfur trioxide radical anion ($\dot{S}O_3^-$). This anion can react with molecular oxygen to form a peroxyl radical (Mechanism 3). This peroxyl radical should be capable of epoxidizing BP-diols to diolepoxides. Thus, the addition of bisulfite to incubations with PHS should significantly enhance diolepoxide formation in a manner similar to that shown for phenylbutazone. Laskin et al. (1976) and Pauluhn et al. (1985) reported that exposure of rats to SO_2 was co-carcinogenic for the development of benzo(a)pyrene-induced pulmonary carcinoma. Enhanced formation of anti-BP-diol epoxide in pulmonary tissue after SO_2 exposure could occur by the substrate-derived peroxyl radical mechanism. In this paper, we present our characterization of the epoxidation of BP-7,8-diol catalyzed by both autooxidation and peroxidase-mediated bisulfite oxidation.

MATERIALS AND METHODS

Materials

[^{3}H]- and [^{14}C]-7,8-dihydroxy-7,8-dihydrobenzo(a)pyrene (BP-7,8-diol) were obtained from Chemsyn Science Lab (Kansas City, MO) as was the unlabeled compound. Horseradish peroxidase (HRP), H_2O_2, butylated hydroxyanisole (BHA), and diethylenetriaminepentaacetic acid (DETAPAC) were obtained from Sigma Chemical Co. (St. Louis, MO). Arachidonic acid was purchased from Nu-Chek Prep (Elysian, MN). Polyethylene (20) sorbitan monolaurate (tween 20), a detergent used to help solubilize the BP-7,8-diol was purchased from Matheson, Coleman and Bell Manufacturing Chemists (Norwood, OH). Ram seminal vesicles (RSV) were obtained from a local slaughterhouse. The RSV microsomal fraction which contains PHS was prepared by a standard procedure (Boyd et al., 1983).

Reaction conditions

The incubation mixture generally contained 10 μM [^{3}H]-BP-7,8-diol, 0.1 M phosphate buffer pH 7.0, 1 mM DETAPAC (when used), and either HRP or RSV microsomes. The reaction was initiated with addition of either H_2O_2 or arachidonic acid. After incubation the reaction was terminated by the addition of BHA and the incubation mixture was extracted twice with ethyl acetate. The organic extractable metabolites were analyzed by HPLC (Eling et al., 1986). The reader should consult figure legends for details of the incubation mixtures.

RESULTS AND DISCUSSION

Initial efforts focused on establishing that bisulfite autoxidation in the presence of BP-7,8-diol resulted in the formation of diolepoxide. Oxygen consumption was observed immediately upon the addition of bisulfite. Addition of the transition metal chelator DETAPAC inhibited oxygen consumption. HPLC analysis of the organic extracts of incubations containing BP-7,8-diol and bisulfite indicated extensive oxidation of BP-7,8-diol. Some products eluted near the solvent front but the major metabolites co-eluted with authentic BP tetrols, the hydrolysis products of BP-diolepoxides. The ratio of tetrols derived from anti- and syn-diolepoxides was approximately 6:1. Other data also indicated the intermediacy of the diolepoxide (Reed et al., 1986).

Rapid epoxidation of BP-7,8-diol occurred during bisulfite autoxidation and was dependent on bisulfite concentration (Table 1).

Table 1. Epoxidation of BP-7,8-diol during bisulfite autoxidation.

Incubation	μM Tetrols
complete system[a]	0.1
complete system + 0.1 mM bisulfite	0.3
complete system + 0.5 mM bisulfite	3.2
complete system + 1.0 mM bisulfite	3.5
complete system + 1.0 mM bisulfite + 0.5 mM DETAPAC	1.3
complete system + 1.0 mM bisulfite + 10 μM BHA	0.1

[a]10 μM [^{3}H]-BP-7,8-diol was incubated in 0.1 M phosphate buffer pH 7.0 at 37°C for 60 min in the presence or absence of bisulfite, DETAPAC or BHA where indicated. Tetrols were measured by HPLC analysis of the organic extract.

Significant enhancement of tetrol formation occurred at a bisulfite concentration of 0.5 mM concentration. The addition of DETAPAC inhibited bisulfite dependent tetrol formation. The phenolic antioxidant BHA, which readily reacts with peroxyl radicals was a very potent inhibitor of tetrol formation. A concentration of 10 μM BHA completely blocked the epoxidation of BP-7,8-diol by the bisulfite peroxyl radical. In addition to the tetrols, water soluble metabolites of BP-7,8-diol were observed. The formation of these metabolites was dependent on bisulfite and was inhibited by the addition of DETAPAC or BHA. The ability of DETAPAC and BHA to block both bisulfite autoxidation and tetrol formation supports this conclusion. Peroxyl radical epoxidation is stereoselective producing predominantly anti-BP-diolepoxide. We observed the stereochemical preference, supporting the hypothesis that a bisulfite-derived peroxyl radical is formed.

HRP and RSV microsomes, which contains PHS, were used to determine if enhanced BP-tetrol formation can occur enzymatically in the presence of bisulfite. DETAPAC (1 mM) was included in the incubations to suppress bisulfite autoxidation. Both HRP and PHS catalyzed the oxidation of bisulfite to $\dot{S}O_3^-$ (Mottley et al., 1982) as detected by electron spin resonance techniques. Oxygen consumption in solution was also observed suggesting the formation of a sulfur containing peroxyl radical. Table 2 shows HRP and PHS catalyzed BP-tetrol formation.

Table 2. Peroxidase catalyzed bisulfite-dependent epoxidation of BP-7,8-diol

Incubation	μM Tetrols
complete system[a] - bisulfite	0.4
complete system[a] + 3 μM bisulfite	0.55
+ 10 μM bisulfite	0.78
+ 30 μM bisulfite	1.1
+ 100 μM bisulfite	1.5
+ 300 μM bisulfite	1.6
complete system[a] + 1 mM bisulfite + 10 μM BHA	0.70
complete system[b] - bisulfite	0.26
complete system[b] + 10 μM bisulfite	0.34
+ 30 μM bisulfite	0.32
+ 100 μM bisulfite	0.41
+ 300 μM bisulfite	0.51

[a]10 μM [^{3}H]-BP-7,8-diol was incubated in phosphate buffer containing DETAPAC (1 mM), pH 7.0 for 60 min at 37°C in the presence of 50 μg/ml HRP and 100 μM H_2O_2.

[b]10 μM [^{3}H]-BP-7,8-diol was incubated in phosphate buffer pH 7.0 containing DETAPAC (1 mM) for 60 min with 0.5 mg/ml RSV microsomes, and 100 μM arachidonic acid.

HPLC analysis of the organic extractable metabolites indicated the formation of BP tetrols. The rate of anti- to syn-diolepoxide derived tetrols was approximately 7:1, which is in agreement with peroxyl radical dependent epoxidation. The addition of bisulfite significantly enhanced tetrol formation with either enzyme system. In contrast to the autoxidation system, significant enhancement of tetrol formation occurred at bisulfite concentration as low as 10 µM. This low concentration suggests the possibility that enhancement of BP-diolepoxide formation by bisulfite could occur at a physiological concentration. As observed with bisulfite autoxidation, significant water soluble metabolites were present in these incubations. At higher bisulfite concentration, particularly at concentrations where the oxygen in the incubation was depleted, predominant formation of water soluble metabolites was observed (data not shown). The addition of BHA, or absence of H_2O_2 inhibited tetrol formation (data not shown). These data indicate that enhancement of BP-diolepoxide formation can be catalyzed by peroxidase as also observed during bisulfite autoxidation.

In preliminary experiments, we have investigated bisulfite enhancement of [^{14}C]-BP-diolepoxide formation catalyzed by PHS present in mouse keratinocytes. Mouse keratinocytes were chosen since they have a high PHS content, as well as inducible cytochrome P-450 enzymes. The data also can be extrapolated to the mouse skin carcinogenesis model. A 1.5-2 fold enhancement of BP-tetrol formation was observed at a bisulfite concentration of 1 mM (data not shown). Further studies are underway with these keratinocytes and are clearly necessary to understand the enhancement of BP-diolepoxide formation by bisulfite. Additional experimental evidence is required to support our hypothesis that peroxidase mediated bisulfite oxidation is responsible, in part, for the co-carcinogenesis between exposure to SO_2 and benzo(a)pyrene.

ACKNOWLEDGEMENTS

The authors wish to thank Robin Sorrell for her editorial assistance in preparing this manuscript. MFH is supported by ES-07126.

REFERENCES

Boyd, J.A., Harvan, D.J. & Eling, T.E., 1983, The Journal of Biological Chemistry, 258, 8246.

Dix, T.A., Fontana, R., Panthani, A. & Marnett, L.J., 1985, The Journal of Biological Chemistry, 260, 5358.

Krauss, R.S. & Eling, T.E., 1984, Biochemical Pharmacology, 33, 3319.

Laskin, S., Kuschner, M., Sellakuman, A. & Katz, G.V., 1976 Combined carcinogen-irritant animal inhalation studies. In Air Pollution and the Lung, edited by E.F. Aharonson, A. Ben-David & M.A. Klingberg (Halsted), p. 190.

Markey, C.M., Alward, A., Weller, P.E. & Marnett, L.J., 1987, The Journal of Biological Chemistry, 262, 6266.

Marnett, L.J. & Eling, T.E., 1983 Cooxidation during prostaglandin biosynthesis: a pathway for the metabolic activation of xenobiotics. In Reviews in Biochemical Toxicology, Vol. 5, edited by E. Hodgson, J.R. Bend & R.M. Philpot (Elseview Biomedical), p. 135.

Mottley, C., Mason, R.P., Chignell, C.F., Sivarajah, K. & Eling, T.E., The Journal of Biological Chemistry, 257, 5050.

Pauluhn, J., Thyssen, J., Althoff, J., Kimmerle, G. & Mohr, U., 1985, Experimental Pathology, 28, 31.

Reed, G.A., Brooks, E.A. & Eling, T.E., 1984, The Journal of Biological Chemistry, 259, 5591.

Role of Microsomal Cytochromes P-450 and Prostaglandin H-Synthase in 4-Aminobiphenyl-DNA Adduct Formation

FRED F. KADLUBAR [1], MARY ANN BUTLER [1], BARBARA E. HAYES [1], FREDERICK A. BELAND [1] AND F. PETER GUENGERICH [2]

[1] National Center for Toxicological Research, Jefferson, AK 72079, USA and
[2] Department of Biochemistry, Vanderbilt University, Nashville, TN 37232, USA

INTRODUCTION

Oxidative metabolism of carcinogenic aromatic amines represents a critical activation step leading to the formation of carcinogen-DNA adducts (reviewed in Beland & Kadlubar, 1985). In the liver, specific cytochromes P-450 have recently been shown to possess high catalytic activity toward the N-oxidation and aryl ring-oxidation of several primary arylamines, including 2-aminofluorene, 3-amino-1-methyl-5H-pyrido[4,3-_b_]indole (Trp-P-2), 2-amino-3-methylimidazo[4,5-_f_]quinoline (IQ), 2-amino-6-methyldipyrido[1,2-_a_:3',2'-_d_]imidazole (Glu-P-1), and 2-naphthylamine (2-NA; reviewed in Kadlubar & Hammons, 1987; Kato, 1986). In rat liver, 2-aminofluorene is metabolized by several P-450's, Trp-P-2 is effectively N-oxidized by P-450/BNF-B and P-450/ISF-G, while N-oxidation of 2-NA, Glu-P-1, and IQ appears to be catalyzed selectively by cytochrome P-450/ISF-G. Metabolic N-oxidation of 4-aminobiphenyl (4-ABP) by purified cytochromes P-450 has not been extensively studied but catalytic activity has been reported for purified enzymes corresponding to P-450/ISF-G and P-450/BNF-B (Kamataki et al., 1983). N-Hydroxy-4-ABP is then subject to a variety of conjugation pathways that can result in its direct reaction with DNA or in its transport and subsequent macromolecular binding in extrahepatic tissues. The predominant DNA adduct formed by these pathways is N-(deoxyguanosin-8-yl)-4-ABP (C8-dG), with minor amounts of N-(deoxyguanosin-N^2-yl)-4-ABP (N^2-dG) and N-(deoxyadenosin-8-yl)-4-ABP (C8-dA). Each of these adducts is persistent in the urinary bladder epithelium, a target tissue for 4-ABP carcinogenesis (Beland & Kadlubar, 1985).

In extrahepatic tissues, evidence has been obtained that prostaglandin _H_ synthase (PHS) can mediate the peroxidative metabolism of 4-ABP and several other carcinogenic arylamines to DNA-bound products (Kadlubar et al., 1982; Morton et al., 1983). Characterization of these adducts has suggested that the major reactive intermediates of benzidine, 2-NA, and 2-aminofluorene formed in the PHS activation system are benzidinediimine (Kadlubar et al., 1986), 2-imino-1-naphthoquinone (Yamazoe et al., 1985), and 2-imino-7-fluorenoquinone (Krauss et al., in preparation), respectively. These DNA adducts are structurally distinct from those derived from the cytochrome P-450 and conjugation pathways and are found in animal tissues containing high levels of PHS. However, the relative contribution of peroxidative activation to DNA adduct formation varies with the arylamine and the target tissue; for the urothelium, this ranges from a minor role for 2-NA to a major role for benzidine and 2-aminofluorene.

In this study, we have examined the involvement of specific cytochromes P-450 and of PHS in the metabolic activation of 4-ABP and have assessed the role of these pathways in carcinogen-DNA adduct formation. The substrate specificity of purified ram and human bladder microsomal PHS for several carcinogenic arylamines is also reported.

METHODS

[2,2'-^{3}H]4-ABP (55 and 385 mCi/mmol), [2,2'-^{3}H]benzidine (118 mCi/mmol), [5,6,7,8-^{3}H]2-NA (47 mCi/mmol), [^{14}C]Trp-P-2 (46 mCi/mmol), [^{14}C]Glu-P-1 (27 mCi/mmol), [5-^{3}H]IQ (118 mCi/mmol), and [ring-^{3}H]phenacetin (110 mCi/mmol) were obtained from Chemsyn Science Laboratories, Lenexa, KS. Purified PHS from ram seminal vesicles (64,000 units/mg) was purchased from Oxford Biomedical Research, Inc., Oxford, MI. Purified rat liver cytochromes P-450 (Guengerich et al., 1982) and NADPH-cytochrome P-450 reductase (Yasukochi & Masters, 1976) were prepared by the methods cited. Human liver microsomes were prepared from tissues obtained from organ donors immediately after death and were stored at -20°C as described by Guengerich (1982). Human urothelial microsomes were similarly prepared except that the homogenization medium consisted of 0.25 M sucrose containing 20 mM Tris-acetate buffer (pH 7.4), 20 uM butylated hydroxytoluene, 0.1 mM EDTA, and 0.1 mM dithiothreitol. These microsomes were used immediately after preparation.

Enzyme assays were carried out as previously described for liver microsomal (Frederick et al., 1982) and reconstituted cytochromes P-450 (Hammons et al.,1985). The N- and ring-oxidation of 4-ABP and O-deethylation of phenacetin were determined by hplc analyses (Kadlubar et al., 1982). Incubation mixtures for the PHS-dependent activation of arylamines to DNA-bound products contained 50 mM potassium phosphate buffer (pH 7.4), 0.1 mM arachidonic acid, 1 uM hematin (purified PHS only), 2.5 mg/ml DNA, 20 uM radio-labelled arylamine, and 50 ug/ml microsomal PHS (or 100 units purified PHS). The reactions were preincubated at 37°C for 1 min, started by addition of arachidonic acid, incubated for 10 min, and covalent binding to DNA was determined (Yamazoe et al., 1986).

Analysis of 4-ABP-DNA adducts formed in vivo was performed after oral administration of 10 mg/kg [^{3}H]4-ABP to male beagle dogs (7 kg). The dogs were sacrificed at 48 hrs after dosing, urothelial DNA was isolated and hydrolyzed enzymatically, and the carcinogen-deoxyribonucleoside adducts were analyzed by hplc as previously described (Beland et al., 1983).

RESULTS AND DISCUSSION

Oxidative Metabolism of 4-ABP by Cytochromes P-450

The involvement of cytochromes P-450 in the oxidative metabolism of 4-ABP was originally demonstrated by McMahon and coworkers (1980) who showed that liver microsomes from several species catalyzed the formation of both N- and ring-hydroxylated products. N-Hydroxy-4-ABP is a major hepatic metabolite in vivo where it enters the circulation, is transported to the urinary bladder lumen, and becomes bound to urothelial DNA (Kadlubar et al., 1987). In contrast, ring-hydroxylation has been generally regarded as a detoxification pathway. In our previous study (Hammons et al., 1985), 2-NA and 2-aminofluorene were examined as substrates for eight available purified rat liver cytochromes P-450. While the N-oxidation of 2-aminofluorene was catalyzed by six different P-450's, only the isosafrole-induced form, P-450/ISF-G, carried out the N-oxidation of 2-NA. However, P-450/ISF-G, P-450/BNF-B, P-450/PB-B, and P-450 PB-D were very effective for the ring-oxidation of both arylamine substrates. Thus, enzyme specificity was apparent for both activation and detoxification pathways and suggested that the relative distribution of cytochromes P-450 could be a significant factor in tissue, species, and individual susceptibility to arylamine carcinogenesis.

Since 4-ABP is strongly carcinogenic and appears to be a widespread environmental contaminant (Bryant et al., 1987), we sought to determine the relative ability of ten purified hepatic cytochromes P-450 to catalyze its N- and ring-oxidation. As shown in Figure 1, N-oxidation of 4-ABP was similar to that of 2-NA in that it was catalyzed selectively by P-450/ISF-G. However, the rate of N-hydroxy-4-ABP formation by P-450/ISF-G was 13.5 nmol/min/nmol P-450, which

x

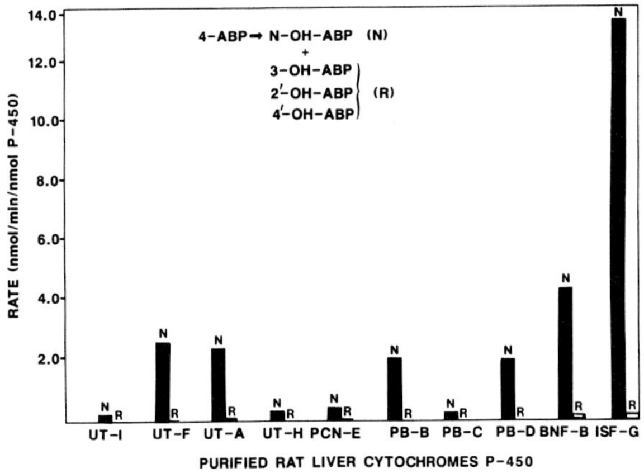

Figure 1. Rates of 4-ABP N-Oxidation (N) and 3,2',4'-Ring-Oxidation (R) by Purified Rat Liver Cytochromes P-450. The P-450's are named according to Guengerich (1987).

was 5-fold faster than the rate of 2-NA N-oxidation and represents one of the highest turnover numbers for this enzyme. Rates of N-oxidation by other P-450's were much lower, ranging from 2% (for P-450/UT-I) to 33% (for P-450/BNF-B) of that observed for P-450/ISF-G. Of interest was the significant rate of 4-ABP N-oxidation by P-450/UT-F (2.7 nmol/min/nmol P-450), since few activities for this isozyme have been reported (Guengerich, 1987). In addition, the rate of 4-ABP ring-oxidation was <2% of the total metabolism for each of cytochromes P-450 examined. This is in marked contrast to the results obtained with 2-NA where 60% of the total metabolites were ring-oxidation products (Hammons et al., 1985). This is consistent with the role of hepatic ring-oxidation as an important detoxification pathway and with the greater relative carcinogenicity of 4-ABP over that of 2-NA (Radomski, 1979).

In view of the immunochemical cross-reactivity and amino-terminal sequence similarity between rat liver P-450/ISF-G and human liver P-450/PA and P-450/HLd (Wrighton et al., 1986), we studied the relative rates of 4-ABP N-oxidation in 19 human liver microsomal preparations and compared these activities with that of phenacetin O-deethylation, which is catalyzed by P-450/PA and antibodies to this protein strongly inhibit human microsomal phenacetin O-deethylase activity (Distlerath et al., 1985). As shown in Figure 2, human liver microsomes varied appreciably in their rates of 4-ABP N-oxidation, ranging from 0.02 to 1.80 nmol/min/mg protein;

while the total cytochrome P-450 levels varied only from 0.14 to 0.74 nmol/mg. As with the rat liver P-450's, 4-ABP ring-oxidation by human liver microsomes was <6% of the total metabolism. However, the most striking comparison was the excellent correlation ($r = 0.91$; $p < 0.001$) observed between 4-ABP N-oxidation and the phenacetin O-deethylase activity of the human microsomal preparations. These data suggest that a single cytochrome P-450 may be primarily responsible for the metabolic activation of 4-ABP and perhaps other carcinogenic arylamines in human liver.

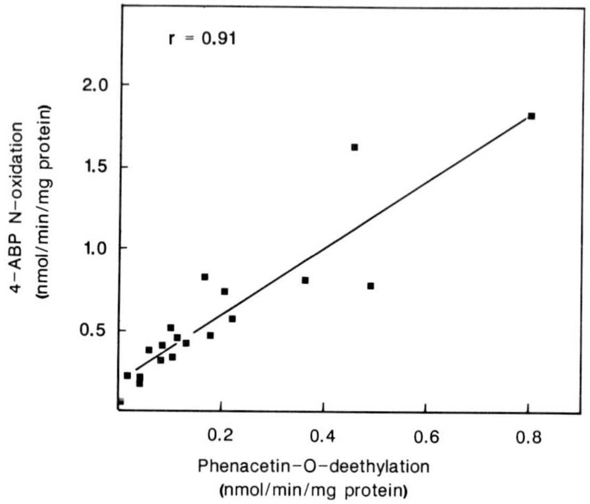

Figure 2. Rates of 4-ABP N-Oxidation and Phenacetin O-Deethylation by Human Liver Microsomes (n = 19).

<u>Peroxidative Metabolism of 4-ABP and Other Arylamines by PHS</u>

The oxidative metabolism of arylamines by microsomal PHS from ram seminal vesicles is known to result in the formation of reactive intermediates that bind covalently to DNA (Kadlubar et al., 1982). Thus, a comparison of the ability of purified ram PHS and human urothelial microsomal PHS to activate 4-ABP and other arylamines to DNA-bound products was undertaken to provide an indicator of the metabolic activation potential in the carcinogen-target tissue (Table 1).

While benzidine was the preferred substrate, significant levels of DNA binding were obtained not only with 4-ABP but with each of the arylamines examined. Considerable variation was observed with the human samples which may be due to the differences in the relative amounts of PHS and endogeneous nucleophiles present in the microsomal preparations.

Nevertheless, these data show that PHS can support the peroxidative activation of arylamines and that the human urothelium has the capacity to metabolize 4-ABP to form DNA adducts.

Table 1. Peroxidative activation of carcinogenic arylamines by purified ram PHS and human urothelial microsomes.

Substrates	Arachidonic Acid-Dependent ram (n = 3)	DNA Binding[a] human (n = 2)[b]
Benzidine	271 ± 23	209, 14.7[c]
2-NA	17.1 ± 2.6	6.4, 3.1
Glu-P-1	14.1 ± 1.0	7.6, 0.9
4-ABP	7.2 ± 0.2	13.6, 2.4
IQ	4.9 ± 0.4	2.7, 0.8
Trp-P-2	1.6 ± 0.2	2.3, 0.4

[a] Incubations were carried out as described in Methods and DNA binding is expressed as pmoles bound/mg DNA.

[b] Inhibition of DNA binding by indomethacin ranged from 65 to 100% for these substrates.

[c] Results with 8 other urothelial preparations gave values ranging from 0.5 to 46 pmoles benzidine bound/mg DNA.

<u>DNA Adducts Formed In Vitro by N-Hydroxy-4-ABP and by the 4-ABP/PHS Activation System.</u>

The reaction of N-hydroxy-ABP with DNA *in vitro* has been shown to result in formation of three 4-ABP-nucleoside adducts, C8-dG, N^2-dG, and C8-dA (Beland & Kadlubar, 1985). To compare these adducts with those DNA-bound products formed by the PHS pathway, [^{3}H]4-ABP was incubated with purified PHS in the presence of calf thymus DNA as shown in Table 1. The modified DNA was reisolated and hydrolyzed enzymatically, combined with the unlabeled adducts derived from N-hydroxy-4-ABP, and analyzed by hplc (Figure 3). The results indicate that the nature of the DNA adducts formed by PHS activation differ considerably from those formed by the cytochrome P-450-dependent N-oxidation of 4-ABP, with putative adducts that elute before and after the C8-dG adduct. Furthermore, preliminary studies on the identity of the major adduct eluting near 17 min are consistent with a product formed by reaction of a 4,3-iminoquinone with a purine or pyrimidine base.

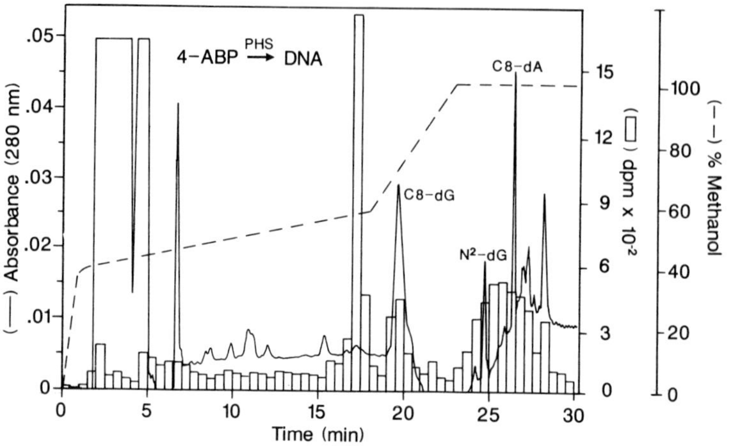

Figure 3. Hplc Profile of the 4-ABP-DNA Adducts Formed *In Vitro* by Reaction of DNA with N-Hydroxy-4-ABP (solid line) and with Reactive Intermediates Formed by the PHS-mediated Activation of [^{3}H]4-ABP (histogram). Elution times for C8-dG, N^2-dG, and C8-dA are shown.

DNA Adducts Formed in Dog Urothelial DNA at 48 Hrs After a Single Dose of [^{3}H]4-ABP.

The dog, which is highly susceptible to 4-ABP-induced urinary bladder carcinogenesis, forms N-hydroxy-ABP as a major metabolite that is transported from the liver to the urinary bladder (Kadlubar et al., 1987). In addition, the dog urothelium contains high levels of microsomal PHS and can similarly activate 4-ABP and other arylamines to DNA-bound products (Wise et al., 1984). Therefore, the relative contribution of these two pathways to carcinogen-DNA adduct formation *in vivo* was assessed by administration of [^{3}H]4-ABP to a male beagle dog and hplc analysis of urothelial DNA (*vide supra*). As shown in Figure 4, only about 10% of the radioactivity bound to the DNA chromatographed with the retention time of the major PHS-dependent, 4-ABP-DNA adduct; while the C8-dG, N^2-dG, and C8-dA accounted for 75%, 9%, and 5% of the total adducts, respectively.

SUMMARY

The metabolic oxidation of the carcinogen 4-ABP is catalyzed both by hepatic cytochromes P-450 and extrahepatic PHS. Of ten purified rat liver cytochromes P-450 examined, P-450/ISF-G exhibits the highest catalytic activity and forms N-hydroxy-4-ABP as the major product. In human liver

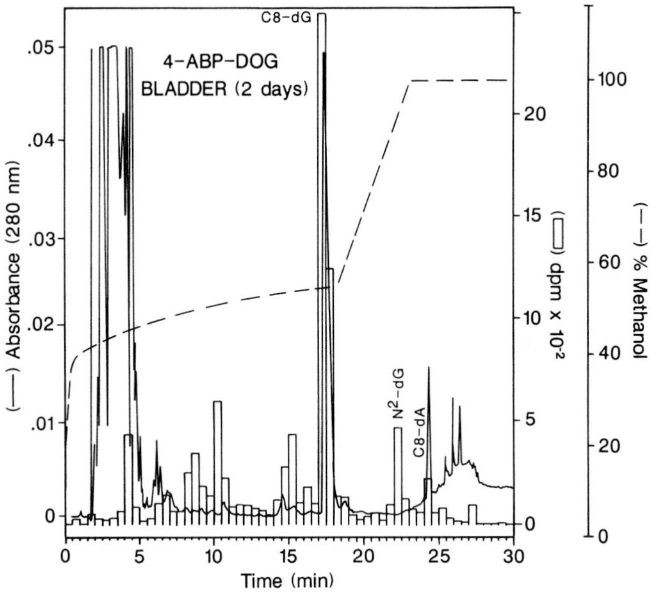

Figure 4. Hplc Profile of the 4-ABP-DNA Adducts Formed <u>In Vitro</u> by Reaction of DNA with N-Hydroxy-4-ABP (solid line) and <u>In Vivo</u> in the Dog Urothelium After Dosing with $[^3H]$4-ABP (histogram). Elution times for C8-dG, N^2-dG, and C8-dA are shown.

microsomes, 4-ABP N-oxidation is strongly correlated (r = 0.91) with levels of cytochrome P-450/PA (phenacetin O-deethylase), a known human drug polymorphism. N-Hydroxy-4-ABP, which is transported to the urinary bladder lumen, reacts with DNA under slightly acidic conditions to form C8-guanine-, N^2-guanine-, and C8-adenine-substituted 4-ABP adducts. PHS, obtained pure from ram seminal vesicles and as a microsomal fraction from human urothelium, mediates the arachidonic acid-dependent cooxidation of 4-ABP and several other carcinogenic arylamines (benzidine, 2-NA, Glu-P-1, Trp-P-2, and IQ) to reactive intermediates that bind to DNA. The major 4-ABP-DNA adduct formed by PHS is structurally distinct from those derived from N-hydroxy-4-ABP and is consistent with formation of a reactive iminoquinone. Analysis of the 4-ABP-DNA adducts formed in the dog urothelium at 48 hrs after dosing with $[^3H]$4-ABP indicates that PHS accounts for only 10% of the total adducts while the remainder are those derived from reaction of N-hydroxy-4-ABP. Thus, cytochrome P-450 appears to play a predominant role in the metabolic activation of this arylamine carcinogen.

REFERENCES

Beland, F.A., Beranek, D.T., Dooley, K.L., Heflich, R.H., and Kadlubar, F.F., 1983, Environ. Health Persp. 48, 12[illegible]

Beland, F.A. and Kadlubar, F.F., 1985, Environ. Health Persp. 62, 19.

Bryant, M.S., Skipper, P.L., Tannenbaum,S.R., Maclure, M., 1987, Cancer Res. 47, 602.

Distlerath, L.M, Reilly, P.E.B., Martin, M.V., Davis, G.G., Wilkinson, G.R., and Guengerich, F.P., 1985, J. Biol. Chem. 260, 9057.

Frederick, C.B., Mays, J.B., Ziegler, D.M., Guengerich, F.P and Kadlubar, F.F., 1982, Cancer Res. 42, 2671.

Guengerich, F.P., Dannan, G.A., Wright, S.T., Martin, M.V., and Kaminsky, L.S., 1982, Biochemistry 21, 6019.

Guengerich, F.P., 1987, Enzymology of rat liver cytochromes P-450. In Mammalian Cytochromes P-450, Vol. I, edited by F.P. Guengerich, CRC Press, Boca Raton, FL, p. 1.

Guengerich, F.P., 1982, Microsomal enzymes involved in toxicology-analysis and separation. In Principles and Methods of Toxicology, edited by A.W. Hayes, Raven Press New York, p. 609.

Hammons, G.J., Guengerich, F.P., Weis, C.C., Beland, F.A., and Kadlubar, F.F., 1985, Cancer Res. 45, 3578.

Kadlubar, F.F., Frederick, C.B., Weis, C.C., and Zenser, T.V., 1982, Biochem. Biophys. Res. Commun. 108, 253.

Kadlubar, F.F., Yamazoe, Y., Lang, N.P., Chu, D.Z.J., and Beland, F.A., 1986, Carcinogen-DNA adduct formation as a predictor of metabolic activation pathways and reactive intermediates in benzidine carcinogenesis. In Biological Reactive Intermediates II, edited by J.J. Kocsis, D.J. Jollow, C.M. Witmer, J.O. Nelson, and R. Snyder, Plenum Press, New York, p. 537.

Kadlubar, F.F. and Hammons, G.J., 1987, The role of cytochrome P-450 in the metabolism of chemical carcinogens. In Mammalian Cytochromes P-450, Vol. II, edited by F.P. Guengerich, CRC Press, Boca Raton, FL, p. 81.

Kadlubar, F.F., Dooley, K.L., Benson, R.W., Roberts, D.W., Butler, M.A., Teitel, C.H., Bailey, J.R. and Young, J.F., 1987, Pharmacokinetic model of aromatic amine-induced urinary bladder carcinogenesis in beagle dogs administered 4-aminobiphenyl. In Third International Conference on Carcinogenic and Mutagenic N-Substituted Aryl Compounds (Proceedings), in press.

Kamataki, T., Maeda, K., Yamazoe, Y., Matsuda, N., Ishii, K. and Kato, R., 1983, Mol. Pharm. 24, 146.

Kato, R., 1986, CRC Crit. Rev. Toxicol., 16, 307.

Krauss, R.S., Kadlubar, F.F., Eling, T.E., in preparation.

Morton, K.C., King, C.M., Vaught, J.B., Wang, C.Y., Lee, M.S., and Marnett, L.J., 1983, Biochem. Biophys. Res. Commun. 111, 96.

McMahon, R.E., Turner, J.C. and Whitaker, G.W., 1980, Xenobiotica 10, 469.

Radomski, J.L., 1979, Ann. Rev. Pharmacol. Toxicol. 19, 129.

Wise, R.W., Zenser, T.V., Kadlubar, F.F. and Davis, B.B., 1984, Cancer Res. 44, 1893.

Wrighton, S.A., Campanile, C., Thomas, P.E., Maines, S.L., Watkins, P.B., Parker, G., Mendez-Picon, G., Haniu, M., Shively, J.E., Levin, W., and Guzelian, P.S., 1986, Mol. Pharmacol. 29, 405.

Yamazoe, Y., Miller, D.W., Weis, C.C., Dooley, K.L., Zenser, T.V., Beland, F.A. and Kadlubar, F.F., 1985, Carcinogenesis 6, 1379.

Yamazoe, Y., Roth, R.W. and Kadlubar, F.F., 1986, Carcinogenesis 7, 179.

Yasukochi, Y. and Masters, B.S.S., 1976, J. Biol. Chem. 251, 5337.

Application of Biochemical and Molecular Techniques to the Epidemiology of Human Lung Cancer

A.WESTON [1], J.C.WILLEY [1], M.J.NEWMAN [4], G.E.HAUGEN [2], D.K.MANCHESTER [1], J.S.CHOI [1], T.KRONTIRIS [3], B.LIGHT [1], D.L.MANN [1] and C.C.HARRIS [1]

[1] Laboratory of Human Carcinogenesis, National Cancer Institute, National Institute of Health, Bethesda, MD 20892, USA. [2] Institute of Occupational Health, Oslo, Norway. [3] Division of Hematology and Oncology, Tufts University, Boston, MA02111, USA. [4] School of Veterinary Medicine, Louisiana State University, Baton Rouge, LA 70803, USA

INTRODUCTION

Biochemical and molecular epidemiology seeks to identify those individuals at risk to disease by virtue of their lifestyle, occupation, or inherited predisposition. With respect to human bronchogenic carcinoma, classical epidemiological studies have identified several chemically complex environmental mixtures: cigarette smoke and coal smoke as etiological agents (Doll 1985 and Mumford et al., 1987). However, not all individuals who are exposed to these carcinogenic agents will develop lung cancer, whereas some less heavily exposed individuals will. This suggests that interindividual variations in xenobiotic metabolic activation, detoxication, DNA repair and other genetically determined host factors have an impact on the role of chemicals in the process of carcinogenesis (Harris 1983). Strategies for the identification of those individuals at high cancer risk are to: obtain evidence of high exposure to carcinogens, detect subsequent alterations in DNA of target cells (putative pathobiological lesions) and examine inherited or acquired host factors. Herein are briefly described the recent scientific advances that have made it possible to obtain evidence for carcinogen exposure, DNA damage and inherited predisposition in humans.

EVIDENCE OF CARCINOGEN EXPOSURE

Measurement of chemical carcinogens in the environment and surveys concerning smoking, drinking or dietary habits are a first step in the determination of human exposure to genotoxic agents. Evidence for bodily exposure to environmental carcinogens can be obtained from an analysis of body fluids, for example; breast milk, seminal fluid, saliva, cervical mucus, urine and blood derived serum (see review, Harris et al., 1987). Indirect evidence of human exposure to putative environmental carcinogens can be gained from testing these body fluids for mutagenic activity (Ames 1979), whereas direct evidence of human exposure to chemical carcinogens can be obtained from specific chemical analysis of these body fluids for the presence of a substance of interest or one or more of its metabolites (Campbell et al., 1970, and Sasson et al., 1985). These types of epidemiological studies provide a tangible basis for human risk assessment, but they do not reveal the extent to which biologically effective macromolecular damage occurs in the individual for the reason that the biological response of the individual to a drug, chemical carcinogen or other xenobiotic agent is a characteristic of the individual.

CARCINOGEN-MACROMOLECULAR ADDUCTS

The metabolic activation of certain environmental contaminants: polycyclic aromatic hydrocarbons (PAHs), hydrazines, N-nitrosamines, aromatic amines, and fungal mycotoxins, have been demonstrated in human tissues and cultured cells (Harris 1987). These biochemical studies, together with experimentation in animals and their cultured cells (Theall-Arce et al., 1987), have led to a better understanding of the genotoxic effects of chemical exposure, which appears to be mediated through the covalent binding of chemical carcinogens to cellular macromolecules (adduct formation) (Weston et al., 1986). In this way, the propensity for metabolic activation of carcinogens in experimental animals broadly correlates with tumor incidence. The formation of qualitatively similar adducts in human tissues treated *in vitro* and observed interindividual variations in levels of metabolic activation have prompted the development of methods suitable for the detection of carcinogen-macromolecular adducts and human biomonitoring following environmental exposures.

Methods that currently prove to be valuable in this respect include the ^{32}P-nucleotide postlabeling assay in combination with thin-layer-chromatography, synchronous scanning spectrophotofluorometry (SFS), mass spectroscopy and immunoassays (Garner 1985 and Harris et al., 1987). Since each method clearly has its own advantages and disadvantages, research in the Laboratory of Human Carcinogenesis has involved the use of more than one corroborative assay. Immunoassays have been used for the detection of PAH-DNA adducts and antibodies directed against these types of adducts and SFS in combination with HPLC has been used for the detection of BPDE-DNA and BPDE-hemoglobin adducts. Recent efforts have involved the development of ^{32}P-postlabeling assays for the detection of both PAH-DNA and alkyl-DNA adducts (Wilson et al., 1987).

Initial studies have focused on attempts to demonstrate the presence of PAH-DNA adducts and antibodies directed against these adducts in people exposured to high levels of PAH. DNA samples that were isolated from the peripheral blood leukocytes of 41 United States coke-oven workers were analysed by SFS, the DNA was hydrolysed under mildly acidic conditions to release PAH residues as tetrol derivatives, then the excitation and emission monochromators of the fluorometer were driven simultaneously with a fixed wavelength difference of 34nm. In the majority of these samples (75%) fluorescent emissions were detected in the region expected for the synchronous peak (374-384nm), however, the signals were primarily of a broad variety that did not permit of quantitation (Harris et al., 1985). Subsequent SFS analysis of 14 of these samples following HPLC has confirmed the presence in 4 of the samples of BPDE-DNA adducts at levels in the order of 1 adduct in 1-2 million nucleotides by comparison with an authentic radiolabeled batch-standard. In addition, DNA from the tissues of lung cancer patients were also analyzed by this method; none of the 5 tumor samples were found to contain BPDE-DNA adducts but 1/4 of the uninvolved lung specimens from the same patients did contain a significant amount of BPDE-DNA adducts.

DNA samples from the peripheral blood leukocytes of the coke-oven workers as described above were also subjected to corroborative immunoassays. Standard curves for the immunoassays (USERIA) were prepared using BPDE-DNA that was diluted with unmodified DNA (20µg or 10µg per microtiter well). Fifty percent inhibition was achieved with a level of DNA modification that fell in the range 8-16 fmol BPDE/µg DNA and levels of antibody binding to test DNA samples corresponding to more than 0.4 fmol BPDE/µg DNA (>20% inhibition) were considered to be positive. Among a group of 27 coke-oven workers, who had been employed for between 5 and 30 years, 18 (67%) were found to be carrying significant levels of adducts in their peripheral blood lymphocytes. Within this group 12 were smokers and 15 were non-smokers; of the smokers 9 (75%) were positive and of the non-smokers 9 (60%) were positive. Seven members of this group were known to work in environments containing high levels of benzene soluble particulates and 6 of them (86%) were found to be positive in the USERIA. In a similar study involving 28 roofers, 20 foundry workers and 9 laboratory workers, 7 (25 and 35%) of the samples from each of the groups subjected to PAH exposure and 2 (22%) of the non-industrially exposed were found to be positive (Shamsuddin et al., 1985).

In a second study using DNA from peripheral blood lymphocytes of a group of Norwegian coke oven workers were analysed by USERIA and SFS (but not using preparative chromatography), and the DNA adduct levels of the immunoassay were 2-6 times higher than those of the fluorescence assay. One possible explanation for this is cross-reactivity of the antiserum with related PAH-DNA adducts that were not detected by the SFS assay. It was found that when five other PAH-DNA adducts formed in vitro from the diol-epoxides of benz[a]anthracene, dibenz[a,c] anthracene, chrysene, and benzo[k]fluoranthene were used in fluid phase antigen competition assays that significant levels of cross-reactivity were observed for the rabbit polyclonal anti-BPDE-DNA antibody (unpublished results). Furthermore it appeared that the chrysene-diol-epoxide-DNA adduct and one of the benz[a]anthracene-diol-epoxide-DNA adducts competed more efficiently for the antibody than BPDE-DNA adducts.

In the corollary to this assay system the presence in human blood derived serum of antibodies directed against synthetic PAH-DNA adducts was used as a measure of molecular exposure to the metabolically activated chemicals. These experiments showed that different antibodies are produced against a range of PAH-DNA adducts in humans. In each of two separate studies, which included 38 Norwegian coke-oven workers (Haugen et al., 1986) and 99 non-industrially exposed individuals (unpublished results), approximately 30% of subjects were found to be positive. Simple antibody binding patterns that were obtained for sera with antibody reactivity against a single adduct suggest that these adducts each possess at least one immunologically unique epitope. However, the complex binding patterns that were obtained when reactivity against more than one adduct was observed in the sera may be the result of the presence in the sera of a range of specific antibodies each of which recognizes a different adduct or the presence of an antibody or antibodies that recognize a common or cross-reactive epitope present on all three adducts. In fact the results of antigen competitive assays indicated the presence in complex sera of both adduct specific and cross-reactive antibodies.

The presence or absence of antibodies to each of the adducts was found not to be correlated with gender or current smoking status of the individual serum donors (unpublished results). However, almost all of the non-smoking contributors of sera to these experiments were reformed smokers, therefore, the possibility that the primary antibody response was produced against adducts formed from PAHs contained in tobacco smoke cannot be precluded. High antibody titers found in coke-oven workers indicates that current exposure is important for the immunological response.

Although carcinogen-hemoglobin or other carcinogen-protein-adducts are not considered to be pathobiological lesions, these complexes do provide a useful resource in molecular dosimetry of exposure to chemical carcinogens. Erythrocytes have a lifespan of about 120 days and the hemoglobin adducts are not subject to repair so that an accumulative exposure is measurable over a period of 3-4 months. Several groups are now using hemoglobin as a dosimeter for exposure to both ethylene-oxide and 4-aminobiphenyl (Tornqvist et al., 1986 and Tannenbaum et al., 1986). In a smoking cessation study that is currently in progress in our laboratory evidence for BPDE-hemoglobin adducts has only been found in 3 out of 18 individuals yet studied by HPLC/SFS, the estimated level of adducts in these cases was one fmol BPDE/µg hemoglobin when compared to a batch standard.

Each of the assay systems used in these studies to evaluate human exposure to PAHs clearly demonstrate the formation of PAH-DNA or PAH-hemoglobin adducts in some individuals. Variations were observed, however, in the levels of PAH-DNA adducts and anti-PAH-DNA-antibodies detected in samples obtained from individuals with ostensibly similar exposure profiles. These variations may be explained on the basis that in vitro studies using normal human tissues have shown wide interindividual differences in the ability to metabolize, detoxify and activate PAHs and other chemical carcinogens (Autrup and Harris 1983). These differences in metabolic capability are now thought to be in part due to genetic polymorphisms within the population (Distlerath and Guengerich 1984). In addition, it may be that some individuals are independently exposed to inducers of P450 enzymes, thereby modifying metabolic activation of PAHs.

Recent experiments in several laboratories (e.g. Randerath et al., 1986) using ^{32}P-postlabeling analysis for the detection of carcinogen-DNA adducts have not provided support for the results of immunoassays or SFS ; and none of the putative chemical-DNA-adducts found in complex biological systems by this method have yet been identified. Specifically, in an analysis of 31 placentas from smokers and non-smokers for the presence of PAH-DNA adducts postlabeling revealed the presence of several unknown but apparently smoking related adducts all of which were found to be different from the PAH-DNA adducts that were formed when seven commonly encountered representative PAHs were activated in a mouse skin model; however, the use of monoclonal and polyclonal anti-BPDE-DNA antisera in these studies revealed high levels of activity, suggesting the presence of PAH-DNA adducts, in the same placental samples (Everson et al.,1986). These divergent results may be explained in part by poor resolution of components of adduct mixtures, since in our laboratory use a modification of the postlabeling technique developed by Gupta et al., (1982) has provided evidence of PAH-DNA adducts in salmon sperm DNA and DNA from human placental syncytiocytotrophoblast (unpublished results).

Based on our observations, and those of others described above, we are designing a system that takes advantage of the crossreactive nature of the polyclonal antibody and specificity conferred by HPLC. This system will use immunoaffinity columns to concentrate PAH-DNA adducts, HPLC to separate complex mixtures and ^{32}P-postlabelling and SFS as end-point assays.

The techniques that have been described above are all directed toward the measurement of covalently bound chemical carcinogens, however, more subtle changes to DNA may occur as the result of the action of free-radicals and radiation. Methods, including immunoassays and ^{32}P-postlabeling (both of which are combined with HPLC and immunocolumns to provide the requisite specificity), are being developed for the detection of these types of DNA damage (Leadon and Hanawalt 1983, Frenkel and Chrzan 1987, and Wilson et al., 1986 & 1987)

INHERITED AND ACQUIRED HOST FACTORS

The results of the studies described above and experiments described in the literature clearly indicate that genetic polymorphisms mediate the pharmacokinetics of drug and xenobiotic metabolism (Smith 1985). N-Acetylation of isoniazid, hydralazine, procainamide, and dapsone is controlled by two autosomal alleles at a single locus. Slow acetylation, which is a recessive trait appears to be associated with risk to bladder cancer in individuals who are occupationally exposed to aromatic amines (Cartwright et al., 1982); in contrast, fast acetylation appears to be associated with risk of colon cancer (Lang et al.,1987 and Ilett et al.,1987). Debrisoquine hydroxylation phenotyping has been correlated with the incidence of bronchogenic and hepatocellular carcinoma, and it has been shown that together with antipyrine and dapsone that this might prove to be a useful indicator drug (Ayesh and Idle 1985). With the advent of advanced techniques in molecular cloning and DNA-restriction fragment length polymorphisms (RFLPs) human genotyping for metabolic capabilities will be possible.

The use of RFLPs in conjunction with epidemiological data has lead to substantial support for the hypothesis suggesting that cancer is the result of a two-hit mutational process (Knudsen 1971) and now predicts the existence of 'tumor suppressor' genes (Knudsen 1985). The best known examples of this mechanisms are in the childhood cancers retinoblastoma and Wilms'tumor (Gallie and Worton 1986). In these diseases it appears that the loss of one of a pair of homologous loci, through either translocation, non-disjunction, mitotic recombination, rearrangement or deletion, predisposes to cancer. Further evidence for this mechanism has been gained from RFLP studies on retinoblastoma (Lundberg et al., 1987),osteosarcoma (Hansen et al., 1985), bladder cancer (Fearon et al., 1985), renal cell carcinoma (Zbar et al., 1987), breast cancer (Theillet et al., 1986) and acoustic neuroma (Seizinger et al., 1986). Karyotype analysis can only detect a small proportion of cases where the deleted region is at least microscopic. However, the use of molecular probes homologous to polymorphic genetic loci are now being used in an attempt to localize the putative 'tumor suppressor' genes. These probes are generally of two types and are characterized either by differences in restriction site location (site polymorphism) or by the presence of a variable tandem repeat (VTR) sequence (insertion/deletion polymorphism). The probes that have a VTR are potentially more informative since an individual is more likely to be heterozygous at these loci than at loci that have a site polymorphism only (Nakamura et al., 1987).

In our laboratory we are examining matched DNA samples of normal and tumor tissues from more than fifty non-small cell lung cancer patients by Southern hybridization with probes homologous to polymorphic genetic loci on chromosomes 3,11,13,15,17,18 and 20. These experiments have been designed to test the hypothesis that the loss of as yet undefined chromosomal genetic elements that may be candidates for 'tumor suppressor' genes are associated with human bronchogenic carcinoma. We have employed probes to six polymorphic genetic loci in order to detect loss of DNA sequences from chromosome 11 and our preliminary data show that deletions of chromosome 11 are frequent in non-small cell carcinoma of the lung (Weston et al., 1987). The pattern of allelic loss that was observed in eighteen cases where three or more of the chromosome 11 probes were informative appears to show that when restriction fragments were lost at the Int-2 locus (11q13) this was always accompanied by loss of heterozygosity at a locus or loci on the short-arm but that the reverse was not necessarily true and loss of sequences from the short-arm was only accompanied by loss of heterozygosity at Int-2 in 5 out of 10 cases.

In a polymorphism analysis of germline DNA, which did not require matched pair samples, analysis of DNA samples from more than 600 individuals Krontiris et al.,(1987) showed that an increased frequency of rare alleles of the cHa ras gene could be found in certain cancer patients. The cHa ras gene is located towards the terminus of the short-arm of chromosome 11 and the RFLP at this locus is of the insertion/deletion type. A VTR region that is tightly spanned by MspI/HpaII restriction sites can be found 3' to the structural gene. The VTR region is composed of reiterated consensus sequences of approximately 28 base pairs and the total number of these sequence repeats determines the overall size of the gene. Consequently the whole of the cHa ras gene region that is bounded by two Bam H1 restriction sites varies in size between approximately 6.5 and 8.3 kilobases. Subtle differences in the length of the VTR may, however, only be revealed by the use of restriction enzymes that tightly span this VTR region. Despite the fact that this type of RFLP may lead to the generation of many different allelotypes, only four allele sizes have hitherto been found to be prevalent in a normal or cancer-free population. We have now analysed DNA samples from 44 lung cancer patients for the presence of both common and rare allelic types and our preliminary data indicate that there is a high frequency of rare alleles of the cHa ras gene in bronchogenic carcinoma.

Restriction fragment length polymorphism analyses are important tools in the localization of genes involved in common malignancies, and development of well characterize polymorphic probes (especially those containing VTRs) wil considerably advance research efforts in this field. The current development of innovative techniques, for example field inversion electrophoresis and specialized cloning protocols will further assist the characterization and isolation of as yet undefined cancer associated genes.

REFERENCES

AMES,B.N., 1979 Science, 204, 587.

AUTRUP, H. & HARRIS,C.C., 1983 metabolism of chemical carcinogens by cultured human tissues. In Human Carcinogenesis, edited by C.C.Harris and H.Autrup (Academic Press, New York), p.169.

AYESH, R. & IDLE, J.R., 1985 evaluation of drug oxidation phenotypes in the biochemical epidemiology of lung cancer risk. In Microsomes and Drug Oxidations, edited by A.R.Boobis, J.Caldwell, F.DeMatteis and C.R.Elcombe (Taylor and Francis Inc., 242 Cherry St, Philadelphia, PA 19106-1906), p.340.

CAMPBELL,T.C., CACEDO,J.P., SALAMAT,L.A. & ENGEL,R.W.,1970 Nature, 227, 403.

CARTWRIGHT, R.A., GLASHAN,R.W., ROGERS,H.J., AHMAD,R.A., BARHAMHALL,D., HIGGINS,E. & KHAN,M.A. 1982, Lancet, 2, 842.

DISLERATH,L.M. & GUENGERICH,F.P. 1984, Proceedings of the National Academy of Sciences, USA, 81, 7348.

DOLL,R. 1985, cancer: a world-wide perspective. In Biochemical and Molecular Epidemiology of Cancer, edited by C.C.Harris (Alan R. Liss, Inc., New York),p.111.

EVERSON,R.B., RANDERATH,E., SANTELLA,R.M., CEFALO,R.C., AVITTS,T.A., & RANDERATH,K., 1986 Science, 231, 54.

FEARON,E.R., FEINBERG,A.P., HAMILTON,S.H. & VOGELSTEIN,B., 1985, Nature, 318, 377.

FRENKEL,K. & CHRZAN,K., 1987, Carcinogenesis, 8, 455.

GARNER,R.C. 1985, Carcinogenesis, 6, 1071.

GALLIE,B.L. & WORTON,R.G. 1986, Journal of Cellular Biochemistry, 32, 215.

GUPTA,R.C., REDDY,M.V. & RANDERATH,K. 1982, Carcinogenesis, 3, 1081.

HARRIS,C.C. 1983, role of carcinogens, cocarcinogens, and host factors in human cancer risk. In Human Carcinogenesis, edited by C.C.Harris (Academic Press, New York), p.941.

HARRIS,C.C., VAHAKANGAS,K., NEWMAN,M.J., TRIVERS,G.E., SHAMSUDDIN,A., SINOPOLI,N., MANN,D.L. & WRIGHT,W.E., 1985, Proceedings of the National Academy of Sciences, USA, 82, 6672.

HARRIS,C.C. 1987, Cancer Rearch, 47, 1.

HARRIS,C.C., WESTON,A., WILLEY,J.C., TRIVERS,G.E. & MANN,D.L., 1987, biochemical and molecular epidemiology of human cancer: indicators of carcinogen exposure, DNA damage, and genetic predisposition. In Environmental Health Perspectives, In press.

HANSEN,M.F., KOUFOS,A., GALLIE,B.L., PHILLIPS,R.A., FODSTAD,O., BROGGER,A., GEDDE-DAHL,T. & CAVENEE,W.K., 1985, Proceedings of the National Academy of Sciences, USA., 82, 6216.

ILETT,K.F., DAVID,B.M., DETCHON,P., CASTLEDON,W.M. & KWA,R., 1987, Cancer Research, 47, 1466.

KNUDSEN,A.G. 1971, Proceedings of the National Academy of Sciences, USA, 68, 820.

KNUDSEN A.G., 1985, Cancer Research, 45, 1437.

KRONTIRIS,T.G., DiMARTINO,N.A., COLB,M., MITCHESON,D.H. & PARKINSON,D.R. 1986, human restriction fragment length polymorphism and cancer risk. In Biochemical and Molecular Epidemiology of Cancer, edited by C.C.Harris (Alan R. Liss, Inc., New York), p.99.

LANG,N.P., CHU,D.Z.J., HUNTER,C.F., KENDALL,M.S.,

FLAMMANG,T.J. & KADLUBAR,F.F. 1987, Archives of Surgery, 121, 1259.

LEADON,S.A. & HANAWALT,P.C. 1983, Mutation Research, 112, 191.

LUNDBERG,C., SKOOG,L., CAVENEE,W.K. & NORDENSKJOLD,M. 1987, Proceedings of the National Academy of Sciences USA, 84, 2372.

MUMFORD,J.L., HE,X.Z., CHAPMAN,R.S., CAO,S.R., HARRIS,D.B., LI,X.M., XIAN,Y.L., JIANG,W.Z., XU,C.W., CHUANG,J.C., WILSON,W.E. & COOKE,M. 1987, Science, 235, 217.

NAKAMURA,Y., LEPPERT,M., O'CONNELL,P., WOLFF,R., HOLM,T., CULVER,M., MARTIN,C., FUJIMOTO,E., HOFF,M., KULMIN,E. & WHITE,R. 1987, Science, 235, 1616.

RANDERATH,E., AVITTS,T.A., REDDY,M.V., MILLER,R.H., EVERSON,R.B. & RANDERATH,K., 1986, Cancer Research, 46, 5869.

SASSON,I.M., HALEY,N.J., HOFFMANN,D., WYNDER,E.L., HELLBERG,D. & NILSSON,S., 1985, New England Journal of Medicine, 312, 315.

SEIZINGER,B.R., MARTUZA,R.L. & COUTELLA,J.F., 1986, Nature, 322, 644.

SHAMSUDDIN, A.K.M., SINOPOLI,N.T., HEMMINKI,K., BOESCH,R.R. & HARRIS,C.C. 1985, Cancer Research, 45, 66.

SMITH,R.L. 1985, genetic polymorphisms of drug oxidation in man. In Microsomes and Drug Oxidations, edited by A.R.Boobis, J.Caldwell, F.Dematteis and C.R.Elcombe (Taylor and Francis Inc., 242 Cherry St, Philadelphia, PA 19106-1906), p.349.

TANNENBAUM,S.T., BRYANT,M.S., SKIPPER,P.L. & MACLURE,M. 1986, hemoglobin adducts of tobacco-related aromatic amines. In Mechanisms in Tobacco Carcinogenesis, edited by D.Hoffman. and C.C.Harris. (Banbury Report #23) (Cold Spring Harbor Laboratories), p.63.

THEALL-ACRE,G., ALLEN,J.W., DOERR,C.L., ELMORE,E., HATCH,G.G., MOORE,M.M., SHARIEFF,Y., GRUNBERGER,D. & NESNOW,S. 1987, Cancer Research, 47, 3388.

THEILLET,C., LIDEREAU,R., ESCOT,C., HUTZELL,P., BRUNET,M., GEST,J., SCHLOM,J. & CALLAHAN,R., 1986, Cancer Research, 46, 4776.

TORNQVIST,M., OSTERMAN-GOLKAR,S., KAURIAINEN,A., JENSEN,S., FARMER,P.B. & EHRENBERG,L. 1986, Carcinogenesis, 7,1519.

WESTON,A., PLUMMER,S.M., GRAFSTROM,R.C., TRUMP,B.F. & HARRIS,C.C. 1986, Food and Chemical Toxicology, 24, 675.

WESTON,A., WILLEY,J.C., HAUGEN,A., KRONTIRIS,T., McDOWELL,E., RESAU,J., TRUMP,B.F. & HARRIS,C.C. 1987. Proceedings of the American Association for Cancer Research, 28, 251.

WILSON, V.L., SMITH, R.A., AUTRUP, H., KROKAN, H., MUSCI, D.E., LE, N-N-T., LONGORIA, J., ZISKA,D. and HARRIS, C.C. 1987,Analytical Biochemistry, 152,275.

WILSON,V.L., BASU,A.K.,ESSIGMANN,J.M. & HARRIS,C.C. 1987, Proceedings of the American Association of Cancer Research, 28, 112.

ZBAR,B., BRAUCH,H., TALMADGE,C. & LINEHAN,M. 1987, Nature, 327, 721.

Post Oxidative Events in Chemical Carcinogenesis: Acquisition of Multidrug Resistance

SNORRI S. THORGEIRSSON

Laboratory of Experimental Carcinogenesis, Division of Cancer Etiology, National Cancer Institute, Bethesda, MD 20892, USA

ABSTRACT

Possible similarities exist between the development of multidrug resistance in tumor cells and early steps in chemical carcinogenesis. We have studied this phenomenon by measuring the expression of the multidrug-resistance (mdr) gene in normal liver cells, in preneoplastic and neoplastic liver lesions and in regenerating liver. Levels of mRNA for mdr gene, which encodes P-glycoprotein, were elevated in both preneoplastic and neoplastic nodules. Expression of the mdr gene also reached very high levels in regenerating rat liver 24-72 hours after partial hepatectomy. These results show that the expression of the mdr gene can be regulated in liver and is likely to be responsible for part of the multidrug-resistance phenotype of carcinogen initiated hepatocytes and regenerating liver cells.

INTRODUCTION

The concept that neoplastic development is a multistep process is now generally accepted and there are currently a number of in vivo models in which this process can be studied (for review see Farber & Sarma 1987). Since the first demonstration of chemically induced hepatocarcinogenesis in 1935 (Sasaki & Yoshida 1987), the liver has been and continues to be a major focus of investigation of many experimental oncologists. The existence of the initiation-promotion-progression phenomenon during chemical hepatocarcinogenesis in the rat has now be clearly established (Peraino et al. 1984; Pitot & Sirica 1980), and consequently modulation of gene expression can be critically examined at different stages in the neoplastic development.

One of the central characteristics of the early preneoplastic cell population in the liver is the acquisition of resistance to the known toxic effects of carcinogens that are exhibited by cells of the normal tissue (Farber & Sarma 1987). The exact time point or stage in the neoplastic development that correspond to the formation of this resistance to toxic effects of carcinogens has not been precisely defined, but all available evidence indicates that the acquisition of the resistant phenotype occurs very early in the neoplastic process, possibly coinciding with initiation (Farber & Sarma 1987). Normal rat hepatocytes in monolayer cultures are highly sensitive to the cytotoxic effects of methotrexate, adriamycin, cycloheximide, and aflatoxin B1, whereas liver cells isolated from carcinogen-treated rats are resistant to the toxic effect of these agents (Farber 1976; Judah et al. 1977; Laishes et al. 1977; Carr & Laishes 1981). Although the mechanisms of this pleiotropic resistance are not known, some data suggest that the intracellular concentration of both the carcinogens and their metabolites is lower in the resistant cell than in the normal hepatocyte (Spiewak Rinauda & Farber 1986). One of the mechanisms which is known to contribute to the development of the multidrug resistance during chemotherapy of neoplastic diseases is the expression of the _mdr_ gene (Roninson et al. 1986). Full-length cDNA for the human and mouse _mdr_ genes have been cloned and used to demonstrate that expression of the _mdr_ gene, which encodes a 170K membrane P-glycoprotein, is responsible for multidrug resistance (Gros et al. 1986; Shen et al. 1986). Sequence analysis indicate that P-glycoprotein has 12 hydrophobic, membrane-spanning regions and two nucleotide binding domains (Chen et al. 1986). Taken together with direct biochemical evidence that P-glycoprotein binds drugs (Cornwell et al. 1986), these results are consistent with its function as an energy-dependent drug efflux pump. Because the P-glycoprotein efflux pump can lower intracellular drug concentration, and because the _mdr_ gene is known to be expressed at somewhat increased levels in normal human and rodent liver (Fojo et al. 1987), we have investigated the possible association between the development of carcinogen-resistance during chemical carcinogenesis and in regenerating liver and the increased expression of the _mdr_ gene.

MATERIALS AND METHODS

Male Fischer rats were used in all experiments. Fetal livers were obtained at 18 days of gestational age. Preneoplastic and neoplastic liver lesions were produced by means of the protocol of Solt & Farber (1976) with minor modifications (Evarts et al. 1985). Briefly, animals were treated with a single intraperitoneal injection of the initiating agent, diethylnitrosamine (200 mg/kg) and after a recovery period of 14 days, animals were administered 2-acetylaminofluorene (AAF) by gavage (1 mg/day) 5 times per week for 2 weeks. On day 5 of AAF feeding, animals were subjected to partial hepatectomy according to Higgins & Anderson (1931), and then continued on AAF for an additional week. Hyperplastic nodules (preneoplastic lesions) isolated 6-8 weeks after initiation with diethylnitrosamine were pooled (8-10 nodules per pool) whereas neoplastic nodules (neoplastic lesions representing primary hepatocellular carcinomas) isolated 6-8 months after initiation were individually isolated and processed.

RNA was isolated with guanithine thiocyanate by the method of Schweizer & Goerttler (1980), and enriched for poly(A)RNA by oligo(dT)-cellulose chromatography (Aviv & Leder 1972). Electrophoresis of poly(A)RNA samples on horizontal denaturing formaldehyde agarose gels with subsequent transfer onto nitrocellulose membranes was performed as described (Huber et al. 1986). Evidence that RNA was not degraded and was equally loaded and transferred to nitrocellulose was obtained by staining the RNA-containing gel with ethidium bromide and observing the gel before and after transfer. The construction of the RNA probe used was as described by Fojo et al. (1987).

RESULTS AND DISCUSSION

Although considerably less is known about the mechanism(s) of carcinogen induced resistance during chemical hepatocarcinogenesis than is known about multidrug-resistance involving chemotherapeutic agents, a number of observations, as pointed out earlier, support the notion that common and/or similar process may be operating in both cases. This hypothesis has brought into focus the possible similarities of two fundamental problems in cancer biology, namely the development of multidrug resistance in tumor cells and the mechanism of initiation, the first step in the multistep process of chemical carcinogenesis (Thorgeirsson et al. 1987).

FIG. 1. mdr Transcripts in normal and fetal rat liver and during chemical hepatocarcinogenesis. (A) Northern blot analysis of poly(A)RNA isolated from neoplastic liver nodules (lanes 1-4); preneoplastic liver nodules (lane 5 and 6) and normal liver (lane 7). All lanes contained 5 ug of poly(A)RNA. (B) Northern blot analysis of poly(A)RNA isolated from fetal liver (lanes 1,2), and adult rat liver (lanes 3,4) and preneoplastic liver nodules (lanes 5,6). All lanes contained 5 ug of poly(A)RNA.

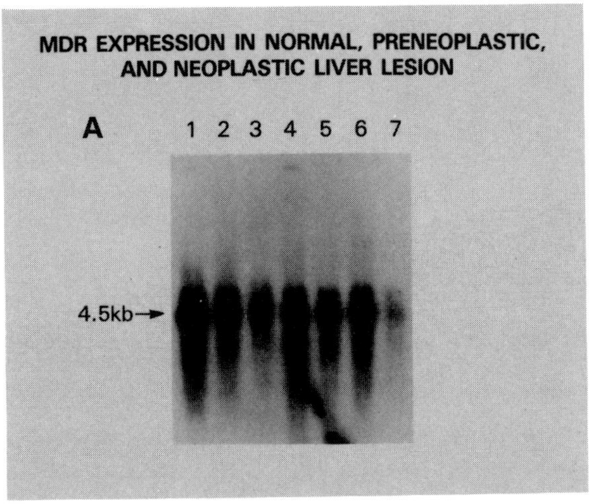

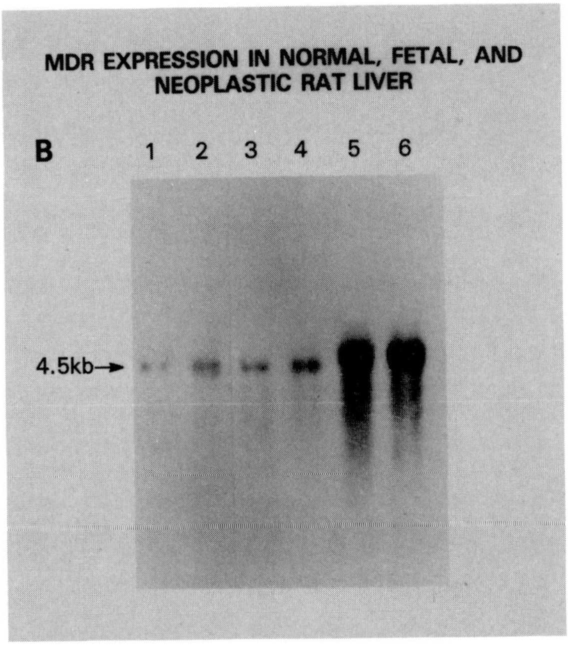

FIG. 2. mdr Transcripts in regenerating rat liver. Poly(A)RNA was isolated at various time points after partial hepatectomies of adult male Fischer rats (for details see legend to Fig. 1). Lanes 1-7 contain 5 ug of poly(A)RNA isolated at 1, 6, 12, 24, 48, 72, and 120 hours after partial hepatectomy, respectively.

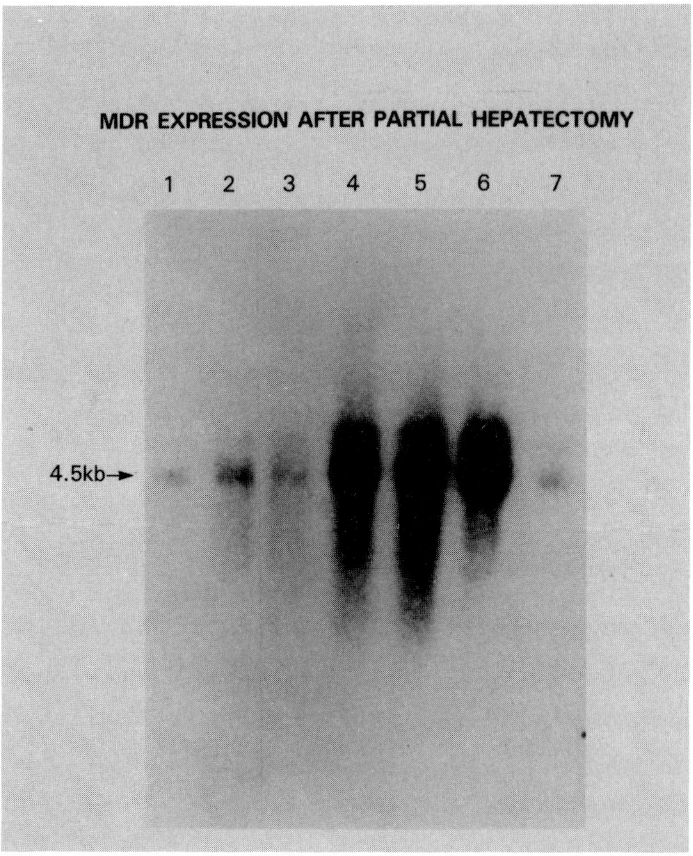

As has been previously shown in normal human liver (Fojo et al. 1987), mdr RNA is readily detectable in normal rat liver (Fig. 1A, lane 7). The expression of mdr gene was increased several fold in both preneoplastic and neoplastic liver lesions produced by the Solt-Farber method (Fig. 1A, lanes 1-6). No apparent difference in mdr expression was observed between early preneoplastic lesions (nodules isolated 6-8 weeks after initiation,

Fig. 1A and B, lanes 5,6), and neoplastic lesions (nodules isolated 6-8 months after initiation; Fig. 1A, lanes 1-4).

Both fetal and adult rat livers showed lower levels of mdr RNA (Fig. 1A, lane 7, Fig. 1B, lanes 1-4). The Solt-Farber protocol involves only a single initiating dose of a carcinogen (in our case diethylnitrosamine 200 mg/kg i.p.) followed by a short period of exposure to a low dose of another carcinogen (e.g. AAF) during which time a partial hepatectomy is performed. Consequently, the preneoplastic and neoplastic cell populations have only been exposed to carcinogens for a brief period of time. The contribution of AAF to the development of the initiated cell population is thought to be the result of growth inhibitory effects of AAF on normal hepatocytes during regeneration after partial hepatectomy (Farber & Sarma 1987).

The expression of mdr RNA was also examined after partial hepatectomy (Fig. 2). No increase in mdr expression was observed for the first 12 hours after the operation (Fig. 2, lanes 1-3). However, mdr RNA rose to a very high level from 24 to 72 hours post-hepatectomy (Fig. 2, lanes 4-6), and falls at 120 hours after hepatectomy (Fig. 2, lane 7). The level of mdr expression seen in these studies is comparable to the high level seen in human adrenal gland or in multidrug-resistant cell lines showing 100-fold increased resistance to colchicine, vinblastine, and adriamycin (Shen et al. 1986). The decline in mdr mRNA levels between 72 and 120 hours post-hepatectomy (Fig. 2, lanes 6 and 7) is consistent with a relatively short half-live for the mdr mRNA under some conditions.

Our results demonstrate, for the first time, modulation of the mdr expression under conditions other than selection of multidrug-resistant cell lines after exposure to chemotherapeutic agents and/or chemical carcinogens. Moreover, an experimental model system now exists that offers exciting possibilities for defining the role of the mdr gene(s) in normal cell biology.

As to the role of the mdr gene in the initiation stage of chemical carcinogenesis, a number of questions need to be answered. It is unlikely that the increased expression of the mdr gene observed both in preneoplastic and neoplastic cell populations can totally explain the resistance to chemical carcinogens observed in early initiated cells, since many of these carcinogens are compounds not affected by the multidrug-resistant phenotype. In addition, if increased expression of the *mdr* gene during the time period of maximal DNA synthesis following partial hepatectomy made liver cells totally resistant to the cytotoxic and mitoinhibitory effects of AAF, then the Solt-Farber protocol would not result in formation of liver nodules. Therefore, other factors, such as overexpression of the glutathione transferases and other unknown mechanisms, may be contributing significantly to the development of the resistant phenotype (Cowan et al. 1986). If this is the case, the regulated expression of the *mdr* gene may be a part of a programmed set of responses of the liver cell to certain toxic injuries.

REFERENCES

Aviv, H. & Leder, P., 1972, *Proceedings of the National Academy of Sciences USA*, 69, 1408.

Carr, B. I. & Laishes, B. A., 1981, *Cancer Research*, 41, 1715.

Chen, C. J., Chin, J. E., Ueda, K., Clark, D. P., Pastan, I., Gottesman, M. M. & Roninson, I. B., 1986, *Cell*, 47, 381.

Cornwell, M. M., Gottesman, M. M. & Pastan, I., 1986, *Journal of Biological Chemistry*, 261, 7921.

Cowan, K. H., Batist, G., Tulpule, A., Sinha, B. K., & Meyers, C. E., 1986, *Proceedings of the National Academy of Sciences USA*, 83, 9328.

Evarts, R. P., Marsden, E. & Thorgeirsson, S. S., 1985, *Carcinogenesis*, 6, 1767.

Farber, E., Parker, S. & Gruenstein, M., 1976, *Cancer Research*, 35, 3279.

Farber, E. & Sarma, D. S. R., 1987, *Laboratory Investigation*, 56, 1.

Fojo, A. T., Ueda, K., Slamon, D. M., Poplack, D. G., Gottesman, M. M. & Pastan, I., 1987, *Proceedings of the National Academy of Sciences USA*, 84, 262.

Gros, P., Neriah, Y. B., Croop, J. M. & Housman, D. E., 1986, *Nature*, 323, 728.

Higgins, G. M. & Anderson, R. M., 1931, *Archives of Pathology* 12, 186.

Huber, B. E., Glowinski, I. B. & Thorgeirsson, S. S., 1986, *Journal of Biological Chemistry*, 261, 12400.

Judah, D. J., Legg, R. F. & Neal, G. E., 1977, *Nature*, 365, 343.

Laishes, B. A., Roberts, E. & Farber, E., 1977, *International Journal of Cancer*, 21, 186.

Peraino, C., Richards, W. L., & Stevens, F. J., 1984, In Mechanisms of Tumor Promotion: Tumor Promotion in the Internal Organs, edited by T.J. Slaga, ed. (CRC Press, New York) p. 1.

Pitot, H. C. & Sirica, A. E., 1980, Biochimica et Biophysica Acta, 605, 191.

Roninson, I. B., Chin, J. E., Choi, K., Gros, P., Housman, D. E., Fojo, A., Shen, D. W., Gottesman, M. M. & Pastan, I., 1986, Proceedings of the National Academy of Sciences USA, 83, 4538.

Sasaki, T. & Yoshida, T., 1987, Archives of Pathology, Anatomy and Physiology, 295, 175.

Schweizer, J. & Goerttler, K., 1980, European Journal of Biochemistry, 112, 243.

Shen, D. W., Fojo, A. T., Chin, J. E., Roninson, I. B., Richert, N., Pastan, I. & Gottesman, M. M., 1986, Science 232, 643.

Shen, D. W., Fojo, A., Roninson, I. B., Chin, J. E., Soffir, R., Pastan, I. & Gottesman, M. M., 1986, Molecular and Cellular Biology, 6, 4039.

Solt, D. & Farber, E., 1976, Nature, 263, 701.

Spiewak Rinaud, J. A. & Farber, E., 1986, Carcinogenesis, 7, 523.

Thorgeirsson, S. S., Huber, B. E., Sorrell, S., Fojo, A. T., Gottesman, M. M. & Pastan, I., 1987, Science, 236, 1120.

INDEX